DEVELOPMENTS IN HYDRAULIC CONDUCTIVITY RESEARCH

Edited by **Oagile Dikinya**

Developments in Hydraulic Conductivity Research
Edited by Oagile Dikinya

Published by InTech
Janeza Trdine 9, 51000 Rijeka, Croatia

Publishing Process Manager Ivana Lorkovic

Technical Editor Teodora Smiljanic

Cover Designer InTech Design Team

Additional hard copies can be obtained from orders@intechopen.com

Developments in Hydraulic Conductivity Research, Edited by Oagile Dikinya
p. cm.
ISBN 978-953-307-470-2

For EU product safety concerns:
IN TECH d.o.o., Prolaz Marije Krucifikse Kozulić 3, 51000 Rijeka, Croatia,
info@intechopen.com or visit our website at intechopen.com.

Meet the editor

Dr. Oagile Dikinya is a Geoscientist/Soil Physicist and is currently a Senior Lecturer at the Department of Environmental Science, University of Botswana. He holds BSc degree (Physics and Environmental Science) from the University of Botswana in 1988; MSc (Soil & Water), University of Wageningen, The Netherlands in 1992 and a PhD degree (Soil Science), University of Western Australia in 2006. As a senior academic, he is involved in the supervision of both undergraduate and postgraduate research projects as well as in research in the area of soil and water management particularly soil permeability investigations. He has numerous research and publications record of 39 publication pieces including 17 refereed journal articles, 1 book chapter, 10 conference articles and 11 technical reports. Most of his refereed articles are extensively published in the so called high impact factor indexed journals. Further, Dr. Dikinya is a reviewer in many international peer reviewed journals.

Contents

Preface

This book provides the state of the art of the investigation and the in-depth analysis of hydraulic conductivity from the theoretical to semi-empirical models to policy development associated with management of land resources emanating from drainage-problem soils. Many international experts contributed to the development of this book. It is envisaged that this thought provoking book of international repute will excite and appeal to academics, researchers and university students who seek to explore the breadth and in-depth knowledge about hydraulic conductivity. Investigations into hydraulic conductivity is important to the understanding of the movement of solutes and water in the terrestrial environment and/or the hydrosphere-biosphere interface. Transport of these fluids has various implications on the ecology and quality of environment and subsequently sustenance of livelihoods of the increasing world population. In particular, water flow in the vadose zone is of fundamental importance to geoscientists, soil scientists, hydrogeologists and hydrologists and is a critical element in assessing environmental implications of soil management. For example, free water at the soil-atmosphere interface is a source of great importance to man. Efficient management of this water will require greater control of hydraulic conductivity in order to solve such wide ranging problems as upland flooding, pollution of surface and ground-waters, and inefficient irrigation of agricultural lands.

It is generally recognized that progress of science depends increasingly on an advanced understanding differences in the methods of investigations and their applicability to solving real problems in the ecological fragile environment. In this book a number of approaches were employed in assessing hydraulic conductivity including theoretically and quasi-semi empirical models and conclude with applied policy considerations. Of particular importance is the analysis of hydraulic conductivity at the macro-scale to pore scale i.e theoretical conceptions from the geological structures (rock pore space) to broken rock mass or saprolite (soil-microscopic level) in order to understand the transport phenomena in underground aquifers and porous media in soils. Water flow and solute transport through soil are directly related to the geometry of the available pore space. The role of macroporosity in groundwater movement cannot be overemphasized, for example, recharging the groundwater by the rapid movement of water through soil macropores may affect simultaneous movement of undesirable constituents and resultant rapid contamination of the subsurface water resources. This has accentuated the need to come up with robust modeling approaches to analyzing the hydraulic conductivity. For instance, in the last two decades, models have been explored in the field of soil physics to study flow processes at the pore scale.

While the measurements of hydraulic conductivity is usually tedious and a difficult, it is imperative that different approaches be employed to derive or predict and accurately estimate hydraulic conductivity. For this reason, the book is covered in 4 sections including: Part 1- *Mechanistic and Geotechnical Modelling Approaches*; Part 2-*Empirical Modeling Approaches to Estimating Hydraulic Conductivity*; Part 3- *Laboratory Hydraulic Conductivity Measurements* and Part 4- *Implications of Hydraulic Conductivity on Land Management and Policy Development.*

Part 1 explores the robust mechanical and geotechnical theoretical methods to analyse hydraulic conductivity and includes topics such as i) Influence of degree of saturation in the electric resistivity-hydraulic conductivity relationship, ii) Stress/strain-dependent properties of hydraulic conductivity for fractured rocks and Hydraulic conductivity and water retention curve of highly comprehensible materials iii). On the other hand ***part 2*** covers in-depth analysis of the empirical models to estimating hydraulic conductivity and covered salient topic features including i) Correlations between hydraulic conductivity and selected hydrogeological properties of rocks, ii) Hydraulic Conductivity of layered anisotropic media, and iii) Rock mass hydraulic conductivity estimated by two empirical models. Use of measuring techniques and deterministic laboratory hydraulic conductivity measurements are covered in ***part 3*** and includes topics such as i) Unsaturated hydraulic conductivity for evaporation in heterogeneous soils ii) Determination of hydraulic conductivity of undisturbed soil column: A measurement accomplished with the Gamma Ray Transmission Technique and iii) Hydraulic conductivity of semi-quasi stable soils: Effects of particulate mobility. ***Part 4*** concludes the book with some applications of hydraulic conductivity implications on land management and policy development and covers interesting topics such as i) Saturated hydraulic conductivity and land use change, new insights to the payments for ecosystem services programs: a case study from a tropical montane cloud forest watershed in eastern central Mexico and ii) Hydraulic conductivity and landfill construction.

In conclusion, this book is structured in a way as to overview the state of the art analyses of hydraulic conductivity and I hope it will serve the interests of the stakeholders involved in the applications of science of transport of water and solutes to understanding the dynamics of fluids flow in porous media. In particular the development of policies and strategies pertinent to water availability and management is critical to the improved livelihoods of the nations of the world.

Thank you

Dr. Oagile.Dikinya
Senior Lecturer,
University of Botswana,
Botswana

Part 1

Mechanistic and Geotechnical Modelling Approaches

1

Stress/Strain-Dependent Properties of Hydraulic Conductivity for Fractured Rocks

Yifeng Chen and Chuangbing Zhou
State Key Laboratory of Water Resources and Hydropower Engineering Science, Key Laboratory of Rock Mechanics in Hydraulic Structural Engineering, Wuhan University, P. R. China

1. Introduction

In the last two decades there has seen an increasing interest in the coupling analysis between fluid flow and stress/deformation in fractured rocks, mainly due to the modeling requirements for design and performance assessment of underground radioactive waste repositories, natural gas/oil recovery, seepage flow through dam foundations, reservoir induced earthquakes, etc. Characterization of hydraulic conductivity for fractured rock masses, however, is one of the most challenging problems that are faced by geotechnical engineers. This difficulty largely comes from the fact that rock is a heterogeneous geological material that contains various natural fractures of different scales (Jing, 2003). When engineering works are constructed on or in a rock mass, deformation of both the fractures and intact rock will usually occur as a result of the stress changes. Due to the stiffer rock matrix, most deformation occurs in the fractures, in the form of normal and shear displacement. As a result, the existing fractures may close, open, grow and new fractures may be induced, which in turn changes the structure of the rock mass concerned and alters its fluid flow behaviours and properties. Therefore, the fractures often play a dominant role in understanding the flow-stress/deformation coupling behavior of a rock system, and their mechanical and hydraulic properties have to be properly established (Jing, 2003).

Traditionally, fluid flow through rock fractures has been described by the cubic law, which follows the assumption that the fractures consist of two smooth parallel plates. Real rock fractures, however, have rough walls, variable aperture and asperity areas where the two opposing surfaces of the fracture walls are in contact with each other (Olsson & Barton, 2001). To simplify the problem, a single, average value (or together with its stochastic characteristics) is commonly used to describe the mechanical aperture of an individual fracture. A great amount of work (Lomize, 1951; Louis, 1971; Patir & Cheng, 1978; Barton et al., 1985; Zhou & Xiong, 1996) has been done to find an equivalent, smooth wall hydraulic aperture out of the real mechanical aperture such that when Darcy's law or its modified version is applied, the equivalent smooth fracture yields the same water conducting capacity with its original rough fracture. It is worth noting that clear distinction manifests between the geometrically measured mechanical aperture (denoted by b in the context) and the theoretical smooth wall hydraulic aperture (denoted by b^*), and the former is usually larger in magnitude than the latter due to the roughness of and filling materials in rock fractures (Olsson & Barton, 2001).

The ubiquity of fractures significantly complicates the flow behaviour in a discontinuous rock mass. The primary problem here is how to model the flow system and how to determine its corresponding hydraulic properties for flow analysis. Theoretically, the representative elementary volume (*REV*) of a rock mass can serve as a criterion for selecting a reasonable hydromechanical model. This statement relates to the fact that *REV* is a fundamental concept that bridges the micro-macro, discrete-continuous and stochastic-determinate behaviours of the fractured rock mass and reflects the size effect of its hydraulic and mechanical properties. The *REV* size for the hydraulic or mechanical behaviour is a macroscopic measurement for which the fractured medium can be seen as a continuum. It is defined as the size beyond which the rock mass includes a large enough population of fractures and the properties (such as hydraulic conductivity tensor and elastic compliance tensor) basically remain the same (Bear, 1972; Min & Jing, 2003; Zhou & Yu, 1999; Wang & Kulatilake, 2002). Owing to high heterogeneity of fractured rock masses, however, the *REV* can be very large or in some situations may not exist. If the *REV* does not exist, or is larger than the scale of the flow region of interest, it is no longer appropriate to use the equivalent continuum approach. Instead, the discrete fracture flow approach may be applied to investigate and capture the hydraulic behaviour of the fractured rock masses. However, due to the limited available information on fracture geometry and their connectivity, it is not a trivial task to make a detailed flow path model. Thus, in practice, the equivalent continuum model is still the primary choice to approximate the hydraulic behaviour of discontinuous rocks.

The hydraulic conductivity tensor is a fundamental quantity to characterizing the hydromechanical behaviour of a fractured rock. Various techniques have been proposed to quantify the hydraulic conductivity tensor, based on results from field tests, numerical simulations, and back analysis techniques, etc. Earlier investigations focused on using field measurements (e.g. aquifer pumping test or packer test (Hsieh & Neuman, 1985)) to estimate the three-dimensional hydraulic conductivity tensor. This approach, however, is generally time-consuming, expensive and needs well controlled experimental conditions. Numerical and analytical methods are also used to estimate the hydraulic properties of complex rock masses due to its flexibility in handling variations of fracture system geometry and ranges of material properties for sensitivity or uncertainty estimations. In the literature, both the equivalent continuum approach (Snow, 1969; Long et al., 1982; Oda, 1985; Oda, 1986; Liu et al., 1999; Chen et al., 2007; Zhou et al., 2008) and the discrete approach (Wang & Kulatilake, 2002; Min et al., 2004) are widely applied. In this chapter, however, only the equivalent continuum approach is focused for its capability of representing the overall behaviour of fractured rock masses at large scales.

Among many others, Snow (1969) developed a mathematical expression for the permeability tensor of a single fracture of arbitrary orientation and aperture and considered that the permeability tensor for a network of such fractures can be formed by adding the respective components of the permeability tensors for each individual fracture. Oda (1985, 1986) formulated the permeability tensor of rock masses based on the geometrical statistics of related fractures. Liu et al. (1999) proposed an analytical solution that links changes in effective porosity and hydraulic conductivity to the redistribution of stresses and strains in disturbed rock masses. Zhou et al. (2008) suggested an analytical model to determine the permeability tensor for fractured rock masses based on the superposition principle of liquid dissipation energy. Although slight discrepancy exists between the permeability tensor and the hydraulic conductivity tensor (the former is an intrinsic property determined by fracture geometry of the rock mass, while the latter also considers the effects of fluid viscosity and gravity), when taking into account the flow-stress coupling effect, the above models presented, respectively, by Snow (1969), Oda (1985) and Zhou et al. (2008) were proved to be functionally equivalent for a certain fluid (Zhou et al., 2008). A common limitation with the above models lies in the fact that the hydraulic conductivity tensor of a fractured rock mass is all formulated to be either stress-dependent or elastic strain-dependent. Consequently, material nonlinearity and post-peak dilatancy are not considered in the formulation of the hydraulic conductivity tensor for disturbed rock masses. To address this problem, Chen et al. (2007) extended the above work and proposed a numerical model to establish the hydraulic conductivity for fractured rock masses under complex loading conditions.

Based on the observation that natural fractures in a rock mass are most often clustered in certain critical orientations resulting from their geological modes and history of formation (Jing, 2003), characterizing the rock mass as an equivalent continuum containing one or multiple sets of planar and parallel fractures with various critical orientations, scales and densities turns out to be a desirable approximation. Starting from this point of view, the deformation patterns of the fracture network can be first characterized by establishing an equivalent elastic or elasto-plastic constitutive model for the homogenized medium. On this basis, a stress-dependent hydraulic conductivity tensor may be formulated for the former for describing the hydraulic behaviour of the rock mass at low stress level and with overall elastic response; and a strain-dependent hydraulic conductivity tensor for the latter for demonstrating the influences of material non-linearity and shear dilatancy on the hydraulic properties after post-peak loading. This chapter mainly presents the research results on the stress/strain-dependent hydraulic properties of fractured rock masses under mechanical loading or engineering disturbance achieved by Chen et al. (2006), Zhou et al. (2006), Chen et al. (2007) and Zhou et al. (2008).

The stress-dependent hydraulic conductivity model (Zhou et al., 2008) was proposed for estimation of the hydraulic properties of fractured rock masses at relatively lower stress level based on the superposition principle of flow dissipation energy. It was shown that the model is equivalent to Snow's model (Snow, 1969) and Oda's model (Oda, 1986) not only in form but also in function when considering the effects of mechanical loading process on the evolution of hydraulic properties. This model relies on the geometrical characteristics of rock fractures and the corresponding fracture network, and demonstrates the coupling effect between fluid flow and deformation. In this model, the pre-peak dilation and contraction effect of the fractures under shear loading is also empirically considered. It was applied to estimate the hydraulic properties of the rock mass in the dam site of the Laxiwa Hydropower Project located in the upstream of the Yellow River, China, and the model predictions have a good agreement with the site observations from a large number of single-hole packer tests.

The strain-dependent hydraulic conductivity model (Chen et al., 2007), on the other hand, was established by an equivalent non-associative elastic-perfectly plastic constitutive model with mobilized dilatancy to characterize the nonlinear mechanical behaviour of fractured rock masses under complex loading conditions and to separate the deformation of weaker fractures from the overall deformation response of the homogenized rock masses. The major advantages of the model lie in the facts that the proposed hydraulic conductivity tensor is related to strains rather than stresses, hence enabling hydro-mechanical coupling analysis to include the effect of material nonlinearity and post-peak dilatancy, and the proposed model is easy to be included in a FEM code, particularly suitable for numerical analysis of hydromechanical problems in rock engineering with large scales. Numerical simulations were performed to investigate the changes in hydraulic conductivities of a cube of fractured rock mass under triaxial compression and shear loading as well as an underground circular excavation in biaxial stress field at the Stripa mine (Kelsall et al., 1984; Pusch, 1989), and the simulation results are justified by in-situ experimental observations and compared with Liu's elastic strain-dependent analytical solution (Liu et al., 1999).

Unless otherwise noted, continuum mechanics convention is adopted in this chapter, i.e., tensile stresses are positive while compressive stresses are negative. The symbol (:) denotes an inner product of two second-order tensors (e.g., $\boldsymbol{a}:\boldsymbol{b}=a_{ij}b_{ij}$) or a double contraction of adjacent indices of tensors of rank two and higher (e.g., $\boldsymbol{c}:\boldsymbol{d}=c_{ijkl}d_{kl}$), and ($\otimes$) denotes a dyadic product of two vectors (e.g., $\boldsymbol{a}\otimes\boldsymbol{b}=a_ib_j$) or two second-order tensors (e.g., $\boldsymbol{c}\otimes\boldsymbol{d}=c_{ij}d_{kl}$).

2. Stress-dependent hydraulic conductivity of rock fractures

In this section, the elastic deformation behaviour of rock fractures at the pre-peak loading region will be first presented, and then a stress-dependent hydraulic conductivity model will be formulated. The deformation model (or indirectly the hydraulic conductivity model) is validated by the laboratory shear-flow coupling test data obtained by Liu et al. (2002). The main purpose of this section is to provide a theory for developing a stress-dependent hydraulic conductivity tensor for fractured rock masses that will be presented later in Section 4.

2.1 Characterization of rock fractures

One of the major factors that govern the flow behaviour through fractured rocks is the void geometry, which can be described by several geometrical parameters, such as aperture, orientation, location, size, frequency distribution, spatial correlation, connectivity, and contact area, etc. (Olsson & Barton, 2001; Zhou et al., 1997; Zhou & Xiong, 1997). Real fractures are neither so solid as intact rocks nor void only. They have complex surfaces and variable apertures, but to make the flow analysis tractable, the geometrical description is usually simplified. It is common to assume that individual fractures lie in a single plane and have a constant hydraulic aperture. When the fractures are subjected to normal and shear loadings, the fracture aperture, the contact area and the matching between the two opposing surfaces will be altered. As a result, the equivalent hydraulic aperture of the fractures varies with their normal and shear stresses/displacements, which demonstrates the apparent coupling mechanism between fluid flow and stress/deformation (Min et al., 2004).

The aperture of rock fractures tends to be closed under applied normal compressive stress. The asperities of the surfaces will be crushed when their localized compressive stresses exceed their compressive strength. As a large number of asperities are crushed under high compressive stress, the contact area between the fracture walls increases remarkably and the crushed rock particles partially or fully fill the nearby void, which decreases the effective flow area, reduces the hydraulic conductivity of the fracture, and even changes the flow paths through fracture plane. Fig. 1 depicts the increase in contact area of fractures under increasing compressive stresses modelled by boundary element method (Zimmerman et al., 1991).

The coupling process between fluid flow and shear deformation is more related to the roughness of fractures and the matching of the constituent walls. Fig. 2 shows the impact of the fracture structure on the shear stress-deformation coupling mechanism. In Fig. 2(a), the opposing walls of the fracture are well matched so that the fracture always dilates and the hydraulic conductivity increases under shear loading as long as the applied normal stress is not high enough for the asperities to be crushed. For the state shown in Fig. 2(c), shear loading will result in the closure of the fracture and the reduction in hydraulic conductivity. Fig. 2(b) illustrates a middle state between (a) and (c), and its shearing effect depends on the direction of shear stress. When the matching of a fracture changes from (a) to (b) then to (c) under shear loading, shear dilation occurs. On the other hand, shear contraction takes place from the movement of the matching from (c) to (b) then to (a). In a more complex scenario, shear dilation and shear contraction may happen alternately, resulting in the fluctuation of the hydraulic behaviour of the fractures.

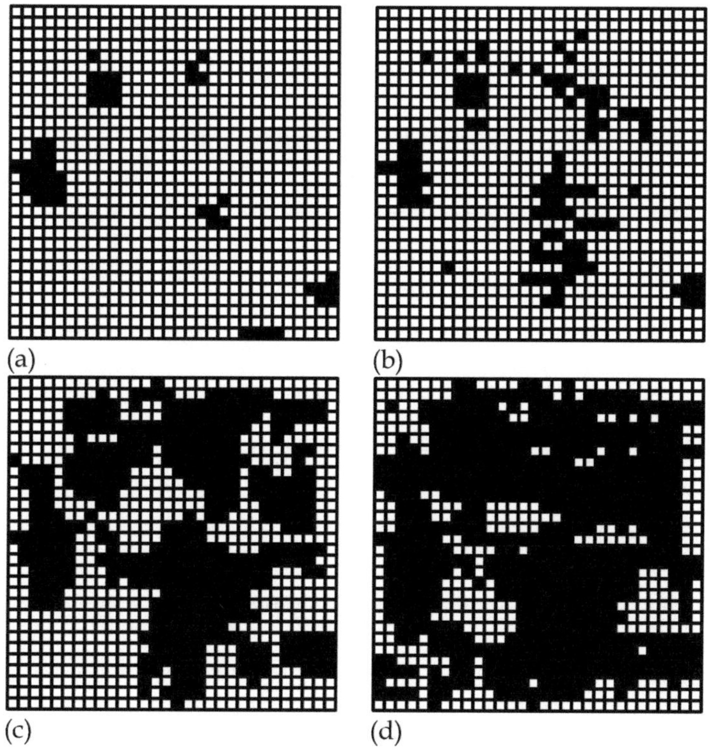

Fig. 1. Variation of contact surface of fractures under increasing compressive stresses (after Zimmerman et al., (1991): (a) P=0 MPa; (b) P=20 MPa; (c) P=40 MPa and (d) P=60 MPa

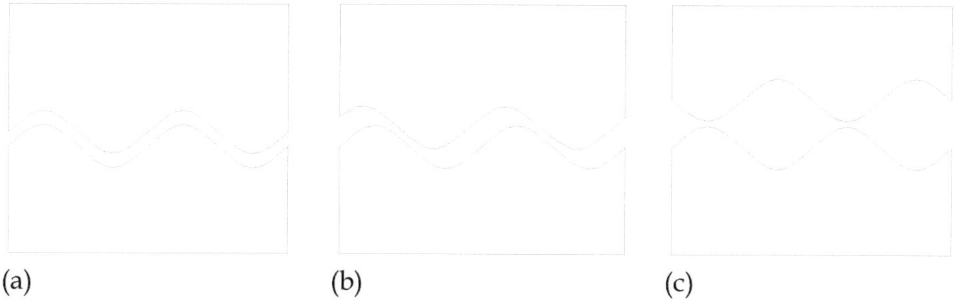

Fig. 2. Shear dilation and shear contraction of fractures: (a) well-matched; (b) fair-matched; and (c) bad-matched

2.2 An elastic constitutive model for rock fractures

To formulate the stress-dependent hydraulic conductivity for rock fractures, we model the fractures by an interfacial layer, as shown in Fig. 3. The interfacial layer is a thin layer with complex constituents and textures (depending on the fillings, asperities and the contact area between its two opposing walls). Assumption is made here that the apparent mechanical response of the interfacial layer can be described by Lame's constant λ and shear modulus μ. Because the thickness of the interfacial layer (i.e., the initial mechanical aperture of the fracture) is generally rather small comparing to the size of rock matrix, it is reasonable to assume that $\varepsilon_x=\varepsilon_y=0$ and $\gamma_{xy}=\gamma_{yx}=0$ within the interfacial layer. Then according to the Hooke's law of elasticity, the elastic constitutive relation for the interfacial layer under normal stress σ_n and shear stress τ can be written in the following incremental form:

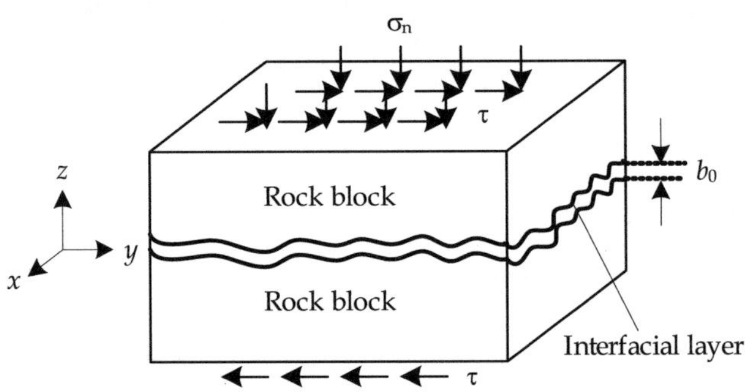

Fig. 3. The interfacial layer model for rock fractures

$$\begin{Bmatrix} d\sigma'_n \\ d\tau \end{Bmatrix} = \begin{bmatrix} \lambda+2\mu & 0 \\ 0 & \mu \end{bmatrix} \begin{Bmatrix} d\varepsilon_n \\ d\gamma \end{Bmatrix} \tag{1}$$

For convenience, we use u_1 to denote the relative normal displacement of the interfacial layer caused by the effective normal stress σ'_n, δ to denote the relative tangential displacement caused by the shear stress τ, and u_2 to denote the relative normal displacement caused by shear dilation or contraction (positive for dilatant shear, negative for contractive shear). Hence, the total normal relative displacement u is represented as

$$u = u_1 + u_2 \tag{2}$$

The increments of strains, $d\varepsilon_n$ and $d\gamma$, can be expressed in terms of the increments of relative displacements, du_1 and $d\delta$, as follows:

$$\begin{cases} d\varepsilon_n = du_1 / (b_0 + u) \\ d\gamma = d\delta / (b_0 + u) \end{cases} \tag{3}$$

where b_0 is the thickness of the interfacial layer or the initial mechanical aperture of the fracture. Substituting Eq. (3) in Eq. (1) yields:

$$\begin{Bmatrix} d\sigma'_n \\ d\tau \end{Bmatrix} = \begin{bmatrix} k_n & 0 \\ 0 & k_s \end{bmatrix} \begin{Bmatrix} du_1 \\ d\delta \end{Bmatrix} \tag{4}$$

where k_n and k_s denote the tangential normal stiffness and tangential shear stiffness of the interfacial layer, respectively.

$$k_n = (\lambda + 2\mu) / (b_0 + u), \quad k_s = \mu / (b_0 + u) \tag{5}$$

Interestingly, k_n and k_s show a hyperbolic relation with normal deformation and characterize the deformation response of the interfacial layer under the idealized conditions that each fracture is replaced by two smooth parallel planar plates connected by two springs with stiffness values k_n and k_s. As can be seen from Eq. (5), as long as the initial normal stiffness and shear stiffness with zero normal displacement, k_{n0} and k_{s0}, are known, they can be used as substitutes for λ and μ.

Substituting Eq. (2) in Eq. (4) results in:

$$\mathrm{d}\sigma'_n = \frac{(\lambda + 2\mu)\mathrm{d}u_1}{b_0 + u_1 + u_2} \tag{6}$$

$$\mathrm{d}\tau = \frac{\mu \mathrm{d}\delta}{b_0 + u_1 + u_2} \tag{7}$$

Suppose normal stress σ_n is firstly applied before the loading of shear stress, u_1 can be obtained by directly integrating Eq. (6):

$$u_1 = (b_0 + u_2)\left[\exp\left(\frac{\sigma'_n}{\lambda + 2\mu}\right) - 1\right] \tag{8}$$

Here, it is to be noted that the elastic constitutive model for the rock fracture leads to an exponential relationship between the fracture closure and the applied normal stress, which has been widely revealed in the literature, e.g., in Min et al. (2004).

On the other hand, the shear expansion caused by $\mathrm{d}\delta$ can be estimated from shear dilation angle d_m:

$$\mathrm{d}u_2 = \tan d_m \, \mathrm{d}\delta \tag{9}$$

By introducing two parameters, s and φ, pertinent to normal stress σ_n, we represent the dilation angle d_m under normal stress σ_n in the form of Barton's strength criterion for joints (Barton, 1976) ($\tau = \sigma_n \tan(2d_m + \varphi_b)$, where φ_b is the basic frictional angle of joints):

$$\tan d_m = \frac{1}{2}\left[\arctan\left(\frac{\tau}{s}\right) - \varphi\right] \tag{10}$$

Obviously, s is a normal stress-like parameter, and φ is a frictional angle-like parameter. But to make the above formulation still valid into pre-dilation state (i.e., shear contraction state), s and φ differ from their initial implications. Later, we will show how they can be back calculated from shear experimental data.

Substituting Eqs. (9) and (10) into (7) yields:

$$\frac{\mathrm{d}u_2}{b_0 + u_1 + u_2} = \frac{1}{2\mu}\left[\arctan\left(\frac{\tau}{s}\right) - \varphi\right]\mathrm{d}\tau \tag{11}$$

By integrating Eq. (11), we have:

$$u_2 = (b_0 + u_1)\left\{\exp\left[\frac{|\tau|}{2\mu}\left(\arctan\frac{|\tau|}{s} - \varphi\right) - \frac{s}{4\mu}\ln\left(1 + \frac{\tau^2}{s^2}\right)\right] - 1\right\} \tag{12}$$

By solving the simultaneous equations, Eqs. (8) and (12), we have:

$$\begin{cases} u_1 = \dfrac{A(1+B)}{1-AB} b_0 \\ u_2 = \dfrac{B(1+A)}{1-AB} b_0 \end{cases} \tag{13}$$

where

$$A = exp\left(\frac{\sigma_n'}{\lambda + 2\mu}\right) - 1 \tag{14}$$

$$B = exp\left[\frac{|\tau|}{2\mu}\left(arctan\frac{|\tau|}{s} - \varphi\right) - \frac{s}{4\mu}ln\left(1 + \frac{\tau^2}{s^2}\right)\right] - 1 \tag{15}$$

Thus, the total normal deformation under normal and shear loading can be obtained,

$$u = u_1 + u_2 = \frac{A + B + 2AB}{1 - AB} b_0 \tag{16}$$

The actual aperture of the fracture, $b = b_0 + u$, is given by:

$$b = b_0 + u = (1 + \chi)b_0 \tag{17}$$

where

$$\chi = \frac{A + B + 2AB}{1 - AB} \tag{18}$$

2.3 Stress-dependent hydraulic conductivity for rock fractures

Since natural fractures have rough walls and asperity areas, it is not appropriate to directly use the aperture derived by Eq. (17) for describing the hydraulic conductivity of the fractures. Instead, an equivalent hydraulic aperture is usually taken to represent the percolation property of the fractures, as demonstrated in Section 1. Based on experimental data, the relationship between the equivalent hydraulic aperture and the mechanical aperture has been widely examined in the literature, and the empirical relations proposed by Lomize (1951), Louis (1971), Patir & Cheng (1978), Barton el al. (1985) and Olsson & Barton (2001) are listed in Table 1. For example, if Patir and Cheng's model is used to estimate the equivalent hydraulic aperture that accounts for the flow-deformation coupling effect in pre-peak shearing stage, then there is

$$b^* = (1 + \chi)b_0[1 - 0.9\exp(-0.56 / C_v)]^{1/3} \tag{19}$$

where C_v is the variation coefficient of the mechanical aperture of the discontinuities, which is mathematically defined as the ratio of the root mean squared deviation to the arithmetic mean of the aperture. For convenience, Eq. (19) is rewritten as:

$$b^* = b_0 f(\beta) \tag{20}$$

Obviously, $f(\beta)$ is a function of the normal and shear loadings, the mechanical characteristics and the aperture statistics of the fractures.

Thus, the hydraulic conductivity of the fractures subjected to normal and shear loadings is approximated by the hydraulic conductivity of the laminar flow through a pair of smooth parallel plates with infinite dimensions:

$$k = \frac{gb^{*2}}{12\nu} \tag{21}$$

where k is the hydraulic conductivity, g is the gravitational acceleration, and ν is the kinematic viscosity of the fluid.

An alternative approach to account for the deviation of the real fractures from the ideal conditions assumed in the parallel smooth plate theory is to adopt a dimensionless constant, ς, to replace the constant multiplier, 1/12, in Eq. (21), where $0<\varsigma\leq 1/12$ (Oda, 1986). In this manner, the hydraulic conductivity of the fractures is estimated by

$$k = \varsigma \frac{gb^2}{\nu} \tag{22}$$

Clearly, the constant, ς, approaches 1/12 with increasing scale and decreasing roughness of the fractures.

Eqs. (21) and (22) show that the hydraulic conductivity of a rock fracture varies quadratically with its mechanical aperture. The latter depends, by Eq. (18), on the normal and shear stresses applied on the fracture. Hence, we call the established model, Eq. (21) or (22), the stress-dependent hydraulic conductivity model, and it is suitable to describe the hydraulic behaviour of the fractures subjected to mechanical loading in the pre-peak stage.

Authors	Expressions	Descriptions
Lomize (1951)	$b^* = b\left[1.0 + 6.0(e/b)^{1.5}\right]^{-1/3}$	b^* is the equivalent hydraulic aperture of fractures, b the mechanical aperture, e the absolute asperity height, e_m the average asperity height, D_H the hydraulic radius, C_v the variation coefficient of the mechanical aperture, JRC the joint roughness coefficient, JRC_0 the initial value of JRC, JRC_{mob} the mobilized JRC, δ the shear displacement and δ_p the peak shear displacement.
Louis (1971)	$b^* = b\left[1.0 + 8.8(e_m/D_H)^{1.5}\right]^{-1/3}$	
Patir & Cheng (1978)	$b^* = b\left[1 - 0.9\exp(-0.56/C_v)\right]^{1/3}$	
Barton, et al. (1985)	$b^* = b^2 JRC^{-2.5}$	
Olsson & Barton (2001)	$\begin{cases} b^* = b^2 JRC_0^{-2.5} & \delta \leq 0.75\delta_p \\ b^* = b^{1/2} JRC_{mob} & \delta \geq \delta_p \end{cases}$	

Table 1. Empirical relations between equivalent hydraulic aperture and mechanical aperture

2.4 Validation of the elastic constitutive model

The key point of the stress-dependent hydraulic conductivity model is whether the established elastic constitutive model can properly describe the variation of mechanical aperture under normal and shear loadings at low stress level. Here, we use the results of the laboratory test performed by Liu et al. (2002) to validate the mechanical model. The test was conducted to study shear-flow coupling properties for a marble fracture with fillings of sand under low normal stresses and small shear displacements.

The marble specimen for shear-flow coupling test is illustrated in Fig. 4, which was collected from the Daye Iron Mine in China. The uniaxial compressive strength and density of the rock sample are 52.4 MPa and 2.66×10^3 kg/m^3, respectively. The specimen was cut into round shape and the fracture surfaces were polished, with its size of 290 mm in diameter and 200 mm in height. The opposite walls of the fracture were cemented with a layer of filtered sands with their diameters ranged from 0.5 to 0.69 mm, and the fracture was further filled with the same sands. The initial aperture of the fracture, b_0, is about 1.31 mm.

The coupled shear-flow test were conducted by first applying a prescribed normal stress ranging between 0.1 and 0.5 MPa and then applying shear displacement in steps until a maximum displacement of about 0.4 mm was reached. During tests, steady-state fluid flow rate and normal displacement were continuously recorded.

With such low normal stresses and small shear displacements, it is reasonable to consider that the fracture behaves elastic during the coupled shear-flow test. According to the experimental results, the elastic parameters, λ and μ, of the fracture with fillings are estimated as λ=1.81 MPa and μ=3.62 MPa. In order to enable Eq. (16) to predict the mechanical aperture of the facture under normal and shear loads, the normal stress-like parameter, s, and the frictional angle-like parameter, φ, should be further determined. Fortunately, both of them can be derived by fitting the experimental curve between normal displacement and shear displacement, as plotted in Fig. 5, using Eq. (16) such that the least square error is minimized. With this approach, we obtain that for σ_n=0.1 MPa, s=0.062, φ=1.324, and for σ_n=0.4 MPa, s=0.046, φ=1.310.

Fig. 5 plots the experimental results as well as the model predictions of the relation between mechanical aperture and shear displacement of the fracture under constant normal stresses. Generally, the proposed elastic constitutive model manifests the behaviour of the fracture with fillings during the shear-flow coupling test with low normal and shear loads. Shear contraction is observed in the initial 0.06-0.08 mm of shear displacement, which is followed by shear dilation in the remaining of the shear displacement. This property, which is actually ensured by the empirical relation assumed in Eq. (9), suggests that the resultant model is suitable for phenomenologically describing the pre-peak shear-flow coupling effect of fractures.

Fig. 6 further depicts the sensitivity of s and φ on the behaviour of the fracture. In Fig. 6(a), φ is fixed to 1.324, while s varies from 0.02 to 0.08. As s increases, shear contraction more apparently manifests, and the mechanical aperture versus shear displacement curves become lower as a whole. On the other hand, the effect of varying φ from 0.524 to 1.222 but fixing s to 0.062 is plotted in Fig. 6(b). For small value of φ, shear contraction is trivial and the curve extends with a larger slope. As φ increases, however, shear contraction becomes relatively remarkable and the curve turns relatively flat. Thus, by adjusting s and φ, the mechanical and hydraulic behaviours of the fracture can be appropriately established.

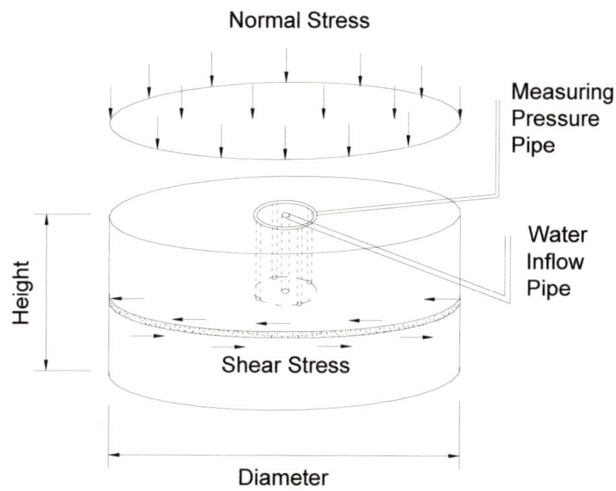

Fig. 4. Sketch of the marble specimen for shear-flow coupling test

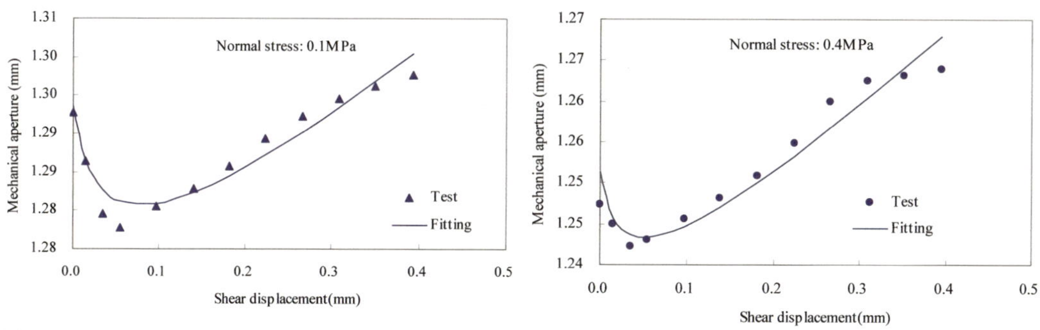

(a) (b)

Fig. 5. Mechanical aperture versus shear displacement curve under constant normal stress: (a) Normal stress: 0.1 MPa and (b) Normal stress: 0.4 MPa.

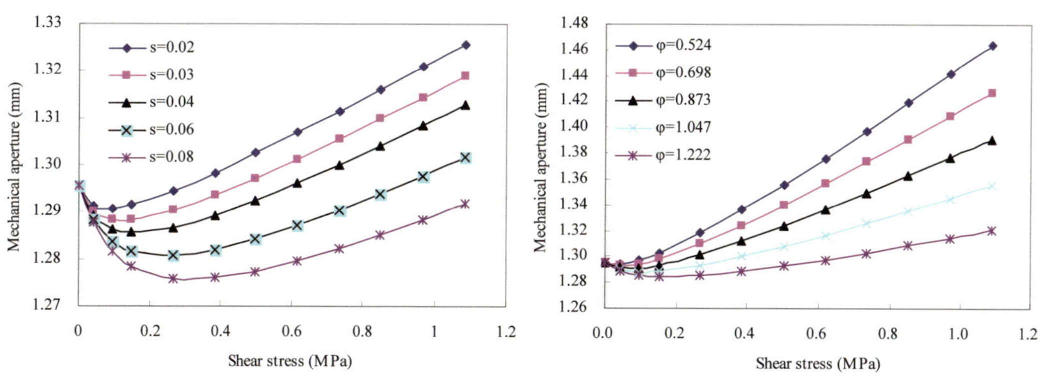

(a) (b)

Fig. 6. The sensitivity of s and φ on the behavior of the fracture: (a) φ=1.324 and (b) s=0.062

3. Strain-dependent hydraulic conductivity of rock fractures

In this section, we develop an elasto-plastic constitutive model for single hard rock fractures with consideration of nonlinear normal deformation and post-peak shear dilatancy, and then formulate the strain-dependent hydraulic conductivity for the fractures under dilated shear loading. Compared with the stress-dependent model presented in Section 2, one major difference is that the strain-dependent model is capable of describing the influence of post-peak mechanical response on the hydraulic properties of the fractures. This work is of paramount importance in the sense that the theoretical results are directly comparable with the experimental data of coupled shear-flow test, e.g. in Esaki et al. (1999). The strain-dependent hydraulic conductivity tensor can then be developed on this basis, which will be presented later in Section 5.

3.1 An elasto-plastic constitutive model for rock fractures

The underlying physical model considered is the same with the model plotted in Fig. 3, in which a fracture of hard rock is located at the mid-height of a specimen between two intact rock blocks. The height of the specimen is denoted by s, and the initial aperture of the fracture is b_0. When constant normal stress σ_n and increasing shear displacement δ are applied on the specimen, typical and idealized curves of shear displacement versus shear stress and shear displacement versus normal displacement (i.e. $\delta\sim\tau$ curve and $\delta\sim u$ curve) are plotted in Fig. 7. The shear stress increases linearly with the shear displacement (linked by the initial shear stiffness of the fracture, k_{s0}) until the shear stress approaches the peak, τ_p, which is then followed by a shear softening process in which the shear stress decreases to a residual level at a decreasing gradient with increasing shear displacement. For the purpose of deriving the hydraulic property of the fracture in post-peak loading section, however, an elastic-perfectly plastic $\delta\sim\tau$ relationship can be assumed, as shown in Fig. 7(a).

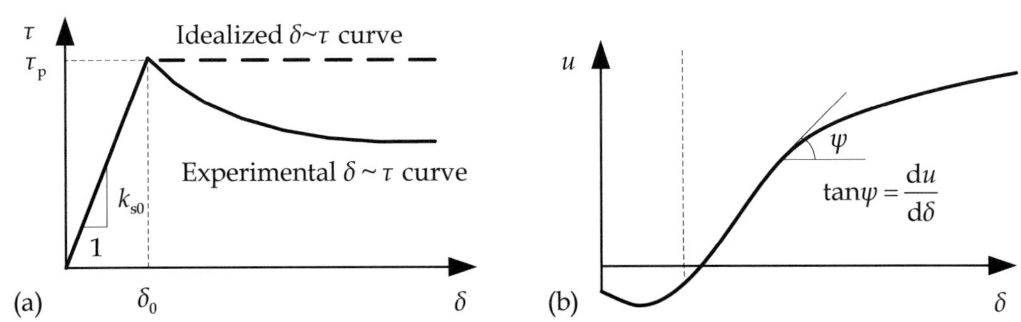

Fig. 7. Typical and idealized curves of shear displacement versus shear stress and shear displacement versus normal displacement of a fracture subjected to normal and shear loads

The deformation response of a rock fracture subjected to normal and shear loadings includes two components: one is the nonlinear closure of the fracture due to normal compression, and the other is the opening of the fracture due to shear dilation. Experimental results in Esaki et al. (1999) show that in the shearing process under constant normal loading, dilatancy will start when the shear stress approaches the peak and it increases at a decreasing gradient with increasing shear displacement, as illustrated in Fig. 7(b). As a result, the aperture of the fracture and then the hydraulic conductivity vary with increasing shear displacement.

Therefore, we may consider that shear dilatancy as well as the change in hydraulic conductivity accompanies normal and plastic shear deformations of the fracture. To deduce the hydraulic conductivity of the fracture with an averaging method, which will be further used later for deriving the hydraulic conductivity tensor for fractured rocks, we view the specimen with fracture as an equivalent continuous medium, i.e. the hydromechanical properties of the fracture are averaged into the whole specimen. As can be seen later, such a treatment does not affect our final solution to a single fracture, but it renders valid the small strain assumption on the fractures in the presence of post-sliding plasticity.

For a one-dimensional problem with a single rock fracture, the elasto-plastic constitutive model can be represented in the following forms:

$$\gamma_p = \gamma - \gamma_e = \frac{\delta}{s} - \frac{\delta_0}{s} = \frac{\delta}{s} - \frac{\tau_p}{sk_{s0}} \tag{23}$$

$$\varepsilon_n = \frac{\sigma'_n}{sk_n} + \int \tan\psi \, d\gamma_p \tag{24}$$

where γ, γ_e and γ_p are the total shear strain, the elastic shear strain and the plastic shear strain of the fracture, respectively; ε_n is the normal strain of the fracture; τ_p is the peak shear stress of the fracture under effective normal stress σ'_n; k_n and k_{s0} are, respectively, the normal stiffness and the initial shear stiffness of the fracture; δ_0 is the maximum elastic shear displacement upon shear yielding, with $\delta_0 = \tau_p/k_{s0}$, as shown in Fig. 7(a); and ψ is the mobilized dilatancy angle of the fracture. Note that in Eq. (24), the first term on the right hand side denotes the nonlinear closure of the fracture subjected to effective normal stress σ'_n, while the second term denotes the opening of the fracture due to shear dilatancy.

Existing studies have indicated that shear dilatancy is highly dependent on the plasticity already experienced by the fractures and normal stress, and non-negligibly dependent on scale (Barton & Bandis, 1982; Yuan & Harrison, 2004; Alejano & Alonso, 2005). The decaying process of the dilatancy angle in line with plasticity can be described by the following negative exponential expression through the plastic shear strain, γ_p, or indirectly through the plastic shear displacement, δ, on the basis of Eq. (23):

$$\psi = \psi_{peak} \exp\left[-r(\delta - \delta_0)\right] \tag{25}$$

where r is a parameter for modelling the rate of decay that ψ undergoes as the plastic shear strain evolves. If r=0, then a constant dilatancy angle is recovered. As $r \to \infty$, the dilatancy angle quickly decays to zero. ψ_{peak} is the peak dilatancy angle of the fracture in the form of (Barton & Bandis, 1982)

$$\psi_{peak} = JRC \cdot \log_{10} \frac{JCS}{-\sigma'_n} \tag{26}$$

where *JRC* and *JCS* are the roughness coefficient and the wall compressive strength of fractures, respectively, and the actual values of them should be scale-corrected (Barton & Bandis, 1982). Thus, the dependencies of fracture dilatancy on plasticity, normal stress and scale are established through Eqs. (25) and (26).

Note that Eq. (25) shares the same shape with the asperity angle proposed for the description of shear dilatancy and surface degradation (Plesha, 1987), but the latter is represented as a function of the plastic tangential work. With the assumption of elastic-perfectly plasticity, they are fully equivalent for monotonic loading (Jing et al., 1993). Cyclic loading is not a concern in this simple model, but when cyclic loading is involved, another independent function can be associated to the reverse loading that starts from the original point, just as the suggestion given in Plesha (1987) for asperity angles in two opposite directions, in order to satisfy the thermodynamic restriction condition presented in Jing et al. (1993).
Using the Mohr-Coulomb criteria, the peak shear stress τ_p of the fracture under effective normal stress σ'_n satisfies

$$\tau_p = -\sigma'_n \tan\varphi + c \tag{27}$$

where φ and c are the frictional angle and the cohesion of the fracture.
Differentiating Eq. (23) yields

$$d\gamma_p = d\gamma = \frac{1}{s} d\delta \tag{28}$$

Combining Eqs. (24) and (28) results in

$$\Delta b \approx s\varepsilon_n = \frac{\sigma'_n}{k_n} + \int_{\delta_0}^{\delta} \tan\psi(\delta) d\delta \tag{29}$$

An interesting phenomenon in Eq. (29) is, as described before, the change in the aperture of the fracture, Δb, is irrelevant to the height of the specimen, s. To conveniently use this formulation, two remedies can be further made:
First, suppose that the hyperbolic variation of k_n with the increase of aperture can be considered in the following (Huang et al., 2002):

$$k_n = \frac{-\sigma'_n + b_0 k_{n0}}{b_0} \tag{30}$$

where k_{n0} is the initial normal stiffness of the fracture.
Second, by employing the Taylor series expansion (truncated at the third order term), $\tan\psi$ can be adequately approximated by $\psi + \psi^3/3$ in radians for a rather large ψ_{peak}, e.g. 30°.
From Eq. (29) and the above two remedies, we have

$$\Delta b = \chi b_0 \tag{31}$$

$$b = b_0 + \Delta b = (1+\chi) b_0 \tag{32}$$

with the parameter, χ, in the following form

$$\chi = \frac{\sigma'_n}{-\sigma'_n + b_0 k_{n0}} + \frac{1}{b_0} \left\{ \frac{\psi_{peak}}{r} \left[1 - e^{-r(\delta-\delta_0)} \right] + \frac{\psi_{peak}^3}{9r} \left[1 - e^{-3r(\delta-\delta_0)} \right] \right\} \tag{33}$$

3.2 Strain-dependent hydraulic conductivity for rock fractures

Rewrite from Eq. (22) the initial hydraulic conductivity of the fracture, k_0, in the following form:

$$k_0 = \varsigma \frac{g b_0^2}{\nu} \tag{34}$$

Then, the hydraulic conductivity of the fracture under effective normal stress σ'_n and shear displacement δ can be described by

$$k = \varsigma \frac{g b^2}{\nu} = k_0 (1 + \chi)^2 \tag{35}$$

Hence, a theoretical model of the hydraulic conductivity for a single rock fracture is finally formulated, which is totally determined by the effective normal stress σ'_n and the shear displacement δ, as well as a set of parameters characterizing the behaviour of the fracture (i.e. b_0, ς, k_{n0}, k_{s0}, φ, c, JRC, JCS and r, which all can be deduced or back-calculated from experimental data).

Note that by Eqs. (35) and (33), the proposed hydraulic conductivity model for rock fractures subjected to normal and shear loadings with mobilized dilatancy behaviour depends in form on the plastic shear displacement, but from Eq. (23), one observes that the model depends indirectly on the plastic shear strain. Thus, we classify the established model into the stain-dependent hydraulic conductivity model.

3.3 Validation of the proposed model

Esaki et al. (1999) systematically investigated the coupled effect of shear deformation and dilatancy on hydraulic conductivity of rock fractures by developing a new laboratory technique for coupled shear-flow tests of rock fractures. In this section, we validate the theory proposed in Section 3.2 using the experimental data reported in Esaki et al. (1999). For this purpose, we first briefly introduce the experiments, and then predict our analytical results through Eqs. (31) and (35) by directly comparing with the experimental data.

3.3.1 The coupled shear-flow tests

The coupled shear-flow tests were conducted with an artificially created granite fracture sample under various constant normal loads and up to a residual shear displacement of 20 mm (Esaki et al., 1999). The underlying specimen for coupled shear-flow tests is sketched in Fig. 3, with its size of 120 mm in length, 100 mm in width and 80 mm in height. The initial aperture of the created fracture, b_0, is about 0.15 mm. The value of JRC is 9, and the value of JCS is 162 MPa, respectively.

The coupled shear-flow tests were conducted by first applying a prescribed normal stress ranging between 1 MPa and 20 MPa and then applying shear displacement in steps at a rate of 0.1 mm/s until a maximum shear displacement of 20 mm was reached. During tests, steady-state fluid flow rate, shear loading and dilatancy were all continuously recorded. The hydraulic aperture and conductivity were back-calculated by applying the cubic law, with the flow equations solved by using a finite difference method.

3.3.2 Determination of the parameters for the proposed model

Some of the experimental values of the mechanical parameters of the fracture specimen during the coupled shear-flow tests are listed in Table 2 (taken from Table 1 in Esaki et al. (1999)). Using the data as listed in Table 2, we plot the peak shear stress versus normal stress curve in Fig. 8, which can be fitted by a linear equation τ_p=1.058σ_n+0.993 with a high correlation coefficient of 0.9999. Therefore, the shear strength of the specimen can be derived as φ=46.6° and c=0.99 MPa, respectively.

σ_n (MPa)	τ_p (MPa)	k_{s0} (MPa/mm)
1	2.06	3.37
5	6.16	10.65
10	11.74	11.97
20	22.10	17.97

Table 2. Mechanical parameters of the artificial fracture (After Esaki et al. (1999))

The initial normal stiffness of the fracture of the specimen, k_{n0}, has to be estimated from the recorded initial normal displacement with zero shear displacement under different normal stresses. From the data plotted in Fig. 9 (which is taken from Fig. 7b in Esaki et al. (1999)), k_{n0} can be estimated as k_{n0}=100 MPa/mm by considering the possible deformation of the intact rock under high normal stresses. It is to be noted that in the remainder of this section, the hard intact rock deformation of the small specimen is neglected, meaning that the normal displacement of the specimen mainly occurs in the fracture of the specimen and it is approximately equal to the increment of the mechanical aperture of the fracture.

Theoretically, the decay coefficient of the fracture dilatancy angle, r, can be directly measured from the normal displacement versus shear displacement curves as plotted in Fig. 9. A better alternative, however, is to fit the experimental curves using Eq. (31) such that the least square error is minimized. By this approach, we obtain that r=0.13 with a correlation coefficient of 0.9538.

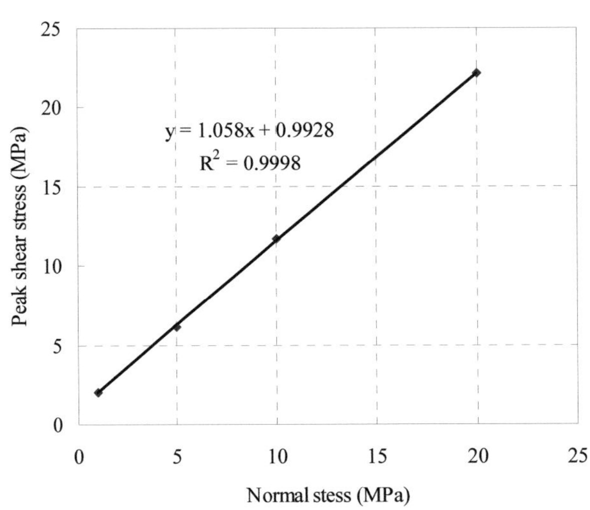

Fig. 8. Peak shear stress versus normal stress curve of the fracture.

To obtain the dimensionless constant, ς, in Eq. (35) that relates the mechanical aperture to the hydraulic conductivity of the fracture under testing, further efforts are needed. A simple approach is to back-calculate ς directly using Eq. (34) with initial hydraulic conductivity, k_0. But similarly, the better alternative is to fit the hydraulic conductivity versus shear displacement curves, as plotted in Fig. 11 (which is taken from Fig. 7c-f in Esaki et al. (1999)), using Eq. (35) such that the least square error is minimized. With such a method, we obtain that ς=0.00875. This means that the mechanical aperture, b, and the hydraulic aperture, b^*, are linked with $b^*=0.324b$, which is very close to the experimental result shown in Fig. 8 in Esaki et al. (1999).

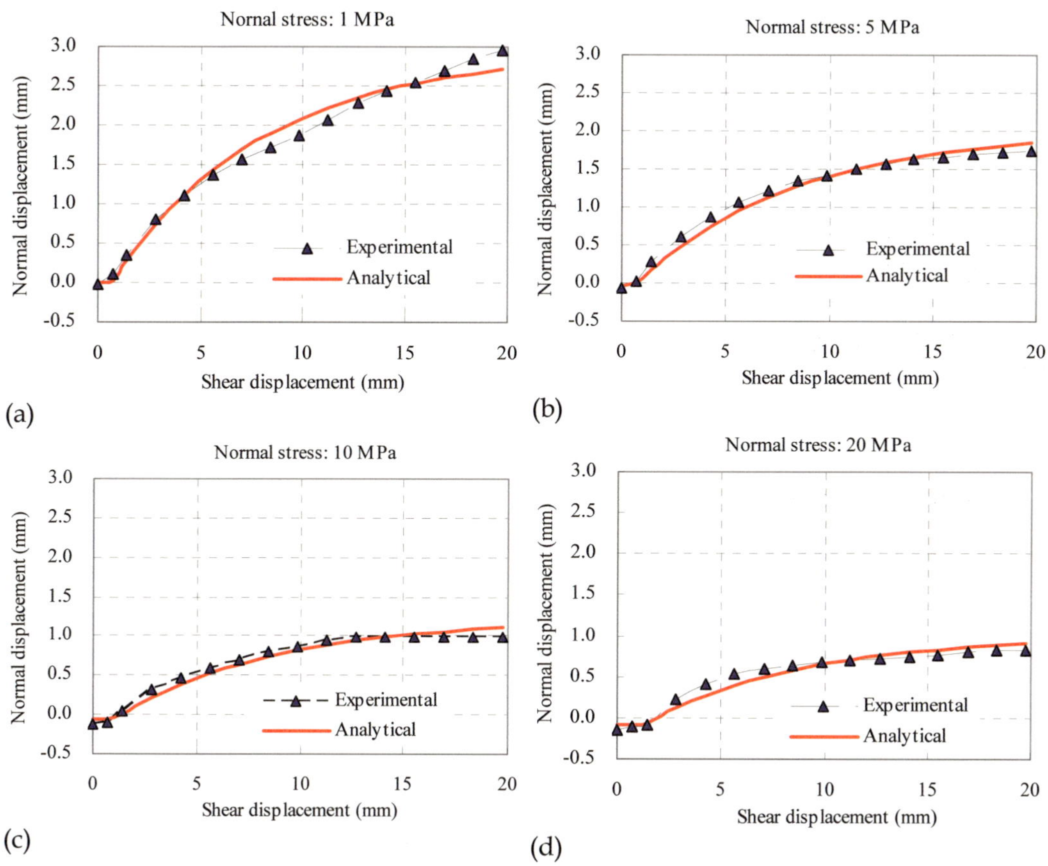

Fig. 9. Comparison of the fracture aperture analytically predicted by Eq. (31) with that measured in coupled shear-flow tests.

3.3.3 Validation of the proposed theory

With the necessary parameters obtained in Section 3.3.2, we are now ready to compare the proposed model in Eqs. (31) and (35) with the experimental data presented in Esaki et al. (1999). Note that although the experimental data are available for one cycle of forward and reverse shearing, only the results for the forward shearing part are considered. The reverse shearing process, however, can be similarly modelled.

Fig. 9 depicts the relations between the mechanical aperture and shear displacement that were measured from the coupled shear-flow tests presented in Esaki et al. (1999) and predicted by using the proposed model given in Eq. (31) under different normal stresses applied during the testing. It can be observed from Fig. 9 that our proposed analytical model is able to describe the shear dilatancy behaviour of a real fracture under wide range of normal stresses between 1 MPa and 20 MPa by feeding appropriate parameters. Even the fracture aperture increases by one order of magnitude due to shear dilation, the analytical model still fitted the experimental results well. For practical uses, the slight discrepancies between the analytical results and the experimental data are negligible and the proposed model is accurate enough to characterize the significant dilatancy behaviour of a real fracture.

This performance is largely attributed to the dilatancy model introduced through Eqs. (25) and (26). The dilatancy angles of the fracture evolving with the plastic shear displacement under different normal stresses are illustrated in Fig. 10. The high dependencies of the dilatancy angle of the fracture on normal stress and plasticity are clearly demonstrated in the curves. The peak dilatancy angle, which can be rather accurately modelled by Barton's peak dilatancy relation (Barton & Bandis, 1982), decreases logarithmically with the increase of the applied normal stress. For normal stresses of 1 MPa, 5 MPa, 10 MPa and 20 MPa, the peak dilatancy angles are 19.9°, 13.6°, 10.9° and 8.2°, respectively. On the other hand, the dilatancy angle undergoes negative exponential decay with increasing plastic shear displacement, a process related to surface degradation of rough fractures.

Fig. 11 shows the hydraulic conductivity versus shear displacement relations that were back-calculated from fluid flow results using the finite difference method from the coupled shear-flow tests presented in Esaki et al. (1999) and that are predicted by the proposed model given in Eq. (35) under different normal stresses during testing. As shown in the semi-logarithmic graphs in Fig. 11, the proposed analytical model can well predict the evolution of hydraulic conductivity of the tested rock fracture, with the change in the magnitude of 2 orders, during coupled shear-flow tests under different normal stresses. The ratios of the predicted hydraulic conductivities to the corresponding experimental results all fall in between 0.3 and 3.0, indicating that they are rather close in orders of magnitude and the predicted results are suitable for practical use.

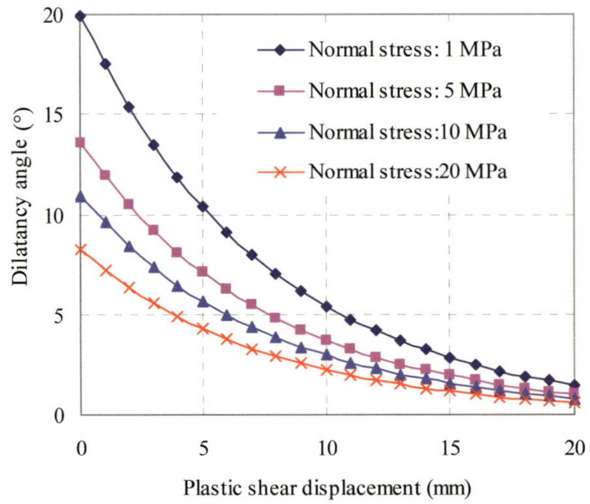

Fig. 10. Dilatancy angles of the fracture evolving with the plastic shear displacement under different normal stresses.

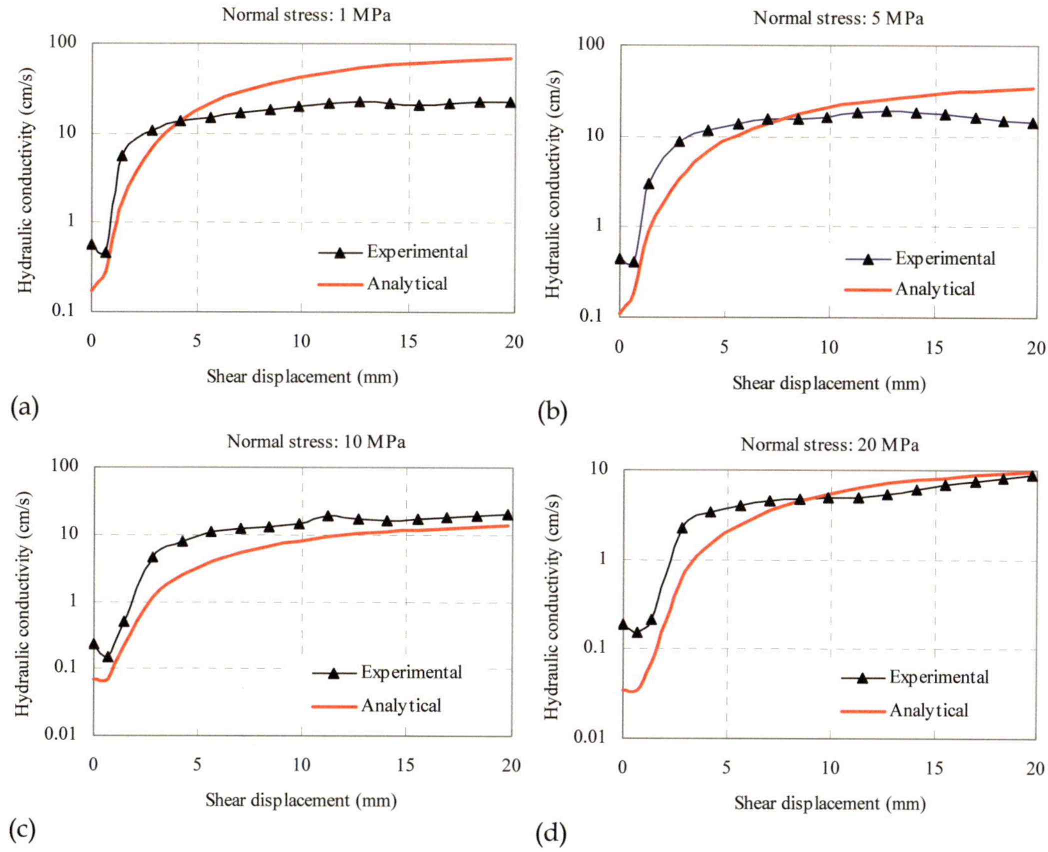

Fig. 11. Comparison of the hydraulic conductivity analytically predicted by Eq. (35) with that calculated from coupled shear-flow tests with finite difference method.

4. Stress-dependent hydraulic conductivity tensor of fractured rocks

When the response of each fracture under normal and shear loading is understood (see Section 2), the remaining problem is how to formulate the hydraulic conductivity for fractured rock mass based on the geometry of the underlying fracture network. Fig. 12 depicts a two-dimensional fracture network (taken after Min et al. (2004)) in a biaxial stress field. As shown in Fig. 12, each fracture plays a role in the hydraulic conductivity of the rock mass, and its contribution primarily depends on its stress state, its occurrence, as well as its connectivity with other fractures. Also shown in Fig. 12 is the scale effect of the rock mass on hydraulic properties. When the size of the rock mass is small, only a few number of fractures are included and heterogeneity of the hydraulic conductivity of the rock mass may dominate. As the population of factures grows with the increasing size, an upscaling scheme may be available to derive a representative hydraulic conductivity tensor for the rock mass at the macroscopic scale.

Based on the above observations, in this section, we formulate an equivalent hydraulic conductivity tensor for fractured rock mass based on the superposition principle of liquid dissipation energy, in which the concept of *REV* is integrated and the applicability of an equivalent continuum approach is able to be validated.

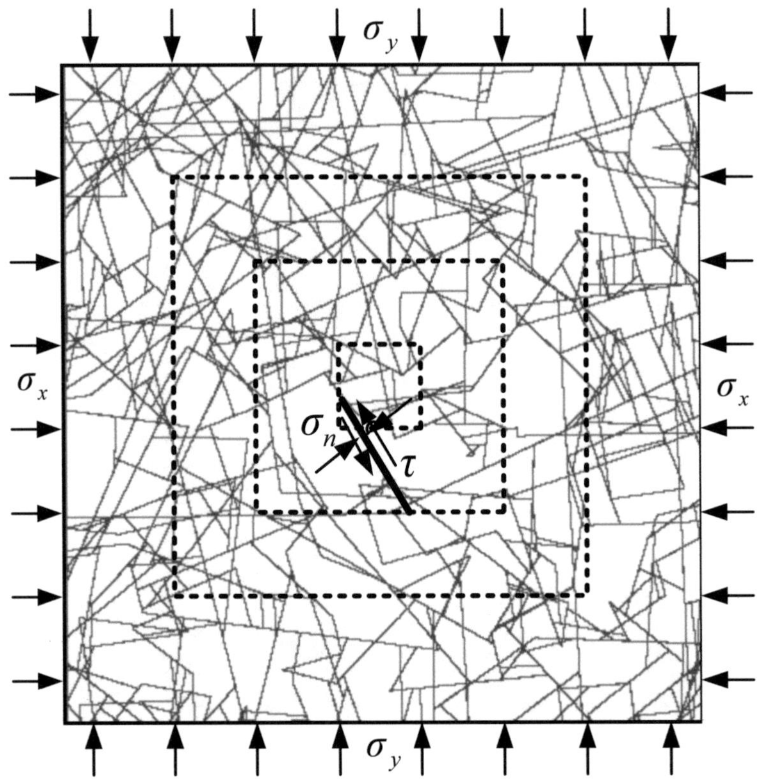

Fig. 12. A fracture network (taken after Min et al. (2004)) in biaxial stress field and the scale effect of the rock mass

4.1 Computational model

Without loss of generality, the global coordinate system $X_1X_2X_3$ is established in such a way that its X_1-axis points towards the East, X_2-axis toward the North and X_3-axis vertically upward. A local coordinate system $x_1^f x_2^f x_3^f$ is associated with the *f*th set of fractures such that the x_1^f -axis is along the main dip direction, the x_2^f -axis is in the strike, and the x_3^f -axis is normal to the fractures, as shown in Fig. 13.

In order to formulate the stress-dependent hydraulic conductivity tensor for fractured rock masses using the aforementioned elastic constitutive model for rock fractures, the following assumptions, similar to Oda (1986), are made in this section:

1. A cube of volume, V_p, is considered as the flow region of interest, which is cut by n sets of fractures. The orientation of each set of fractures is indicated by a mean azimuth angle β and a mean dip angle α. Other geometrical statistics of the fractures are assumed to be available through field measurements or empirical estimations.
2. Even though the geometry of real fractures is complex, generally it can be simplified as a thin interfacial layer with radius r and aperture b^*.
3. The rock mass is regarded as an equivalent continuum medium, which means the representative elementary volume (*REV*) exists in the rock mass and its size is smaller than or equal to V_p.

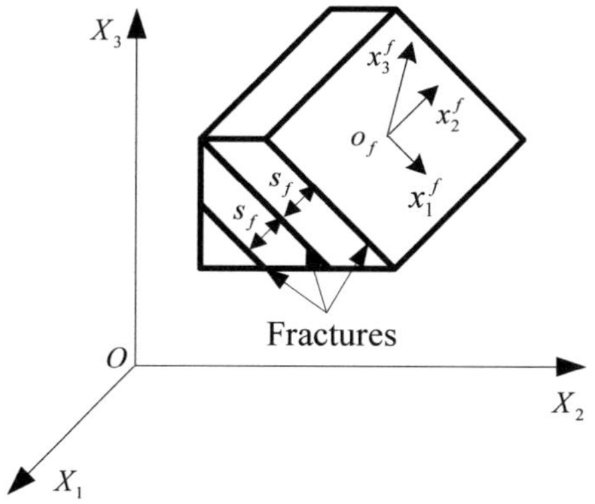

Fig. 13. Coordinate systems

4.2 Stress-dependent hydraulic conductivity tensor

Fluid flow through the equivalent continuum media can be described by the generalized 3-D Darcy's law as follows:

$$v = \mathbf{KJ} \tag{36}$$

where v denotes the vector of flow velocities, $\mathbf{J}$ denotes the vector of hydraulic gradients, and $\mathbf{K}$ is the hydraulic conductivity tensor for the rock mass.

For steady-state seepage flow, the dissipation energy density, $e(X_1, X_2, X_3)$, of fluid flow through the media can be represented as (Indelman & Dagan, 1993):

$$e = \frac{1}{2}\mathbf{J}^{\mathrm{T}}\mathbf{KJ} \tag{37}$$

Hence, the total flow dissipation energy, E, in the rock mass V_{p} can be calculated by performing an integration throughout the whole flow domain:

$$E = \int_{V_{\mathrm{p}}} e\mathrm{d}\Omega = \frac{1}{2}\int_{V_{\mathrm{p}}} \mathbf{J}^{\mathrm{T}}\mathbf{KJ}\mathrm{d}\Omega \tag{38}$$

If REV does exist in the rock mass and its size is smaller than or equal to V_{p}, by defining $\overline{\mathbf{J}}$ to be the vector of the average hydraulic gradient within V_{p} and $\overline{\mathbf{K}}$ to be the average hydraulic conductivity tensor, Eq. (38) can be reduced to:

$$E = \frac{1}{2}\overline{\mathbf{J}}^{\mathrm{T}}\overline{\mathbf{K}}\overline{\mathbf{J}}V_{\mathrm{p}} \tag{39}$$

Suppose that the volume density of the ith set of fractures is J_{vi}. The number of this set of fractures can be estimated by $m_i = J_{vi} V_{\mathrm{p}}$.

For permeable rock matrix, the flow dissipation energy shown in Eq. (39) consists of two components, i.e., the flow dissipation energy through rock matrix, E_{r}, and the flow dissipation energy through crack network, E_{c}:

$$E = E_{\mathrm{r}} + E_{\mathrm{c}} \tag{40}$$

E_r can be represented as:

$$E_r = \frac{1}{2}\bar{\mathbf{J}}^T \bar{\mathbf{K}}_r \bar{\mathbf{J}} V_p \tag{41}$$

where $\bar{\mathbf{K}}_r$ denotes the hydraulic conductivity tensor for rock matrix. If rock matrix is impermeable, all elements in $\bar{\mathbf{K}}_r$ vanish.

To estimate E_c, we introduce a weight coefficient W_{ij} to describe the effect of the connectivity of the fracture network on fluid flow:

$$W_{ij} = \xi_{ij} / \bar{\xi}_i \tag{42}$$

where ξ_{ij} is a stochastic variable denoting the number of fractures intersected by the jth fracture belonging to the ith set; and $\bar{\xi}_i$ denotes the maximum number of fractures cut by the ith set of fractures. Obviously, $0 \le W_{ij} \le 1$ and when $\xi_{ij} = 0$, $W_{ij} = 0$. This implies that an entirely isolated fracture which does not intersect any other fracture effectively contributes nothing to the hydraulic conductivity of the total rock mass.

For the jth fracture belonging to the ith set, a void volume equal to $\pi r_{ij}^2 b_{ij}^*$ is associated with it. Then, the flow dissipation energy through it is described as:

$$E_{cij} = W_{ij} e_{ij} \pi r_{ij}^2 b_{ij}^* \tag{43}$$

where e_{ij} is shown as follows:

$$e_{ij} = \frac{1}{2} k_{ij} \bar{\mathbf{J}}_{ci}^T \bar{\mathbf{J}}_{ci} \tag{44}$$

where k_{ij} denotes the hydraulic conductivity of the jth fracture of the ith set, which can be calculated by the stress-dependent hydraulic conductivity model, Eq. (21).

$\bar{\mathbf{J}}_{ci}$ denotes the hydraulic gradient within the ith set of fractures:

$$\bar{\mathbf{J}}_{ci} = (\boldsymbol{\delta} - \mathbf{n}_i \otimes \mathbf{n}_i)\bar{\mathbf{J}} \tag{45}$$

where $\boldsymbol{\delta}$ is the Kronecker delta tensor, and $\boldsymbol{n}_i$ denotes the unit vector normal to the ith set of fractures, with its components $n_1 = \sin\alpha\sin\beta$, $n_2 = \sin\alpha\cos\beta$, and $n_3 = \cos\alpha$.

Thus, E_c can be represented as

$$E_c = \frac{g\pi}{12\nu} \sum_{i=1}^{n} \sum_{j=1}^{m_i} W_{ij} r_{ij}^2 b_{ij}^{*3} \bar{\mathbf{J}}^T (\boldsymbol{\delta} - \mathbf{n}_i \otimes \mathbf{n}_i) \bar{\mathbf{J}} \tag{46}$$

From Eqs. (39)-(41), (46) and (20), it can be referred that

$$\bar{\mathbf{K}} = \bar{\mathbf{K}}_r + \frac{g\pi}{12\nu V_p} \sum_{i=1}^{n} \sum_{j=1}^{m_i} W_{ij} f^3(\beta_{ij}) r_{ij}^2 b_{0ij}^3 (\boldsymbol{\delta} - \mathbf{n}_i \otimes \mathbf{n}_i) \tag{47}$$

In Eq. (47), $\boldsymbol{n}$ is determined by the orientation of the fractures, which reflects the effect of the orientation of the fractures on the fluid flow. r and b_0 represent the size or the scale of the fractures; they retrain the fluid flow through the fractures from their developing magnitude. W is a parameter introduced to show the impact of the connectivity of the fracture network on fluid flow. Finally, $f(\beta)$ is a function used to demonstrate the coupling effect between fluid flow and stress state.

The hydraulic tensor for fractured rock masses given in Eq. (47) is related to the volume of the flow region, V_p, which exactly shows the size effect of the hydraulic properties. Intuitively, the smaller the V_p size is, the less number of fractures is contained within the volume, and thus the poorer the representative of the computed hydraulic conductivity tensor. On the other hand, when V_p is increased up to a certain value, the fractures involved in the cubic volume are dense enough and the hydraulic conductivity tensor for the rock mass does not vary with the size of the volume. This V_p size is exactly the representative elementary volume, REV, of the flow region. The V_p size of the flow region is required to be larger than REV for estimating the hydraulic conductivity tensor for the fractured rock mass. Otherwise, treating the fractured rock mass as an equivalent continuum medium is not appropriate, and the discrete fracture flow approach is preferable.

4.3 Comparison with Snow's and Oda's models

Now we make a comparison between the formulation of the hydraulic conductivity tensor presented in Eq. (47) and the formulation given by Snow (1969) as well as the formulation given by Oda (1986). The Snow's formulation is as follows:

$$\mathbf{K} = \frac{g}{12\nu}\sum_{i=1}^{n}\frac{b_i^3}{s_i}(\boldsymbol{\delta} - \mathbf{n}_i \otimes \mathbf{n}_i) \tag{48}$$

where s_i is the average spacing of the ith set of fractures. If we neglect the hydraulic conductivity of the rock matrix and the connectivity of the factures, and define

$$b_i = \frac{1}{m_i}\sum_{j=1}^{m_i} f(\beta_{ij})b_{0ij} \quad \text{and} \quad s_i^{-1} = \frac{\pi}{V_p}\sum_{j=1}^{m_i} r_{ij}^2 \tag{49}$$

Then, the formulation presented in Eq. (47) is totally equivalent to Snow's formulation, Eq. (48).

On the other hand, the Oda's formulation is described by

$$\mathbf{K} = \varsigma(P_{kk}\boldsymbol{\delta} - \mathbf{P}) \tag{50}$$

where $\boldsymbol{P}$ is the fracture geometry tensor, with $P_{kk} = P_{11}+P_{22}+P_{33}$.

$$\mathbf{P} = \pi\rho\int_0^\infty\int_0^\infty\int_\Omega r^2 b^3 \mathbf{n}\otimes\mathbf{n}E(n,r,b)\mathrm{d}\Omega\mathrm{d}r\mathrm{d}b \tag{51}$$

where $E(n, r, b)$ is a probability density function of the geometry of the fractures, ρ is the number of fracture centers per unit of volume, with $\rho = m_v/V_p$, $m_v = \sum m_i$, and ς is the dimensionless scalar adopted to penalize the permeability of real fractures with roughness and asperities. Assuming that a statistically valid REV exists and being aware that the fracture orientation is a discrete event, the fracture geometry tensor may be empirically constructed by the following direct summation

$$\mathbf{P} = \frac{\pi}{V_{\mathrm{p}}} \sum_{i=1}^{m_{\mathrm{v}}} r_i^2 b_i^3 \mathbf{n}_i \otimes \mathbf{n}_i \tag{52}$$

Following a similar deduction, it can be inferred that all these three formulations are equivalent not only in form but also in function, though they are derived from different approaches and different assumptions. The formulation presented in Eq. (47) can be directly obtained from Snow's formulation by considering the connectivity and roughness of the fractures and integrating the aperture changes under engineering disturbance. The discretized form of the Oda's formulation is much closer to the current formulation, and the latter can also be directly achieved from the former by considering the connectivity of the fracture network. However, the proposed method for formulating an equivalent hydraulic conductivity tensor for complex rock mass based on the superposition principle of liquid dissipation energy is a widely applicable approach not only to equivalent continuum but also to discrete medium.

4.4 A numerical example: hydraulic conductivity of the rock mass in the Laxiwa Hydropower Project

In order to validate the theoretical model presented in Section 4.2, we investigated the hydraulic conductivity of a fractured rock mass at the construction site of the Laxiwa Hydropower Project, the second largest hydropower project on the upstream of the Yellow River. The selected construction site for a double curvature arch dam is a V-shaped valley formed by granite rocks, as shown in Fig. 14. The dam height is 250 m, the top elevation of the dam is 2460 m, the reservoir storage capacity is 1.06 billion m^3 and the total installed capacity is 4200 MW.

A typical section of the Laxiwa dam site is illustrated in Fig. 15. Besides faults, four sets of critically oriented fractures are developed in the rock mass at the construction site. The geological characteristics of the fractures are described by spacing, trace length, aperture, azimuth, dip angle, the joint roughness coefficient, *JRC*, of the fractures as well as the connectivity of the fracture network (i.e., the number of fractures intersected by one fracture). According to site investigation, the statistics (i.e., the averages and the mean squared deviations, as well as the distribution of the characteristics) of the fractured rock mass on the right bank of the valley are listed in Table 3.

Fig. 14. Site photograph of the Laxiwa valley

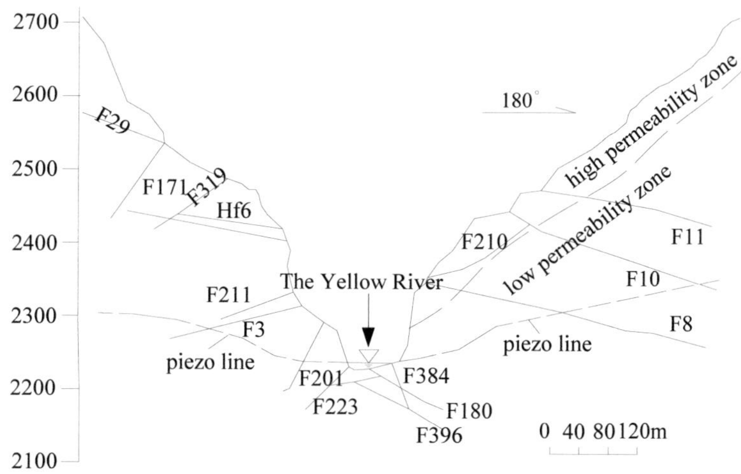

Fig. 15. A typical section of the Laxiwa dam site

Set	Spacing (m)	Length (m)		Aperture (mm)		Azimuth (°)		Dip (°)		Connectivity	
		avg.	dev.	avg.	dev.	avg.	dev.	avg.	dev.	avg.	dev.
1	1.45	5	1.5	0.096	0.02	85.3	10	54.5	10	5	3
2	2.62	3	1.0	0.096	0.02	355.1	20	29.8	5	3	2
3	10.96	3	1.0	0.096	0.02	287.4	20	61.4	10	3	2
4	10.96	3	1.0	0.096	0.02	320.2	20	11.9	5	3	2
Distribution	logarithmic normal	negative exponential		Gama		normal		normal		normal	

*'avg.' denotes arithmetic mean of a variable,
'dev.' represents root mean squared deviation

Table 3. Characteristic variables of the fractured rock mass*

At the construction site of the Laxiwa dam, a total number of 1450 single-hole packer tests were conducted to measure the hydraulic properties of the rock mass, with 113 packer tests for the shallow rock mass on the right bank in 0–80 m horizontal depth and 278 packer tests for the deeper rock mass. The measurements of the hydraulic conductivity range from 10^{-5} cm/s to 10^{-6} cm/s for the shallow rock mass and from 10^{-6} cm/s to 10^{-7} cm/s for the deeper rock mass, with in average 4.94×10^{-5} cm/s for the former and 3.80×10^{-6} cm/s for the latter, respectively (Liu, 1996). On the other hand, in-situ stress tests showed that the geostress in the base of the valley and in deep rock mass has a magnitude of 20–60 MPa, with the direction of the major principal stress pointing towards NNE. As a result of stress release, the release fractures are frequently developed and a high permeability zone of 0–80 m horizontal depth is formed in the bank slope, as shown in Fig. 15. The stress release fractures, however, become infrequent in deeper rock mass, and the measured hydraulic conductivity is generally 1–2 orders of magnitude smaller than the hydraulic conductivity of the rock mass in shallow depth away from the bank slope. Therefore, the hydraulic conductivity of the rock mass at the construction site of the Laxiwa arch dam is mainly controlled by the fracture network and the stress state.

Based on these statistics given in Table 3, fracture networks can be generated and calibrated for the rock mass at the construction site of the Laxiwa Hydropower Project using the Monte-Carlo method by assuming that each fracture is a smooth, planar disc, with its center uniformly distributed in the simulated area. For each set of fractures, the geometrical parameters of any one are sampled by Monte-Carlo method until enough fractures are included in the simulated area. Then, a calibration procedure is invoked to check whether the generated model satisfies the distribution mode of the real fracture network. If doesn't, the fracture network will be regenerated until one matches the distribution mode. With the generated fracture network, the actual connectivity can be computed by spatial operation on the fractures. But for calibrated fracture network, a more convenient approximate approach to determine the connectivity of the fracture network, as it is adopted here, is to directly produce ξ_{ij} in Eq. (42) with the Monte-Carlo method and the characteristics presented in Table 3, then W_{ij} is derived from Eq. (42) with $\overline{\xi}_i$, the maximum number of fractures cut by the ith set of fractures. Field measurements are used to estimate $\overline{\xi}_i$, with $\overline{\xi}_1$=11, $\overline{\xi}_2$=8 and $\overline{\xi}_3 = \overline{\xi}_4$ =6 for the four sets of fractures, respectively. Fig. 16 illustrates a simulated fracture network with size of 20×20×20 m.

On the basis of the fracture network generated above, we compute the hydraulic conductivity tensor for the simulated cubic volume of rock mass with size of 20×20×20 m using the method given by Snow (1969) and the method presented in Section 4.2, respectively. To show the coupling effect of stress/deformation on hydraulic properties, we consider two scenarios for examination. In the first scenario, we consider the fracture network located in the shallow depth away from the bank slope, where the impact of the in-situ stress is negligible. While in the second scenario, the fracture network is situated in larger depth, and a typical stress state with σ_x=σ_z=10 MPa and σ_y=20 MPa is associated with it. Based on laboratory test results, the shear modulus of the fractures is estimated as μ=2 MPa, and then by taking the Poisson's ratio as ν=0.25, the Lame's constant is derived with λ=2 MPa. The kinematic viscosity of underground water is set to be ν_w=1.14×10^{-6} m^2/s and the frictional angle-like parameter and the normal stress-like parameter are taken as φ=0.4363 and s=σ_n/20.

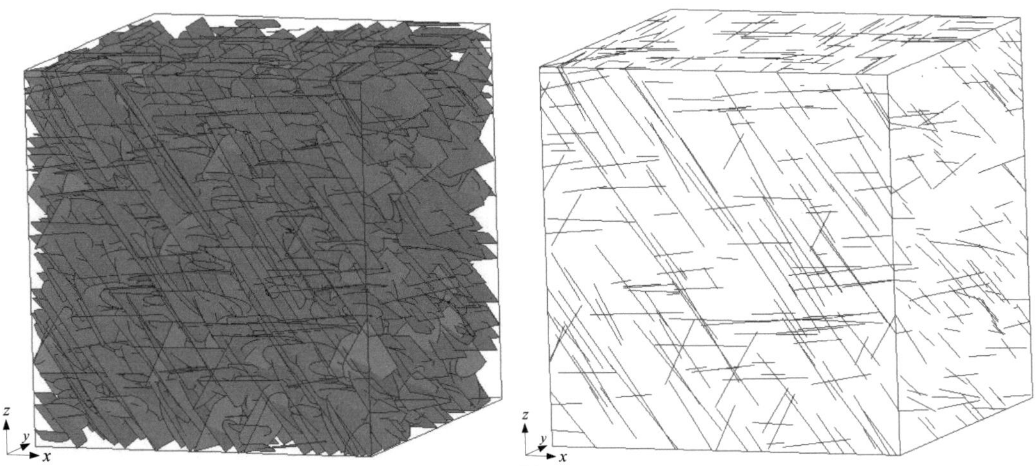

(a) (b)

Fig. 16. A three dimensional fracture network with size of 20×20×20 m generated by using the Monte-Carlo method for the rock mass in the Laxiwa Hydropower Project: (a) fracture network and (b) traces of the fractures on the surfaces of the simulated area

The predicted hydraulic conductivity tensor for the examined rock mass is listed in Table 4. From Table 4, one observes that for shallow rock mass (where the effect of in-situ stress is not considered), the Snow's method and the method presented in Section 4.2 predict similar results and the predicted hydraulic conductivity is in the magnitude of 10^{-5} cm/s and close to in-situ hydraulic observations, but the anisotropy in hydraulic conductivity manifests due to non-uniform distribution of fractures. Compared with the hydraulic conductivity of the shallow rock mass, the predicted hydraulic conductivity for the rock mass in larger depth with the same fracture network decreases in 2 orders of magnitude due to the closure of the fractures applied by the in-situ stresses, but the anisotropic property of the hydraulic conductivity remains, which suggests that the occurrence of the fractures has a significant impact on permeability. Taking into consideration the applied stress level, the reduction of hydraulic conductivity in orders of magnitude is very close to the results achieved in Min et al. (2004) through a discrete element method, and generally agrees with the in-situ hydraulic observations.

Snow's model (for shallow rock mass)		
4.78E–05	–4.76E–07	–1.71E–05
–4.76E–07	7.49E–05	–1.41E–05
–1.71E–05	–1.41E–05	4.08E–05
The proposed model (for shallow rock mass)		
1.93E–05	–1.75E–07	–6.39E–06
–1.75E–07	2.99E–05	–5.81E–06
–6.39E–06	–5.81E-06	1.64E–05
The proposed model (for deep rock mass)		
9.06E–08	–4.81E–09	–6.10E–08
–4.81E–09	1.85E–07	–1.92E–08
–6.10E–08	–1.92E–08	1.10E–07

Table 4. Predicted hydraulic conductivity tensor of the rock mass at the construction site of the Laxiwa dam (cm/s)

Now, we take for example the rock mass in shallow depth to estimate the *REV* size of the rock mass. For this purpose, the scale of the rock mass is increased gradually from 3×3×3 m to 40×40×40 m with an increment of 1 m in each dimension. In each step, a fracture network with prescribed size is generated by using the Monte-Carlo method described above, and it is worth noting that this method is somewhat different from the method used by Min & Jing (2003) and Long et al. (1982). For each fracture network, the hydraulic conductivity tensor is calculated from Eq. (47) and then the principal hydraulic conductivities are further obtained from the hydraulic conductivity tensor. The relationship between the computed principal hydraulic conductivities and the sizes of the rock mass is illustrated in Fig. 17. As we can see from Fig. 17, when the block size of the rock mass is smaller than 18×18×18 m, the population of fractures is not dense enough and the principal hydraulic conductivities fluctuate dramatically. On the other hand, as the size scales up to about 20×20×20 m, the examined rock mass has included enough fractures and the computed principal hydraulic conductivities approach a rather steady state, with k_1, k_2, k_3 estimated to be 2.41×10^{-5} cm/s, 3.59×10^{-5} cm/s, 1.08×10^{-5} cm/s, respectively. This suggests that the *REV* does exist in the rock mass and the rock mass can be regarded as an equivalent continuum medium as long as its size is no less than, e.g., 20×20×20 m or 8000 m^3.

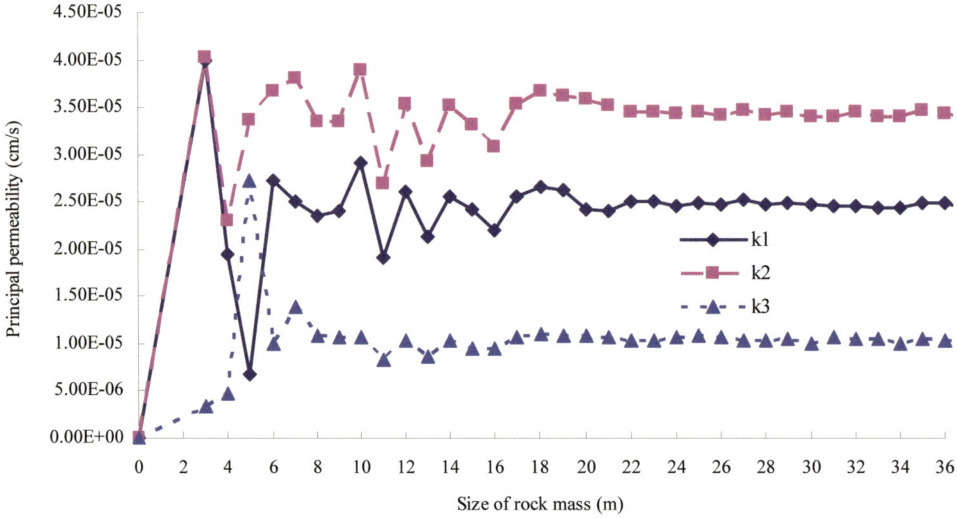

Fig. 17. Hydraulic conductivity versus the volume size of the fractured rock mass

5. Strain-dependent hydraulic conductivity tensor of fractured rocks

On the basis of the strain-dependent model presented in Section 3 for rock fractures, this section formulates the strain-dependent hydraulic conductivity tensor for fractured rock masses cut by one or multiple sets of parallel fractures. The major difference between the model in this section and the stress-dependent model presented in Section 4 is that the former is capable of describing influence of the post-peak mechanical behaviours on the hydraulic properties of the rock masses, and is suited for modelling the coupled processes in rock masses at high stress level and in drastic engineering disturbance condition.

5.1 An equivalent elasto-plastic constitutive model for fractured rocks

Consider a fractured rock mass cut by n sets of planar and parallel fractures of constant apertures with various orientations, scales and densities. The global response of the fractured rock mass under loading comes both from weak fractures and from stronger rock matrix. Based on this observation, an equivalent elasto-plastic constitutive model can be formulated by imposing assumptions on the interaction between fractures and rock matrix.

The coordinate systems are defined in the same way with those defined in Section 4.1 (see Fig. 13). Denote the unit vector along X_i-axis of the global frame as e_i (i=1, 2, 3) and the unit vector along x_i^f -axis of the fth local frame as $\mathbf{e}_i^f$ (i=1, 2, 3). Then, a second order tensor, $\mathbf{l}^f$, can be defined for transforming physical quantities between the frames, with the components in the form of

$$l_{ij}^f = \mathbf{e}_i^f \cdot \mathbf{e}_j \tag{53}$$

Regarding the fractured rock mass as a continuous medium at the macroscopic scale, it is rational to assume that the global strain increment of the fractured rock mass is composed of the strain increments of rock matrix and fractures (Pande & Xiong, 1982; Chen & Egger, 1999), i.e.

$$\mathrm{d}\boldsymbol{\varepsilon} = \mathrm{d}\boldsymbol{\varepsilon}^{\mathrm{R}} + \sum_{\mathrm{F}} \mathrm{d}\boldsymbol{\varepsilon}^{\mathrm{F}} \tag{54}$$

where $\mathrm{d}\boldsymbol{\varepsilon}$, $\mathrm{d}\boldsymbol{\varepsilon}^{\mathrm{R}}$ and $\mathrm{d}\boldsymbol{\varepsilon}^{\mathrm{F}}$ are the total incremental strain tensor, the incremental strain tensor of rock matrix and the incremental strain tensor of *f*th set of fractures measured in the global coordinate system, respectively. Note that a variable with a superscript in upper case (i.e. R or F) means that it is measured in the $X_1X_2X_3$ system, while a variable with a superscript in lower case (i.e. *f*) is measured in $x_1^f x_2^f x_3^f$ system, respectively. Unless otherwise specified, the superscripts F and *f* are not summing indices.

On the other hand, traction continuity has to be ensured across the fracture interfaces. In the global coordinate system, this condition can be strictly represented by (Pande & Xiong, 1982; Chen & Egger, 1999)

$$\mathrm{d}\boldsymbol{\sigma} = \mathrm{d}\boldsymbol{\sigma}'^{\mathrm{R}} = \mathrm{d}\boldsymbol{\sigma}'^{\mathrm{F}} \tag{55}$$

where $\mathrm{d}\boldsymbol{\sigma}'$, $\mathrm{d}\boldsymbol{\sigma}'^{\mathrm{R}}$ and $\mathrm{d}\boldsymbol{\sigma}'^{\mathrm{F}}$ are the effective incremental stress tensor of the fractured rock mass, the effective incremental stress tensor of rock matrix and the effective incremental stress tensor of *f*th set of fractures, respectively. The effective stress tensor $\boldsymbol{\sigma}'$ is defined as

$$\boldsymbol{\sigma}' = \boldsymbol{\sigma} + \alpha p \boldsymbol{\delta} \tag{56}$$

where $\boldsymbol{\sigma}$ is the total stress tensor (positive for tension), p is the pore water pressure (positive for compressive pressure), and α ($\alpha \leq 1$) is an effective stress parameter.

Combining the plastic potential flow theory and the consistency conditions of rock matrix and fractures, an equivalent elasto-plastic constitutive model can be derived from Eqs. (54) and (55):

$$\mathrm{d}\boldsymbol{\varepsilon} = \mathbf{S}^{\mathrm{ep}} : \mathrm{d}\boldsymbol{\sigma}' \tag{57}$$

with

$$\mathbf{S}^{\mathrm{ep}} = (\mathbf{C}^{\mathrm{R,ep}})^{-1} + \sum_{\mathrm{F}} (\mathbf{C}^{\mathrm{F,ep}})^{-1} \tag{58}$$

where $\mathbf{S}^{\mathrm{ep}}$ is the equivalent elasto-plastic compliance tensor of the fractured rock mass.

$\mathbf{C}^{\mathrm{R,ep}}$ in Eq. (58) is the elasto-plastic modulus tensor of rock matrix. Neglecting the degradation of rock strength in the volume close to fracture intersections, $\mathbf{C}^{\mathrm{R,ep}}$ can be written as

$$\mathbf{C}^{\mathrm{R,ep}} = \mathbf{C}^{\mathrm{R}} - \frac{\mathbf{C}^{\mathrm{R}} : \dfrac{\partial Q_{\mathrm{R}}}{\partial \boldsymbol{\sigma}'} \otimes \dfrac{\partial F_{\mathrm{R}}}{\partial \boldsymbol{\sigma}'} : \mathbf{C}^{\mathrm{R}}}{\dfrac{\partial F_{\mathrm{R}}}{\partial \boldsymbol{\sigma}'} : \mathbf{C}^{\mathrm{R}} : \dfrac{\partial Q_{\mathrm{R}}}{\partial \boldsymbol{\sigma}'} + H_{\mathrm{R}}} \tag{59}$$

where $\mathbf{C}^{\mathrm{R}}$ is the fourth-order elastic modulus tensor of rock matrix, which can be represented in terms of the Lame's constants λ and μ:

$$C_{ijkl}^{\mathrm{R}} = \lambda \delta_{ij} \delta_{kl} + \mu(\delta_{ik}\delta_{jl} + \delta_{il}\delta_{jk}) \tag{60}$$

F_R, Q_R and H_R in Eq. (59) are the yield function, the plastic potential function and the hardening modulus of rock matrix, respectively. A non-associative flow rule with elastic-perfectly plasticity (i.e. H_R=0) is adopted for better modeling dilatant behavior of rock matrix by virtue of, for example, the Druker-Prager criterion with its cone fully inscribed by the Mohr-Coulomb hexagon, defined by functions

$$F_R = aI_1' + \sqrt{J_2} - \kappa = 0 \tag{61}$$

$$Q_R = \beta I_1' + \sqrt{J_2} \tag{62}$$

with

$$\alpha = \sin\varphi_R \,/\, \sqrt{3(3+\sin^2\varphi_R)} \tag{63}$$

$$\kappa = 3c_R \cos\varphi_R \,/\, \sqrt{3(3+\sin^2\varphi_R)} \tag{64}$$

$$\beta = \sin\psi_R \,/\, \sqrt{3(3+\sin^2\psi_R)} \tag{65}$$

where c_R and φ_R are the cohesion and the friction angle of rock matrix, respectively. I_1' and J_2 are the first invariant of the effective stress and the second invariant of the deviatoric stress of rock matrix, respectively. ψ_R is the mobilized dilatancy angle of rock matrix.

It should be noted here that in the literature, Drucker-Prager criterion has been used by many authors to model the elasto-plastic behaviour of intact rock matrix, see Pande & Xiong (1982) and Chen & Egger (1999) for example. Although a modified Drucker-Prager yield function may be more suitable for this formulation in order to model plastic deformation properties of intact rock such as pressure dependency, strain hardening, transition from compressibility to dilatancy and stress path dependency (Chiarelli et al., 2003), the criterion given above may keep the formulation compact and does not lose generality. Other yield functions, such as the modified Drucker-Prager criterion (Chiarelli et al., 2003) or the modified Hoek-Brown criterion (Hoek et al., 1992), can also be integrated into the formulation without major mathematical difficulties.

With the researches conducted by Yuan & Harrison (2004) and Alejano & Alonso (2005), the decaying process of the rock dilatancy angle in line with plasticity can be described by the following negative exponential expression through the equivalent plastic strain of rock matrix, $\bar{\varepsilon}_R^p$ (Lai, 2002):

$$\psi_R = \psi_R^{peak} \exp(-r_R \bar{\varepsilon}_R^p) \tag{66}$$

where $r_R \geq 0$ is a parameter for modelling the decaying process of the dilatancy angle, and ψ_R^{peak} is the peak dilatancy angle of rock matrix and the following expression has been proposed by recovering the shape of the peak dilatancy angle of fractures given by Barton & Bandis (1982) and by assuming $\psi_R^{peak} = \varphi_R$ for null confinement pressures (Alejano & Alonso, 2005):

$$\psi_{\mathrm{R}}^{\mathrm{peak}} = \frac{\varphi_{\mathrm{R}}}{1+\log_{10}\sigma_{\mathrm{c}}}\log_{10}\frac{\sigma_{\mathrm{c}}}{-\sigma_3'+0.1} \tag{67}$$

where σ_c is the unconfined compressive strength for intact rock. By Eqs. (66) and (67), the dependencies of rock dilatancy on plasticity, confining stress and scale are produced.

The equivalent plastic strain $\bar{\varepsilon}^{\mathrm{p}}$ is computed by the following:

$$\bar{\varepsilon}^{\mathrm{p}} = \int \mathrm{d}\bar{\varepsilon}^{\mathrm{p}} = \int\sqrt{\frac{2}{3}\mathrm{d}\boldsymbol{\varepsilon}^{\mathrm{p}} : \mathrm{d}\boldsymbol{\varepsilon}^{\mathrm{p}}} \tag{68}$$

Similarly, $\mathbf{C}^{\mathrm{F,ep}}$ in Eq. (58) is the elasto-plastic modulus tensor of fth set of fractures measured in the $X_1X_2X_3$ system, which can be calculated from its corresponding elasto-plastic modulus tensor measured in the $x_1^f x_2^f x_3^f$ system, $\mathbf{C}^{f,\mathrm{ep}}$, with the assumption of small strain and by imposing the following tensor transformation:

$$C_{ijkl}^{\mathrm{F,ep}} = l_{mi}^f l_{nj}^f l_{ok}^f l_{pl}^f C_{mnop}^{f,\mathrm{ep}} \tag{69}$$

with

$$\mathbf{C}^{f,\mathrm{ep}} = \mathbf{C}^f - \frac{\mathbf{C}^f : \dfrac{\partial Q_f}{\partial \boldsymbol{\sigma}'} \otimes \dfrac{\partial F_f}{\partial \boldsymbol{\sigma}'} : \mathbf{C}^f}{\dfrac{\partial F_f}{\partial \boldsymbol{\sigma}'} : \mathbf{C}^f : \dfrac{\partial Q_f}{\partial \boldsymbol{\sigma}'} + H_f} \tag{70}$$

where $\mathbf{C}^f$ is the fourth-order tangential elastic modulus tensor of the fth set of fractures, with $C_{3333}^f = s_f k_{\mathrm{n}f}$, $C_{2323}^f = C_{3131}^f = s_f k_{\mathrm{s}f}$, and with all other elements equal to zero. The symbols $k_{\mathrm{n}f}$, $k_{\mathrm{s}f}$ and s_f are the normal stiffness, the tangential stiffness and the spacing of the fth set of fractures, respectively. The expressions for the elements in $\mathbf{C}^f$ mean that the strain of fractures is evaluated over the fracture spacing, not over the fracture aperture, thus enabling the proposed model to consider the post-sliding plasticity of fractures and nonlinear variations of $k_{\mathrm{n}f}$ and $k_{\mathrm{s}f}$ with dilatancy caused by shear loading, without violating the small strain assumption.

F_f, Q_f and H_f in Eq. (70) are the yield function, the plastic potential function and the hardening modulus of the fth set of fractures, respectively. The elasto-plastic behaviour of the fractures is treated in a similar fashion as that for the rock matrix, with a non-associative Mohr-Coulomb criterion:

$$F_f = \sqrt{\tau_{zxf}^2 + \tau_{zyf}^2} + \sigma_{zf}'\tan\varphi_f - c_f = 0 \tag{71}$$

$$Q_f = \sqrt{\tau_{zxf}^2 + \tau_{zyf}^2} + \sigma_{zf}'\tan\psi_f \tag{72}$$

where σ_{zf}', τ_{zxf} and τ_{zyf} are the effective normal stress and the shear stresses on the fracture surfaces, respectively. c_f, φ_f and ψ_f are the cohesion, the friction angle and the mobilized dilatancy angle of the fth set of fractures, respectively. Similar to Eq. (66), ψ_f is also a shrinking function of the equivalent plastic strain of fractures $\bar{\varepsilon}_f^{\mathrm{p}}$, and depends on normal stress and scale as well, in the following form:

$$\psi_f = \psi_f^{\text{peak}} \exp(-r_f \bar{\varepsilon}_f^{\text{p}}) \tag{73}$$

where r_f is the decaying parameter and ψ_f^{peak} is the peak dilatancy angle of the fth set of fractures, respectively, with the latter calculated by Eq. (26).

Thus at any loading step, as long as the stress increment of the equivalent rock mass, $d\sigma'$, is obtained, the local strain pertinent to fth set of fractures can be derived as follows:

$$d\boldsymbol{\varepsilon}^{\text{F}} = (\mathbf{C}^{\text{F,ep}})^{-1} : d\boldsymbol{\sigma}' \tag{74}$$

and

$$d\boldsymbol{\varepsilon}_{ij}^{f} = l_{im}^{f} l_{jn}^{f} d\boldsymbol{\varepsilon}_{mn}^{\text{F}} \tag{75}$$

The separation of the incremental strain of fractures from that of the rock mass through the proposed equivalent constitutive model plays a significant role in the present study. It enables the formulation of strain-dependent hydraulic conductivity that accounts for the mobilized dilatancy behaviour, which will be demonstrated in the following section.

5.2 Strain-dependent hydraulic conductivity tensor for fractured rocks

Consider a domain of flow that has been discretized into several sub-domains according to rock quality classification. Suppose that each sub-domain contains n sets of fractures, with average initial aperture b_{f0} and spacing s_f for the fth set of fractures. Starting from Eq. (22) and using the averaging concept for the hydraulic conductivity over the whole sub-domain, the equivalent initial hydraulic conductivity of the fth set of fractures, k_{f0}, in the examined sub-domain can be represented as (Castillo, 1972; Liu et al., 1999)

$$k_{f0} = \varsigma \frac{g b_{f0}^3}{\nu s_f} \tag{76}$$

where ς, as pointed out before, is a dimensionless constant introduced to penalize the real water conducting capacity of natural fractures with rough walls, finite scales, asperity areas and filling materials. The validity of using a constant value of ς has been examined by Zhou et al. (2006).

Assuming that the change in spacing s_f during modeling is negligible, under normal and shear stress loadings we have

$$k_f = \varsigma \frac{g b_f^3}{\nu s_f} = \varsigma \frac{g(b_{f0} + \Delta b_f)^3}{\nu s_f} \tag{77}$$

where Δb_f and k_f are the increment of the aperture and the equivalent hydraulic conductivity of the fth set of fractures under loading, respectively. Suppose that strain localization (Lai, 2002; Vajdova, 2003) is not dominantly exhibited in the concerned fractures, it is approximately valid that

$$\Delta b_f = s_f \Delta \varepsilon_{zf} \tag{78}$$

where $\Delta\varepsilon_{zf}$ is the increment of the normal strain of the fth set of fractures, which can be directly obtained from Eq. (75).
Substituting Eq. (78) into Eq. (77) then yields

$$k_f = k_{f0}\left(1+\frac{s_f}{b_{f0}}\Delta\varepsilon_{zf}\right)^3 \tag{79}$$

Following the theory proposed by Snow (1969), a strain-dependent equivalent hydraulic conductivity tensor for fractured rock masses with n sets of fractures is represented by

$$\mathbf{K} = \sum_f k_f(\boldsymbol{\delta} - \mathbf{n}_f \otimes \mathbf{n}_f) = \sum_f k_{f0}\left(1+\frac{s_f}{b_{f0}}\Delta\varepsilon_{zf}\right)^3 (\boldsymbol{\delta} - \mathbf{n}_f \otimes \mathbf{n}_f) \tag{80}$$

where $\boldsymbol{K}$ is the equivalent hydraulic conductivity tensor of the examined rock mass, and $\boldsymbol{n}_f$ is the unit vector normal to the fth set of fractures.
The following significant implications can be observed from the formulation of $\boldsymbol{K}$ in Eq. (80):

1. $\boldsymbol{K}$ is a cubic function of $\Delta\varepsilon_{zf}$, and any variation in ε_{zf} under loading will trigger the change in $\boldsymbol{K}$, even in orders of magnitude. This exactly accounts for the coupling effect of mechanical loading (strain/stress) on hydraulic properties.
2. $\boldsymbol{K}$ depends on incremental strains, rather than on stresses, which makes it possible to integrate various material nonlinearities in hydro-mechanical coupling analysis.
3. In addition to cubic relation, the influence of $\Delta\varepsilon_{zf}$ on $\boldsymbol{K}$ is amplified by s_f/b_{f0}, indicating that $\boldsymbol{K}$ can be rather sensitive to b_{f0} and s_f. Therefore, techniques for estimating b_{f0} and s_f need to be carefully developed, on the basis of laboratory or in-situ hydraulic test data.
4. The orientations of fractures possibly render $\boldsymbol{K}$ highly anisotropic, even if $\boldsymbol{K}$ is initially assumed isotropic, as has been systematically examined, e.g. by Liu et al. (1999).
5. When implemented in a FEM code, a different $\boldsymbol{K}$ can be associated to each geological sub-domain or even to each element, as long as k_{f0}, b_{f0} and s_f for the sub-domains or elements can be estimated in advance.
6. As a nature of the homogenized equivalent continuum approach, the size effect of fractures, especially the size-dependency of aperture, is not fully considered in the formulation of $\boldsymbol{K}$ for simplicity, even though it can be reflected to some degree through ς and scaled *JRC* and *JCS* values. The connectivity and the intersection effect of fractures, on the other hand, may have a more significant influence on $\boldsymbol{K}$, but similarly, they cannot be properly considered in the equivalent continua without explicit representation of fractures. A rough remedy is to process the fracture system in such a way that only the connected fracture populations are included for conducting analyses.

To determine $\boldsymbol{K}$ of a fractured rock under any loading paths, a coupled hydro-mechanical process has to be invoked. With the assumption of incompressible rock matrix and fluid (e.g. groundwater), the governing equations for the coupled process of saturated fluid flow and deformation are given below as balance equation, geometric equation and fluid flow equation, respectively:

$$\sigma'_{ij,j} - \alpha p_{,i} + f_i = 0 \tag{81}$$

$$\varepsilon_{ij} = \frac{1}{2}\left(u_{i,j} + u_{j,i}\right) \tag{82}$$

$$\frac{\partial}{\partial x_i}\left(k_{ij}\frac{\partial h}{\partial x_j}\right) = \frac{\partial \varepsilon_v}{\partial t} \tag{83}$$

where f_i and u_i are the components of the body force and displacement in the ith direction, $h=p/\gamma_w+z$ the water head, z the vertical coordinate, γ_w the unit weight of water, and ε_v the volume strain of the rock mass.

In the coupled process given above, mechanical loading or disturbance to the rock mass results in change in flow properties and flow behaviour through Eqs. (80) and (83), while the change in flow behavior leads to change in mechanical response of the rock mass through Eq. (81). When the coupled process reaches a stable state, the solution to $\boldsymbol{K}$ is also available.

Now we briefly discuss how to determine k_{f0}, b_{f0} and s_f in Eq. (80) based on laboratory or in-situ hydraulic test or site investigation data. Obviously, the initial hydraulic conductivity, k_{f0}, can be determined by in-situ hydraulic tests. Suppose the initial hydraulic conductivity tensor, $\boldsymbol{K}_0$, is known through in-situ hydraulic test, as suggested by Hsieh & Neuman (1985), then $\boldsymbol{K}_0$ can be rewritten, from Eq. (80), in the following form:

$$\mathbf{K}_0 = \sum_f k_{f0}(\boldsymbol{\delta} - \mathbf{n}_f \otimes \mathbf{n}_f) \tag{84}$$

By optimizing Eq. (84), k_{f0} (f=1, …, n) can be estimated if the number of the sets of critically oriented fractures, n, is less than or equal to 6 (i.e. the number of the independent components of $\boldsymbol{K}_0$), regardless $\boldsymbol{K}_0$ is assumed to be isotropic or anisotropic.

The average spacing of the fth set of fractures, s_f, can be roughly estimated from the statistics of drill holes or scanlines. An alterative, however, is to use *RQD* (Rock Quality Designation) for determining s_f, as suggested by Liu et al. (1999), when the value of *RQD* for a specific rock mass is known *a priori*.

After the initial hydraulic conductivity, k_{f0}, and the average spacing, s_f, of the fractures are determined, the mean initial aperture of the fractures, b_{f0}, is ready to be back-calculated from Eq. (76).

5.3 Validation of the proposed model

5.3.1 Hydraulic conductivity of the surrounding rock of a circular tunnel in the Stripa mine

Here we compare the proposed method with results from a previous study as presented by Liu el al. (1999) by applying the method to an excavated circular tunnel with a biaxial stress field, σ_x and σ_z. The physical model is illustrated in Fig. 18, which is actually a manifestation of the reality of the Stripa mine in Sweden (Kelsall et al., 1984; Pusch, 1989). The following description about the tunnel is directly taken from Liu et al. (1999):

A Buffer Mass Test was conducted in Stripa Mine over the period 1981-1985 (Kelsall et al., 1984; Pusch, 1989) to measure the permeability of a large volume of low permeability fractured rock mass by monitoring water flow into a 33 m long section of the tunnel, as a large scale in-situ experiment for the research and development programs of underground geological disposal of nuclear wastes of the participating countries of the Stripa Project. The radius of the tunnel is about 2.5 m with two major sets of fractures striking obliquely to the

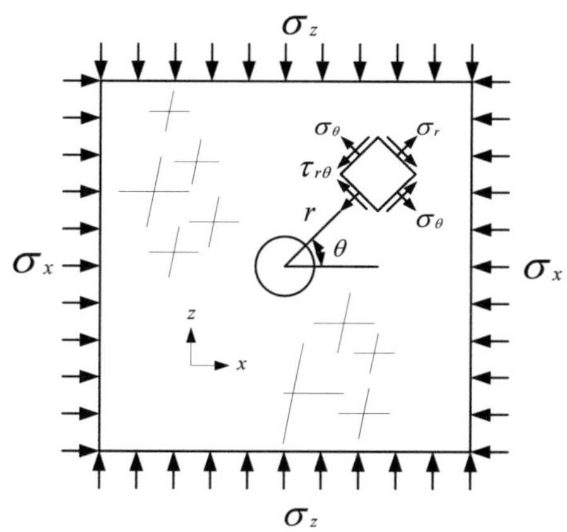

Fig. 18. Sketch of a circular excavation in a biaxial stressed rock mass.

tunnel axis, as shown in Fig. 18. Fracture frequency measured in holes drilled from the tunnel was on average 4.5 fractures/m in inclined holes and 2.9 fractures/m in vertical holes. The initial stress field is anisotropic with high horizontal stress component and the conductivity of the virgin rock is about 10^{-10} m/s. The excavation of the test drift produced a dramatic increase in axial hydraulic conductivity in a narrow zone adjacent to the periphery of the drift. The conductivity increase is estimated to be 3 orders of magnitude.

The following assumptions are made in the calculations, with some of them similar to those in Liu et al. (1999):

1. Statically uniform aperture and spacing distributions exist before excavation;
2. Fracture spacing and continuity are not altered by the excavation;
3. The high obliquity of the two major sets of fractures can be well approximated by two orthogonal sets of fractures;
4. Excavation-induced strain redistribution may be adequately captured by the proposed equivalent elasto-plastic constitutive model.

Some of the parameters are directly taken from Liu et al. (1999), while other unavailable parameters are assumed, as listed in Table 5, in which the initial mechanical aperture of the fractures is back-calculated from Eq. (76) by taking k_0=10^{-10} m/s. Consistent with Liu et al. (1999), the far-field stress components are taken as σ_x=20 MPa and σ_z=10 MPa, respectively.

Category	Parameter	Setting
Intact rock matrix	Elastic modulus, E	37.5 GPa
	Poisson's ratio, ν	0.25
	Cohesion, c_R	5 MPa
	Friction angle, φ_R	46°
Fractures	Initial mechanical aperture, b_0	0.0075 mm
	Spacing, s	0.27 m
	Normal stiffness, k_n	200 GPa/m
	Shear stiffness, k_s	100 GPa/m
	Dimensionless constant, ς	0.0067
	Cohesion, c_f	0.4 MPa
	Friction angle, φ_f	40°

Table 5. Geometrical and mechanical parameters for a circular tunnel

To avoid the difficulty in determining the initial dilatancy angles and the corresponding decay parameters of fractures and intact rock matrix, associative flow rule is used in this simulation. Again for simplicity, both the normal stiffness and the shear stiffness of the fractures are assumed constant during excavation. The finite element mesh of the model is shown in Fig. 19, and the FEM program was run to simulate the excavation effect of the tunnel. Fig. 20 shows the deformation zone and plastic zone of the rock mass after the tunnel excavation. Fig. 21 plots the excavation-induced changes in hydraulic conductivities around the circular tunnel, which are directly compared with the results presented in Liu et al. (1999).

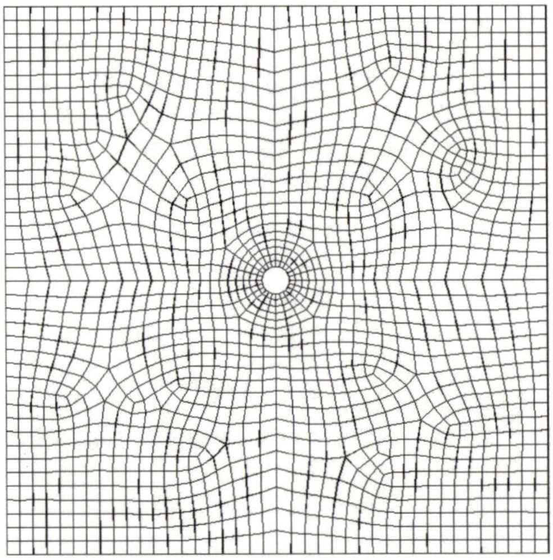

Fig. 19. Finite element mesh for simulation of a tunnel excavation.

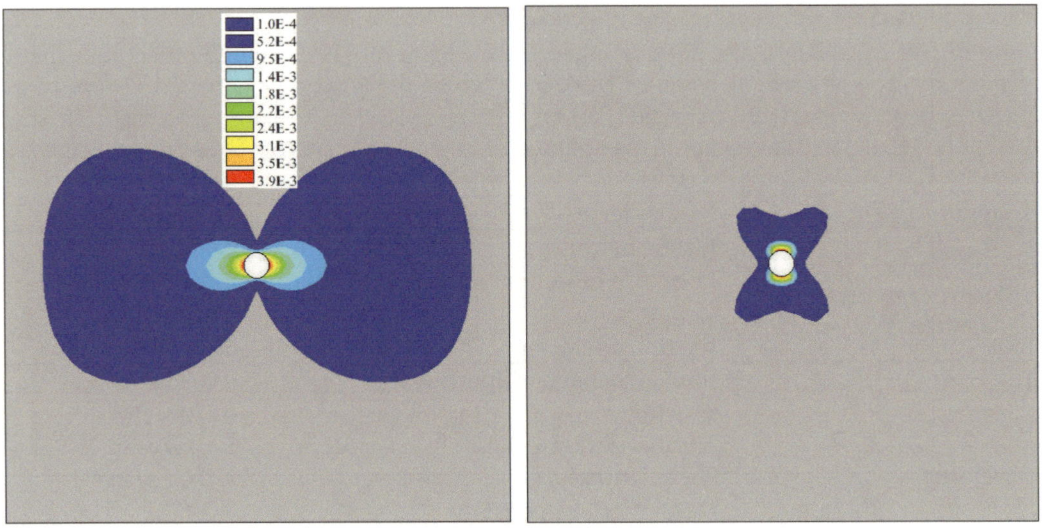

(a) (b)

Fig. 20. Deformation zone and plastic zone induced by the tunnel excavation: (a) deformation zone and (b) plastic zone

(a)

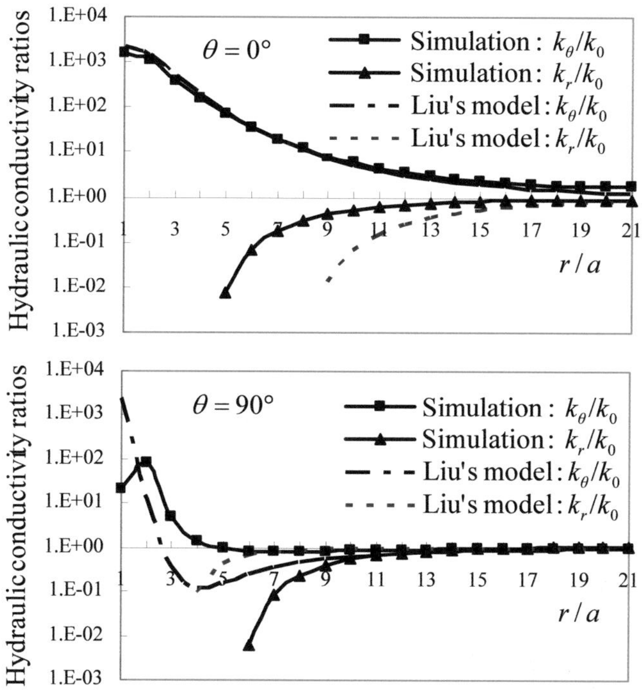

(b)

Fig. 21. Excavation-induced hydraulic conductivity ratios around a circular tunnel in a biaxial stressed rock mass, where a is the radius of the tunnel and r is the distance away from the tunnel center. θ=0° denotes the horizontal direction while θ=90° the vertical direction.

It can be observed from Fig. 21 that generally tangential conductivities are found to increase greatly due to the formation of the excavation disturbed zone around the tunnel, while radial conductivities diminish greatly as a result of closure on related fractures. In the horizontal direction (i.e. θ=0°), the excavation-induced tangential hydraulic conductivity ratios, k_θ/k_0, predicted by our model are very close to the results presented in Liu et al. (1999). For radial hydraulic conductivity ratios, k_r/k_0, however, deviation occurs in the vicinity of the excavation. Such a deviation is also found both for k_θ/k_0 and for k_r/k_0 in the vertical direction (i.e. θ=90°).

Clearly, these deviations are largely resulted from the facts that (1) Different strain distribution patterns are assumed in the elastic model in Liu et al. (1999) and in our elasto-plastic model; (2) Different methods are used to compute the strain increments of fractures. In Liu et al. (1999), normal strains of fractures were separated from rock matrix through a modulus reduction ratio empirically defined as a function of *RMR*, while in this simulation fracture strains were calculated by strain decomposition through an equivalent elasto-plastic constitutive model; (3) Radial and tangential fractures were assumed in Liu et al. (1999), leading to different background fracture networks; and (4) As mentioned above, some of the parameters, such as the shear strength of fractures and rock matrix, the shear stiffness and normal stiffness of the fractures, are unavailable in the literature (Kelsall et al., 1984; Pusch, 1989; Liu et al., 1999) and hence are empirically assumed in the calculations. If these parameters are determined based on in-situ or laboratory experiments, more convincing results may be achieved.

Despite the deviations, the trends of variation of the hydraulic conductivity ratios around the tunnel due to excavation are consistent between the two studies, and basically accord with the in-situ experimental observations, demonstrating the applicability of the present model in this section.

From Fig. 20, one observes that the excavation-induce deformation zone and plastic zone are asymmetric, due to the anisotropic initial stress field. As a result, the predicted hydraulic conductivities are highly anisotropic due to strain redistribution, as shown in Fig. 21. In the horizontal direction (i.e. θ=0°), the deformation zone extends as far as more than 16 times of the tunnel radius and the plastic zone extends 2 times of the tunnel radius, while in the vertical direction (i.e. θ=90°), they are, respectively, within 2 and 5 times of the tunnel radius. The asymmetry of deformation zone and plastic zone demonstrates why the predicted hydraulic conductivities approach k_0 more slowly in the horizontal direction than in the vertical direction. The changes in hydraulic conductivities resulted from strain redistribution in the disturbed rock mass indicate that a different hydraulic conductivity tensor should be associated to each geological sub-domain or even each element of the rock mass, which is important for hydro-mechanical coupling analyses.

5.3.2 Hydraulic conductivity of a cubic block of rock mass with three orthogonal sets of identical fractures

In this section, a numerical simulation is conducted to evaluate hydraulic behaviour of a cubic block of rock mass containing three orthogonal sets of identical fractures under isotropic triaxial compression and shear loading. The primary goal is to investigate the change in the hydraulic conductivity of the rock mass with increasing shear load, which is obviously not achievable through any elastic models considering only the deformation of fractures under normal stresses, e.g. in Liu et al. (1999).

The underlying rock mass block model for examination, with a size of 10×10×10 m (a scale that can represent both the initial mechanical and hydraulic *REV*s (Min et al., 2004)), is assumed to contain three orthogonal sets of identical fractures, as sketched in Fig. 22. The spacing, s, of each set of fractures and the initial aperture, b_0, of each fracture are assumed to be identical, with s=1 m and b_0=1 mm. The mechanical properties of each fracture are also regarded identical and for simplicity, both the normal stiffness and the shear stiffness of the fractures are assumed to be constant during shear loading. All parameters used in this simulation are listed in Table 6, and such parameter settings enable us to demonstrate how the hydraulic conductivity evolves from initial isotropy to anisotropy in the shearing process.

The examined rock mass block model is divided into 1000 brick elements, and the resultant mesh is shown in Fig. 22. The loading condition is as follows. First, triaxial compressive stresses are applied on the surfaces of the cubic block, with $\sigma_x=\sigma_y=\sigma_z$=20 MPa. Then, a shearing load , τ, is applied on the upper and lower surfaces of the block model step by step, increasing at an increment of 1 MPa until a maximum shear load, 20 MPa, is reached. At each step of shear loading, numerical divergence may occur. If numerical divergence does occur, the simulation program terminates after 1000 iterations with a modified Newton-Raphson method.

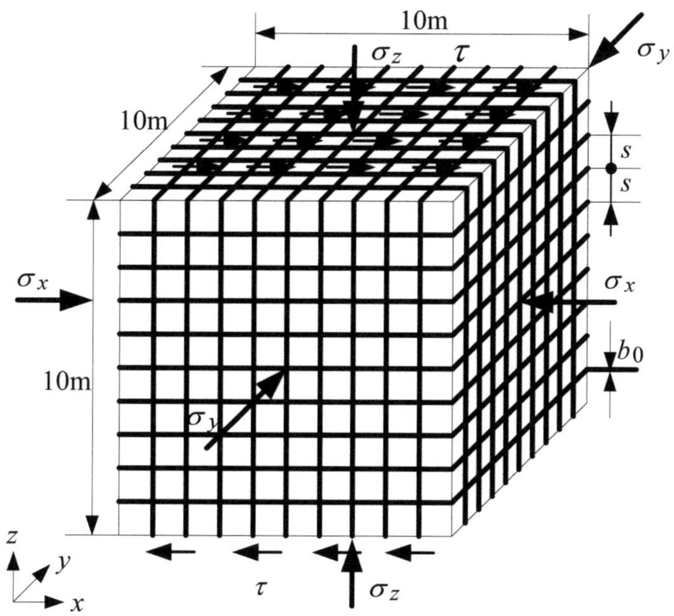

Fig. 22. Sketch of a cubic block of rock mass with three orthogonal sets of identical fractures

Category	Parameter	Setting
Intact rock matrix	Elastic modulus, E	6 GPa
	Poisson's ratio, ν	0.25
	Cohesion, c_R	1 MPa
	Friction angle, φ_R	46°
	Peak dilatancy angle, ψ_R^{peak}	35°
	Decay parameter of dilatancy, r_R	100
Fractures	Initial mechanical aperture, b_0	1 mm
	Spacing, s	1 m
	Normal stiffness, k_n	30 GPa/m
	Shear stiffness, k_s	10 GPa/m
	Dimensionless constant, ς	0.0067
	Cohesion, c_f	0.4 MPa
	Friction angle, φ_f	40°
	Peak dilatancy angle, ψ_f^{peak}	26°
	Decay parameter of dilatancy, r_f	100

Table 6. Geometrical and mechanical parameters for a cubic block of fractured rock mass

Clearly, before the rock mass is loaded, its initial hydraulic properties are isotropic, with $k_{x0}=k_{y0}=k_{z0}=1.30\times10^{-2}$ cm/s by Eq. (84). Under the condition of isotropic compression, the rock mass remains elastic, the isotropic property of hydraulic conductivity is maintained, and the magnitude of the hydraulic conductivity reduces by 2 orders of magnitude due to compression of fractures, with $k_x=k_y=k_z=4.82\times10^{-4}$ cm/s by Eq. (80). When shear stress is added incrementally on the rock mass block model from 0 to 20 MPa, the proposed method

predicts some interesting results, as depicted in Table 7, Figs. 23 and 24, respectively. Table 7 and Fig. 23 show the major hydraulic conductivities of the rock mass and Fig. 24 shows a typical case of mobilized dilatancy angle of a fracture under increasing shear loading.
As can be observed from Fig. 23, shear load has a substantial impact on the evolution of hydraulic conductivity of the rock mass model. Before the shear load reaches 4 MPa, the response of the rock mass model remains elastic, and the hydraulic conductivity components of the rock mass model are basically identical and do not vary with the shear load. When the shear load exceeds 4 MPa, however, hydraulic conductivity of the model becomes anisotropic. Due to shear dilation of fractures in the z-direction, the major hydraulic conductivities parallel to the direction of shear load in x-y plane, k_x and k_y, increase mildly at first when the shear load is smaller than 8 MPa. Afterwards, they increase dramatically, reaching an increase of 3-4 orders of magnitude. They approach a relatively stable state after the shear load increases up to 14 MPa. Obviously, the increase of k_x and k_y is resulted from the dilatancy behavior of the fractures related to equivalent plastic strain, as shown in Fig. 24, where the mobilized dilatancy angle approaches zero as the shear load approaches 14 MPa. When the shear load exceeds 14 MPa, shear dilatancy of the related fractures becomes trivial and hence k_x and k_y become steady. From Table 7 and Fig. 23, we can further see that k_x and k_y are very close to each other in values and they generally have the same varying trend with the increasing shear load.

τ (MPa)	k_x (cm/s)	k_y (cm/s)	k_z (cm/s)	τ (MPa)	k_x (cm/s)	k_y (cm/s)	k_z (cm/s)
-	0.013016	0.013016	0.013016	10	0.279373	0.279350	0.000020
0	0.000482	0.000482	0.000482	11	1.088835	1.088816	0.000056
1	0.000482	0.000482	0.000482	12	2.204162	2.204158	0.000375
2	0.000482	0.000482	0.000482	13	3.171558	3.171559	0.001374
3	0.000483	0.000483	0.000482	14	3.676801	3.697449	0.022811
4	0.000494	0.000486	0.000474	15	3.915193	4.137786	0.224877
5	0.000543	0.000509	0.000444	16	4.063688	4.696511	0.635383
6	0.000657	0.000576	0.000372	17	4.243447	5.407600	1.167070
7	0.000742	0.000643	0.000282	18	4.635512	6.233203	1.600997
8	0.000704	0.000581	0.000207	19	5.390907	7.316177	1.928768
9	0.012562	0.012459	0.000106	20	6.462514	8.618240	2.159053

Table 7. Major hydraulic conductivities of a cubic block of rock mass under isotropic compression and increasing shear loading

With the increase of shear load from 4 to 20 MPa, the change in the major hydraulic conductivity vertical to the direction of shear load, k_z, is even more interesting. Before the shear load reaches 10 MPa, k_z decreases significantly with increasing shear load and manifests a shear contraction-like behavior. When the shear load further increases, shear dilatancy occurs and k_z increases drastically, with changes in as high as 4-5 orders of magnitude. k_z reaches a relatively stable state after the shear load increases up to 17 MPa, which is actually a critical loading point that numerical instability may occur.

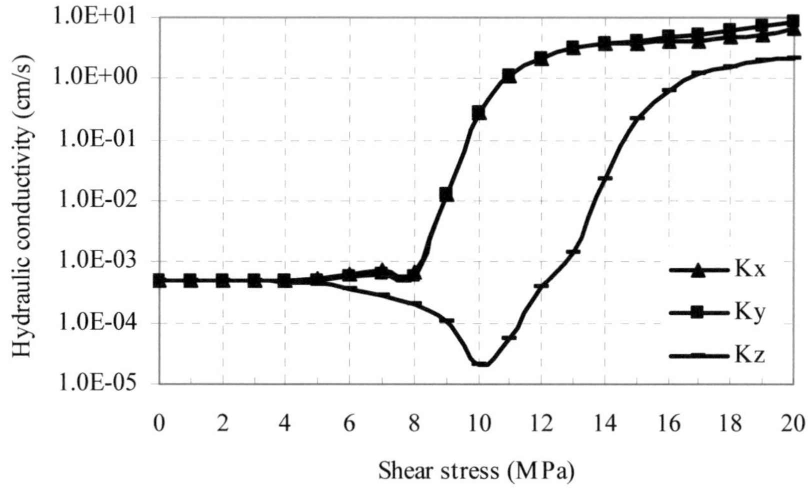

Fig. 23. Major hydraulic conductivities of a cubic block of rock mass with increasing shear load.

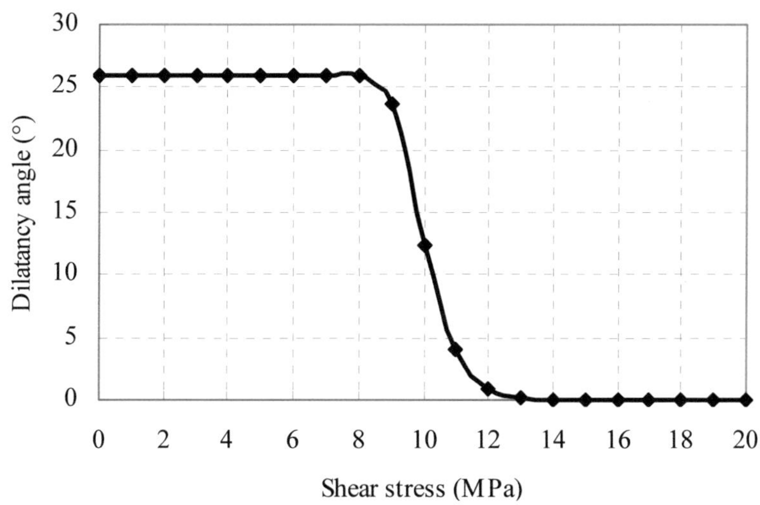

Fig. 24. A typical case of mobilized dilatancy angle of a fracture with increasing shear load.

6. Conclusions

In this chapter, mathematical models were developed to estimate the hydraulic conductivity tensor for fractured rock masses subjected to mechanical loading or engineering disturbance. Emphases are placed on the investigation of the geological characteristics of rock masses as well as the coupling between fluid flow and stress/deformation, especially the effect of shear dilation or shear contraction on the hydraulic behavior of rock fractures.

The stress-dependent hydraulic conductivity tensor was formulated by using the superposition principle of flow dissipation energy on the basis of the concept of representative elementary volume (*REV*) and the assumption that rock masses can be treated as equivalent continuum media. The deformation behaviours of rock fractures subjected to normal and shear loadings are described with an elastic constitutive model, in which the pre-peak shear dilation or contraction of the fractures is empirically modelled. The validity of using the superposition principle of flow dissipation energy for development of the model is supported by the functional equivalence between the current formulation and the Snow's and Oda's models. This model is best suited for estimation of the hydraulic properties of rock masses at low stress level and with overall elastic response, and can be used to determine the applicability of the continuum approach to coupling analysis. The latter is achieved by performing numerical experiments to test the existence of the REV, and if exists, to further estimate the REV by gradually increasing the cubic volume of flow region, V_p, to see whether the hydraulic conductivity of the rock mass can eventually approach a steady point. The hydraulic properties and the REV size of the fractured rock mass at the construction site of the Laxiwa Hydropower Project were evaluated with the proposed model, and the calculation results were compared with the predictions of the Snow's model and validated by in-situ hydraulic tests, hence the feasibility of the proposed model in rock engineering practices is demonstrated.

The strain-dependent hydraulic conductivity tensor, on the other hand, was developed for disturbed rock masses under excavation or loading. In the model, a non-associative elastic-perfectly plastic constitutive model was integrated to describe the deformation behaviours of the rock masses by characterizing them as equivalent continua containing one or multiple sets of parallel fractures. The clear advantages of the formulation are:

- The proposed hydraulic conductivity tensor is related to strains rather than stresses, hence enabling easier hydro-mechanical coupling analysis to include the effect of material nonlinearity of fractured rock masses.
- Beneficial from the equivalent non-associative elastic-perfectly plastic constitutive model, the hydraulic conductivity tensor considers the impact of shear dilatancy of fractures on fluid flow properties via mobilized dilatancy angles.
- When reduced to one dimensional case with a single fracture under normal and shear loadings, a closed-form solution to the hydraulic conductivity can be obtained, enabling validation of the model by laboratory coupled shear-flow tests of rock fractures.
- The proposed model is easy to be implemented in a FEM code, particularly suitable for numerical analysis of coupled hydro-mechanical processes in rock engineering.

The closed-form solution was validated by an existing coupled shear-flow test, and the evaluation results show that the proposed solution can closely describe the hydraulic behavior of a hard rock fracture under a wide range of normal and shear loads. The results of the simulation conducted to predict the excavation-induced hydraulic conductivities around a circular tunnel in a biaxial stress field at the Stripa mine are justified by in-situ experimental observations and compared with an existing elastic strain-dependent model, which show that engineering disturbance such as underground excavations may dramatically alter the hydraulic conductivities of the rock mass surrounding the excavations and change the isotropic pattern of the initial hydraulic conductivities. The numerical simulation on a cubic block model of a rock mass with three orthogonal sets of identical fractures under isotropic triaxial compression and shear loading further demonstrates that shear loading may drastically change the hydraulic properties of fractured rocks, in the magnitude of as high as 4-5 orders, and lead to high anisotropy of the hydraulic properties.

Despite all these efforts, characterizing the hydraulic properties for fractured rock masses remains one of the most difficult research topics in rock mechanics. In the proposed models presented in this chapter, rock masses are assumed with rather regular distribution patterns of fractures, and the existence of a hydraulic conductivity tensor of the rock masses with any distribution of fractures is not discussed. The interaction between the fractures in the rock masses is also out of the scope of this chapter, and its effect on the hydraulic properties remains an open issue. Furthermore, the proposed models are established with a rather intuitive upscaling approach, and more rigorous homogenization schemes should be developed. All of there issues should be addressed in the future research.

7. Acknowledgements

The financial support from the National Natural Science Foundation of China (No. 51079107) and the National Natural Science Fund for Distinguished Young Scholars of China (No. 50725931), and the Program for New Century Excellent Talents in University (No. NCET-09-0610) for this study is gratefully acknowledged.

8. References

Alejano, L. R. & Alonso, E. (2005). Consideration of the dilatancy angle in rocks and rock masses. *International Journal of Rock Mechanics and Mining Sciences*, Vol. 42, No. 4, 481–507

Barton, N. (1976). Rock mechanics review: the shear strength of rock and rock joints. *International Journal of Rock Mechanics and Mining Sciences & Geomechanical abstracts*, Vol. 13, No. 9, 255-279

Barton, N. R. & Bandis, S. C. (1982). Effects of block size on the shear behaviour of jointed rocks. In: *Proc 23rd US Symp Rock Mechanics*, Berkeley

Barton, N.; Bandis, S. & Bakhtar, K. (1985). Strength, deformation and conductivity coupling of rock joints. *International Journal of Rock Mechanics and Mining Sciences & Geomechanical abstracts*, Vol. 22, No. 2, 121-140

Bear, J. (1972). *Dynamics of fluids in porous media*. American Elsevier, New York

Castillo, E. (1972). Mathematical model for two-dimensional percolation through fissured rock. In: *Proc Int Symp Percolation through Fissured Rock*, T1±D1-7, Stuttgart, Germany

Chen, Y. F.; Sheng, Y. Q. & Zhou, C. B. (2006). Strain-dependent permeability tensor for coupled M-H analysis of underground opening. *Proceedings of the 4th Asian Rock Mechanics Symposium*, pp. 271, Singapore, Nov 2006, World Scientific Publishing

Chen, Y. F.; Zhou, C. B. & Sheng, Y. Q. (2007). Formulation of strain-dependent hydraulic conductivity for fractured rock mass. *International Journal of Rock Mechanics and Mining Sciences*, Vol. 44, No. 7, 981-996

Chen, S. H. & Egger, P. (1999). Three dimensional elasto-viscoplastic finite element analysis of reinforced rock masses and its application. *International Journal for Numerical and Analytical Methods in Geomechanics*, Vol. 23, No. 1, 61–78

Chiarelli, A. S.; Shao, J. F. & Hoteit, N. (2003). Modeling of elastoplastic damage behavior of a claystone. *Int J Plasticity*, Vol. 19, 23–45

Esaki, T.; Du, S.; Mitani, Y.; Ikusada, K. & Jing, L. (1999). Development of a shear-flow test apparatus and determination of coupled properties for a single rock joint. *International Journal of Rock Mechanics and Mining Sciences*, Vol. 36, 641–50

Hoek, E.; Wood, D. & Shah, S. (1992). A modified Hoek-Brown criterion for jointed rock masses. In: *Proc Rock Characterization Symp ISRM: Eurock 92*, Hudson, J. A. (Ed.), 209–214, British Geotechnical Society, London

Hsieh, P. A. & Neuman, S. P. (1985). Field determination of the three-dimensional hydraulic conductivity tensor of anisotropic media. Water Resource Research, Vol. 21, No. 11, 1655–1665.

Huang, T. H.; Chang, C. S. & Chao, C. Y. (2002). Experimental and mathematical modeling for fracture of rock joint with regular asperities. Eng Fract Mech, Vol. 69, 1977–1996

Indelman, P. & Dagan, G. (1993). Upscaling of permeability of anisotropic heterogeneous formations. *Water Resources Research*, Vol. 29, No. 4, 917-923

Jing, L. (2003). A review of techniques, advances and outstanding issues in numerical modeling for rock mechanics and rock engineering. *International Journal of Rock Mechanics and Mining Sciences*, Vol. 40, No., 283-353

Jing, L.; Stephansson, O. & Nordlund, E. (1993). Study of rock joints under cyclic loading conditions. Rock Mechanics and Rock Engineering, Vol. 26, No. 3, 215–32

Kelsall, P. C.; Case, J. B. & Chabannes, C. R. (1984). Evaluation of excavation-induced changes in rock permeability. Int J Rock Mech Min Sci & Geomech Abstr, Vol. 21, No. 3, 123–35

Lai, T. Y. (2002). *Multi-scale finite element modeling of strain localization in geomaterials with strong discontinuity*. Ph.D. thesis, Stanford University

Liu, C. H.; Chen, C. X. & Fu, S. L. (2002). Testing study on seepage characteristics of single fracture with sand under shearing displacement. Chinese Journal of Rock Mechanics and Engineering, Vol. 21, No. 10, 1457-1461

Liu, J.; Elsworth, D. & Brady, B. H. (1999). Linking stress-dependent effective porosity and hydraulic conductivity fields to RMR. *International Journal of Rock Mechanics and Mining Sciences*, Vol. 36, 581-596

Liu, S. H. (1996). Generation of flow network and field tests on hydraulic conductivity for fractured rock mass. *Northwestern Hydropower*, Vol. 55, No. 1, 21-27

Lomize, G. M. (1951). *Flow in fractured rocks*. Gosenergoizdat, Moscow

Long, J. C. S.; Remer, J. S.; Wilson, C. R. & Witherspoon, P. A. (1982). Porous media equivalents for networks of discontinuous fractures. *Water Resource Research*, Vol. 18, No. 3, 645–58

Louis, C. (1971). A study of groundwater flow in jointed rock and its influence on the stability of rock masses. *Rock Mechanics Research Report*, No. 10, Imperial College of Science and Technology, London, Maini, YNT

Min, K. B. & Jing, L. (2003). Numerical determination of the equivalent elastic compliance tensor for fractured rock masses using the distinct element method. *International Journal of Rock Mechanics and Mining Sciences*, Vol. 40, No. 6, 795-816

Min, K. B.; Rutqvist, J.; Tsang, C. F. & Jing, L. (2004). Stress-dependent permeability of fractured rock masses: a numerical study. *International Journal of Rock Mechanics and Mining Sciences*, Vol. 41, No. 7, 1191-1210

Oda, M. (1985). Permeability tensor for discontinuous rock masses. *Geotechnique*, Vol. 35, No. 4, 483-195

Oda, M. (1986). An equivalent continuum model for coupled stress and fluid flow analysis in jointed rock masses. *Water Resources Research*, Vol. 22, No. 13, 1845-1856

Olsson, R. & Barton, N. (2001). An improved model for hydromechanical coupling during shearing of rock joints. *International Journal of Rock Mechanics and Mining Sciences*, Vol. 38, No. 3, 317-329

Pande, G. N. & Xiong, W. (1982). An improved multilaminate model of jointed rock masses. In: *Numerical Models in Geomechanics*, Dungar, R.; Pande, G. N. & Studer, J. A. (Ed.), 218–226, Bulkema, Rotterdam

Patir, N. & Cheng, H. S. (1978). An average flow model for determining effects of three-dimensional roughness on hydrodynamic lubrication. *ASME Journal of Lubrication Technology*, Vol. 100, 12-17

Plesha, M. E. (1987). Constitutive models for rock discontinuities with dilatancy and surface degradation. International Journal for Numerical and Analytical Methods in Geomechanics, Vol. 11, 345–62

Pusch, R. (1989). Alteration of the hydraulic conductivity of rock by tunnel excavation. *Int J Rock Mech Min Sci & Geomech Abstr*, Vol. 26, No. 1, 79–83

Snow, D. T. (1969). Anisotropic permeability of fractured media. *Water Resources Research*, Vol. 5, No. 6, 1273-1289

Vajdova, V. (2003). *Failure mode, strain localization and permeability evolution in porous sedimentary rocks*. Ph.D. thesis, Stony Brook University

Wang, M. & Kulatilake, P. H. S. W. (2002). Estimation of REV size and three dimensional hydraulic conductivity tensor for a fractured rock mass through a single well packer test and discrete fracture fluid flow modeling. *International Journal of Rock Mechanics and Mining Sciences*, Vol. 39, 887-904

Yuan, S. C. & Harrison, J. P. (2004). An empirical dilatancy index for the dilatant deformation of rock. *International Journal of Rock Mechanics and Mining Sciences*, Vol. 41, 679–86

Zimmerman, R. W.; Kumar, S. & Bodvarsson, G. S. (1991). Lubrication theory analysis of the permeability of rough-walled fractures. *International Journal of Rock Mechanics and Mining Sciences*, Vol. 28, No. 4, 325-331

Zhou, C. B.; Chen, Y. F. & Sheng, Y. Q. (2006). A generalized cubic law for rock joints considering post-peak mechanical effects. In: Proc GeoProc2006, 188–197, Nanjing, China

Zhou, C. B.; Sharma, R. S.; Chen Y. F. & Rong, G. (2008). Flow-Stress Coupled Permeability Tensor for Fractured Rock Masses. *International Journal for Numerical and Analytical Methods in Geomechanics*, Vol. 32, 1289-1309

Zhou, C. B. & Xiong, W. L. (1996). Permeability tensor for jointed rock masses in coupled seepage and stress fields. *Chinese Journal of Rock Mechanics and Engineering*, Vol. 15, No. 4, 338-344

Zhou, C. B. & Xiong, W. L. (1997). Influence of geostatic stresses on permeability of jointed rock masses. *Acta Seismologica Sinica*, Vol. 10, No. 2, 193-204

Zhou, C. B.; Ye, Z. T. & Han, B. (1997). A study on configuration and hydraulic conductivity of rock joints. *Advances in Water Science*, Vol. 8, No. 3, 233-239

Zhou, C. B. & Yu, S. D. (1999). Representative elementary volume (REV): a fundamental problem for selecting the mechanical parameters of jointed rock mass. *Chinese* Journal of Engineering Geology, Vol. 7, No. 4, 332-336

2

Influence of Degree of Saturation in the Electric Resistivity-Hydraulic Conductivity Relationship

Mohamed Ahmed Khalil[1,2] and Fernando A. Monterio Santos[1]
[1]*Universidade de Lisboa, Centro de Geofísica da Universidade de Lisboa-IDL,*
[2]*National Research Institute of Astronomy and Geophysics,*
[1]*Portugal*
[2]*Egypt*

1. Introduction

The relationship between hydraulic conductivity and electric resistivity is one of the most difficult and challenging approaches in the field of hydrogeophysics. The promising side of this relation is the analogy between electric current flow and water flow, whereas the grand ambiguity is the non-dimensionality between both two quantities. Relationship between hydraulic conductivity and electric resistivity either measured on the ground surface or from resistivity logs, or measured in core samples has been published for different types of aquifers in different locations. Generally, these relationships are empirical and semi-empirical, and confined in few locations. This relation has a positive correlation in some studies and a negative in others. So far, there is no potentially physical law controlling this relation, which is not completely understood. Eelectric current follows the path of least resistance, as do water. Within and around pores, the model of conduction of electricity is ionic and thus the resistivity of the medium is controlled more by porosity and water conductivity than by the resistivity of the rock matrix. Thus, at the pore level, the electrical path is similar to the hydraulic path and the resistivity should reflect hydraulic conductivity. This chapter will discuss the following items:

1. A general revision of the theoretical relation between hydraulic conductivity and electric resistivity and the role of surface conductance as an effective transporting mechanism.
2. A brief revision of different published theoretical and empirical methods to estimate hydraulic conductivity from electric resistivity.
3. Studying the effect of degree of groundwater saturation in the relation between hydraulic conductivity and electric resistivity via a simple numerical analysis of Archie's second law and a simplified Kozeny-Carman equation.

Initially, every hydrogeologic investigation requires an estimate of hydraulic conductivity (K), the parameter used to characterize the ease with which water flows in the subsurface. (J.J. Butler, 2005). Hydraulic conductivity differs significantly from permeability, where hydraulic conductivity of an aquifer depends on the permeability of the hosting rock and viscosity and specific weight of the fluid (Hubbert, 1940), where as permeability is a function of pore space only.

Hydraulic conductivity has been measured long time by traditional hydrogeologic approaches. Such these approaches are: pumping test, slug test, laboratory analysis of core samples, and geophysical well logging.
Pumping tests do produce reliable (K) estimates, but the estimates are large volumetric averages. Laboratory analysis can provide information at a very fine scale, but there are many questions about the reliability of the (K) estimates obtained with those analyses. Although the slug test has the most potential of the traditional approaches for detailed characterization of (K) variations, most sites do not have the extensive well network required for effective application of this approach. (J.J. Butler, 2005). However, these traditional methods are time-consuming and invasive.
Another group of hydrogeological methods are used to measure vertical hydraulic conductivity such as: Dipole- Flow test (DFT), Multilevel slug test (MLST), and Borehole Flow meter test (BFT). These techniques can only be used in wells, which often must be screened across a relatively large portion of the aquifer and provide information about conditions in the immediate vicinity of the well in which they are used.
The ability to reliably predict the hydraulic properties of subsurface formations is one of the most important and challenging goals in hydrogeophysics, since in water-saturated environments, estimation of subsurface porosity and hydraulic conductivity is often the primary objective. (D. P. Lesmes and S. P. Friedman, 2005). Many hydrogeophysical approaches have been used to study the relationship between hydraulic conductivity from surface resistivity measurements.

2. Electric resistivity-hydraulic conductivity relationship

Since the electrical resistivity of most minerals is high (exception: saturated clay, metal ores, and graphite), the electrical current flows mainly through the pore water. According to the famous Archie law (Archie, 1942), the resistivity of water saturated clay-free material can be described as

$$R_o = R_w \,.\, F_i \tag{1}$$

Where,
R_o = specific resistivity of water saturated sand,
R_w = specific resistivity of pore water,
F_i = intrinsic formation factor.
The intrinsic formation factor (F_i) combines all properties of the material influencing electrical current flow like porosity φ, pore shape, and digenetic cementation.

$$F_i = a.\varphi^{-m} \tag{2}$$

Different definitions for the material constant (m) are used like porosity exponent, shape factor, and cementation degree. Factors influencing (m) are, e.g., the geometry of pores, the compaction, the mineral composition, and the insolating properties of cementation. The constant (a) is associated with the medium and its value in many cases departs from the commonly assumed value of one. The quantities (a) and (m) have been reported to vary widely for different formations. The reported ranges are exemplified in table (1), which is based upon separate compilations of different investigators.

Lithology	*a*	*m*	*Author (s)*
Sandstone	0.47-1.8 0.62-1.65 1.0-4.0 0.48-4.31 0.004-17.7	1.64-2.23 1.3-2.15 0.57-1.85 1.2-2.21 0.02-5.67	Hill and Milburn (1956) Carothers (1968) Porter and Carothers (1970) Timur at al. (1972) Gomez-Rivero (1977)
Carbonates	0.73-2.3 0.45-1.25 0.33-78.0 0.35-0.8	1.64-2.1 1.78-2.38 0.39-2.63 1.7-2.3	Hill and Milburn (1956) Carothers (1968) Gomez-Rivero (1977) Schon (1983)

Table 1. Reported ranges of the Archie constants (a) and (m).

Equation (2) is called Archie's first law, where it is valid only in fully saturated clean formations (the grains are perfect insulators).
When the medium is not fully saturated, water saturation plays an important role, where the changing in degree of saturation changes the effective porosity (accessible pore space). It became Archie's second law.

$$F_i = \frac{R_o}{R_w} = a\varphi^{-m} S_w^{-n} \tag{3}$$

Where, R_o is the formation resistivity, R_w is the pore water resistivity, φ is the porosity, S_w is the water saturation, a and m are constants related to the rock type, and n is the saturation index (usually equals 2).
Many studies concluded that Archie's law breaks down in three cases: (1) clay contaminated aquifer (Worthington, 1993, Vinegar and Waxman, 1984, Pfannkuch, 1969), (2) partially saturated aquifer (Börner, et. al., 1996, Martys, 1999), and (3) fresh water aquifer (Alger, 1966, Huntley, 1987).
In Archie condition (fully saturated salt water clean sand), the apparent formation factor equals the intrinsic formation factor (Archie, 1942). Whereas in non-Archie condition the apparent formation factor is no longer equals to the intrinsic formation factor.
Vinegar and Waxman (1984) stated that Archie's empirical equations have provided the basis for the fluid saturation calculations. In shaly sands, however, exchange counter ions associated with clay minerals increase rock conductivity over that of clean sand, and the Archie relations is no longer valid.
Huntley (1986) showed that at low groundwater salinities, surface conduction substantially affects the relation between resistivity and hydraulic conductivity and, with even low clay contents, the relation between hydraulic conductivity and resistivity becomes more a function of clay content and grain size and less dependent (or independent) of porosity.
A large number of empirical relationships between hydraulic conductivity and formation factor have been published. Figure (1), shows some inverse relations between aquifer hydraulic conductivity and formation factor, reported after Heigold, et. al., (1979) using data from Illinois, Plotnikov, et. al.,(1972) using data from Kirgiza in the Soviet Union, Mazac and Landa (1979), Mazac and Landa (1979) analyzing data from Czechoslovakia, and Worthington (1975).

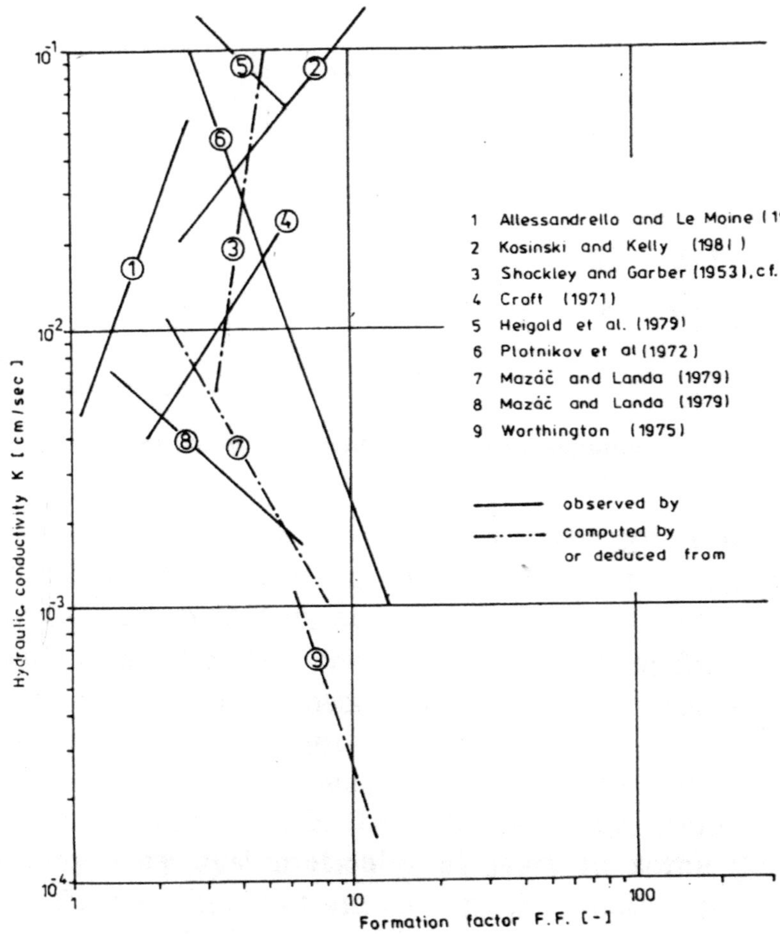

Fig. 1. Reported relation between hydraulic conductivity and aquifer formation factor (after Mazac, et. al., 1985).

Another group of case studies reported the opposite behaviour i.e., the direct relation between aquifer hydraulic conductivity and formation factor, (Allessandrello and Le Moine, 1983, Kosinski and Kelly, 1981, Shockley and Garber, 1953, and Croft, 1971).

In non-Archie conditions, there will be the double-layer phenomenon, which introduces an additional conductivity to the system called surface conductance. Surface conductance is a special form of ionic transport occurs at the interface between the solid and fluid phases of the system (Pfannkuch, 1969). It is found that, the validity of Archie's law depends on the value of the Dukhin number, which is the ratio between surface conductivity at a given frequency to the conductivity of the pore water (Bolève et al. 2007, Crespy et al. 2007). When the Dukhin number is very low with respect to 1, Archie's law is valid.

Theoretical expressions, which include consideration of conductivity in the dispersed (solid) phase and in the continuous (fluid) phase, as well as a grain surface conductivity phase are best represented by an expression in the form of a parallel resistor model (Pfannkuch, 1969). One of the earliest parallel resistor models was proposed by Patnode and Wyllie (1950) to account for the observed effects of clay minerals in shaly sand.

$$\frac{1}{F_a}=\frac{1}{F_i}+\frac{R_w}{R_c} \quad (4)$$

where, R_w is the water resistivity, R_c is the resistivity of clay minerals, F_i is the intrinsic formation factor, and F_a is the apparent formation factor.

Pfannkuch (1969), proposed his parallel resistor model, emphasizing the role that surface conductivity plays in the electrical transport process.

$$\frac{1}{R_e}=\frac{1}{R_f}+\frac{1}{R_d}+\frac{1}{R_s} \quad (5)$$

Or in conductance terms

$$K_e=K_f+K_d+K_s \quad (6)$$

where K_e is the conductance of the combined or bulk phase, K_f is the conductance of the continuous phase (fluid), K_d is the conductance of the dispersed phase (solid), and K_s is the surface conductance.

This model was expressed by Pfannkuch, (1969) in terms of the geometry of the matrix system, incorporating the concept of tortuosity, in the following form:

$$F_a=F_i\left[1+\frac{K_d}{K_f}\frac{1-\varphi}{\varphi}\left(\frac{L_e}{L_d}\right)^2+\frac{K_s}{K_f}S_p\right]^{-1} \quad (7)$$

Where L_e is the tortuous path, L_d is the flow path through the solid material, and S_p is the specific internal pore area (the total interstitial surface area of the pores per unit por volume of the sample).

If the matrix grains consist primarily of non-conducting minerals, such as quartz, the matrix conductivity represented by the second term in the denominator of (7) becomes very small and can be neglected (Urish, 1981). Equation (7) becomes

$$F_a=\frac{F_i}{1+(\frac{K_s}{K_f})S_p} \quad (8)$$

Of particular interest, the term (k_s/k_f) represents the relative magnitude of the surface conductance to pore-water conductance. When (k_f) becomes large due to high molarity concentration of fluid, this term approaches zero. The apparent formation factor (F_a) then approaches the intrinsic formation factor (Fi), which is the case for saline pore-water. But for high-resistivity fresh water sands, the surface conductance effect represented by the term $\left(\frac{k_s}{k_f}\right)S_p$ must be considered (Urish, 1981).

This model is equivalent to Waxman-Smits model (1968) for clayey sediments. It relates the intrinsic formation factor, F_i and the apparent formation factor, F_a (the ratio of bulk resistivity to fluid resistivity), after taking into consideration the shale effect. According to Worthington (1993),

$$F_a = F_i \, (1 + BQ_v \, R_w)^{-1} \tag{9}$$

Waxman and Smits (1968) used two parameters; the first is Q_v, which is the cation exchange capacity (CEC) per unit pore volume of the rock (meq/ml) (Worthington, 1993). It defined as cation concentration (Butler and Knight, 1998), and reflects the specific surface area, which is a constant for a particular rock. It describes also the number of cations available for conduction that are loosely attached to the negatively charged clay surface sites. The ions, which can range in concentration from zero to approximately 1.0 meq/ml, are in addition to those in the bulk pore fluid. Q_v varies with porosity according to the following equation (Worthington, 1993).

$$\log Q_v = -3.56 - 2.74 \log \varphi \tag{10}$$

The second parameter, B, is the equivalent ionic conductance of clay exchange cations (mho-cm^2/meq) as function of *Cw* (specific conductivity of the equilibrating electrolyte solution (mho/cm) (Worthington, 1993). This parameter is called the equivalent electrical conductance, which describes how easily the cations can move along the clay surface (Butler and Knight, 1998). It varies with water resistivity according to the equation

$$B = 3.83\left[1 - 0.83 \exp(-0.5 / R_w)\right] \tag{11}$$

This equation implies that clay conduction will be more important as a mechanism than bulk pore-fluid conduction at low salinities and less important at high salinities.
The product BQ_v has units of conductivity. Comparison between Urish model (1981) (eq.8) and Waxman-Smits model (1968) (eq.9), shows that $Kf = 1/Rw$, and $Ks\ Sp = BQv$.
Equation (9) is modified by (Butler and Knight, 1998) to the following form

$$\sigma_b = \varphi^m \left(\sigma_w + \frac{BQ_v}{S_w} \right) S_w^n \tag{12}$$

Where, the first term in the parentheses represents bulk pore-fluid conduction, while the second represents clay surface conduction. Clay conduction is not as strongly affected by water saturation as is conduction through the bulk pore fluid because the number of clay cations remains constant until very low levels of saturation (Butler and Knight, 1998).
According to Waxman and Smits (1968) model, a shaly formation behaves like a clean formation of the same porosity, tortuosity, and fluid saturation, except the water appears to be more conductive than its bulk salinity. In other words, it says that the increase of apparent water conductivity is dependent on the presence of counter-ion (Kurniawan, 2002). Accordingly, equation (8) could be used also for shaly formations.
Vinegar and Waxman (1984) proposed a complex conductivity form of the Waxman-Smits' model (1968), based on measurements of complex conductivity (σ^*) of shaly sandstone samples as function of pore water conductivity, as shown in equation (13).

$$\sigma^* = \left(\frac{\sigma_w}{F_a} + \frac{BQ_v}{F_i} \right) + i \frac{\lambda Q_v}{Fn} \tag{13}$$

Where the Waxman-Smits' part of the equation is the real component that represents the electrolytic conduction in fluid $\left(\frac{\sigma_w}{F_a}\right)$ and real surface conductivity component $\left(\frac{BQ_v}{F_i}\right)$, which are in-phase with the applied electric field. The imaginary conductivity component $\left(i\frac{\lambda Q_v}{Fn}\right)$ is the conductivity which results from displacement currents that are 90° out of phase with the applied field. Vinegar and Waxman assumed that the displacement currents were caused by the membrane and the counter-ion polarization mechanisms. These two mechanisms were proportional to the effective clay content or specific surface area represented by the parameter (Q_v). The parameter (λ) represents an effective quadrature conductance for these surface polarization mechanisms. (λ) is slightly dependent on salinity.

The low-frequency complex conductivity (σ^*) can be explained by a simple electrical parallel conduction of three components (Vinegar and Waxman 1984, Börner, 1992, Lesmes and Frye (2001)): (1) real electrolytic conductivity (σ_{bulk} ; Archie 1942), (2) real surface conductivity component ($\sigma'_{surf}(\omega)$), and (3) imaginary surface conductivity component ($\sigma''_{surf}(\omega)$) caused by charge polarization.

$$\sigma^* = \left[\sigma_{bulk} + \sigma'_{surf}(\omega)\right] + i\sigma''_{surf}(\omega) \tag{14}$$

The imaginary part of conductivity is widely studied by Börner et al. (1992) and (1996) and Slater and Lesmes (2002). They found a strong relation between surface conductivity components and surface-area-to-porosity ratio (S_{por}), effective grain size (d_{10}), and the product of measured hydraulic conductivity multiplied by true formation factor (K *x* F) as shown in figures (2, a, b, c)

Börner et al. (1992) and (1996) described the imaginary part (σ'') of water-saturated rock as

$$\sigma'' = \frac{lf(\sigma_w)S_p}{F} \tag{15}$$

where *F*, for purposes of simplicity, is the same formation factor for all conductivity components, $f(\sigma_w)$ is a general function concerning salinity dependence of interface conductivity and depending on surface charge density and the ion mobility, and *l* is the ratio between real and imaginary component of interface conductivity that is assumed to be nearly independent of salinity.

Slater and Lesmes (2002) mentioned a power relationship between the saturated hydraulic conductivity and imaginary conductivity as well.

$$K_s = a\left(\frac{lf(\sigma_w)S_p}{F}\right)^b \tag{16}$$

where *a* and *b* are the respective constants. For the dataset they used, they find, $a = 0.0002 \pm 0.0003$ and $b = 1.1 \pm 0.2$ (σ'' in μS/m, *K* in m/s; $R^2 = 0.7$, CI = 95%).

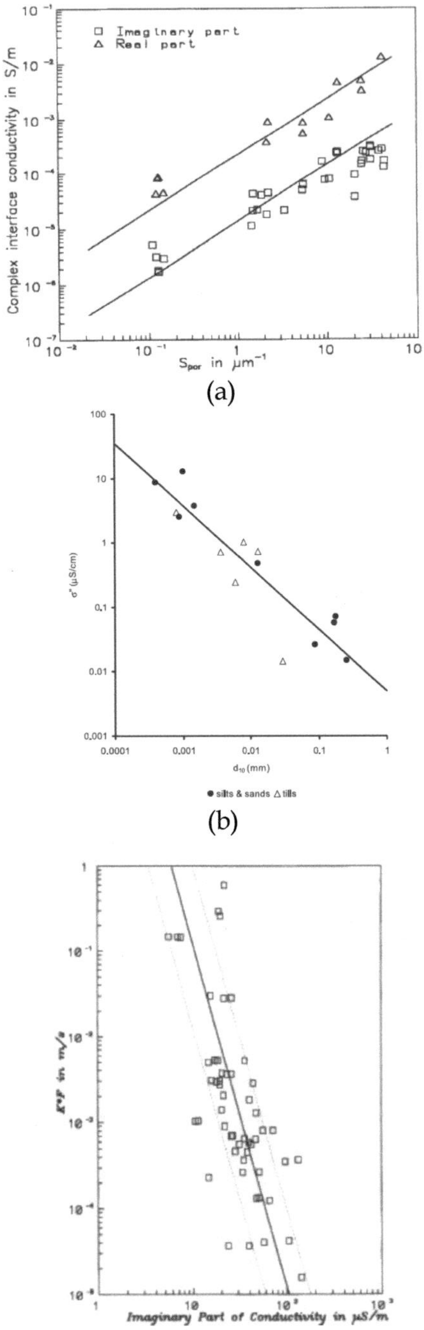

Fig. 2. (a)-Complex interface conductivity components vs. surface-area-to-porosity ratio S_{por} for sandstones (Börner, 1996), (b)- Plot of σ_{surf}'' (1 Hz) versus (d10) (Slater and Lesmes 2002), (c)- Relationship between the imaginary component of complex electrical conductivity and the product of the true formation factor (F) and the permeability (K) (Börner, 1996).

Börner et al (1992, 1996) and Slater and Lesmes (2002) showed that the imaginary (quadrature) surface conductivity resulted in nearly identical numerical values with the geometric hydraulic conductivity. It is proposed also, since there is a large similarity between imaginary surface conductivity component and real surface conductivity component, the later could be used to estimate hydraulic conductivity (Khalil and Fernando, 2010).

3. Estimation of hydraulic conductivity from electric resistivity

Estimation of hydraulic conductivity from electric resistivity measurements can offer the following advantages: (1) It can provide a new and important hydrogeologic trend for the application of resistivity measurements, (2) potential estimation of many hydraulic parameters through hydraulic conductivity, (3) Evaluation of the groundwater potentiality of new reclaimed areas before well drilling. It gives advantage to select the most productive zones for drilling new wells, (4) resistivity data are densely sampled, repetitive, spatially continuous information can be obtained, (5) measurements are indirect and minimally invasive, and (6) the scale of the measurement can be controlled through appropriate field survey design.

In addition to the recently developed method to estimate hydraulic conductivity from imaginary surface conductivity component via complex resistivity or induced polarisation measurements (Börner et. al. 1992, 1996, and Slater and Lesmes 2002), there are many hydrogeophysical approaches that have been used to estimate hydraulic conductivity from surface resistivity measurements. These approaches are classified as follows:

3.1 Combined interpretation of hydrogeologic and geophysical data:

This type of approaches is carried out by S. Niwas and D.C. Singhal (1981). These authors used Vertical electrical sounding and pumping tests to provide analytical relationship to estimate the aquifer transmissivity from transverse resistance in an area of the same geological situation, if hydraulic conductivity of the aquifer at any point therein is known, considering that ($K.\sigma$) is a constant factor. This method was applied at different areas such as Umuahia area of Nigeria (P. D. Mbonu, et al, 1991), Wadi El- Assuity, Egypt (M. A. Khalil, et al, 2005) and in the middle Imo river basin aquifers, south-eastern Nigeria (A.C. Ekwe, et. al., 2006). This method resulted in a fairly good correlation with the measured data.

S. Niwas and D.C. Singhal (1985) introduced normalized aquifer resistivity instead of aquifer resistivity. An Analytical relationship between normalized transverse resistance and aquifer transmissivity has been developed for estimating transmissivity from resistivity sounding data taking into consideration the variation in groundwater quality. This method is applied by Yadav, et al, (1993) and Yadav (1995) for Jayant project, Singrauli coalfields, India. Yadav (1995) found that normalized aquifer resistivity is a very good predictor for transmissivity in this aquifer.

Chandra, S., et al., (2008) developed a similar approach to estimate hydraulic conductivity of Maheshwaram watershed aquifer in hard rock terrain in Hyderabad, India.

Another combined approach was proposed by Soupios, P., et al., (2007); they used groundwater resistivity (Rw) measured from boreholes samples and apparent formation factor (Fa), estimated using formation resistivity from Vertical Electrical Sounding to estimate intrinsic formation factor. Intrinsic formation factor is used to estimate porosity. Estimated porosity is then, used in Kozeny-Carman equation to estimate hydraulic conductivity of Keritis basin in Chania (Crete-Greece).

3.2 Empirical and semi-empirical hydrogeological and geophysical relationship depending on petrophysical relation:

This category is the largest group of approaches in both field and laboratory scale. A) Field scale: P. F. Worthington (1976), correlated between the values of groundwater resistivity (R_w) determined from the chemical analysis of borehole water samples, with the formation resistivity (R_o) as deduced from the interpretation of geoelectric soundings measured nearby boreholes. He concluded that, geoelectric determination of groundwater salinity would be most exact at lower salinities and where porosity is relatively high. W. E. Kelly, (1977), carried out a correlation between resistivity values of six Schlumberger VES and pumping test data of the wells. He got a good direct relation between aquifer resistivity and measured hydraulic conductivity, good direct relation between aquifer resistivity and specific capacity, and good direct relation between formation factor and measured hydraulic conductivity. P. C. Heigold, et al., (1979), used Wenner sounding resistivity and hydraulic conductivity data from pumping test to show an inverse relation between hydraulic conductivity and resistivity due to that poorly sorted sediments are responsible for reduced porosity and thus less hydraulic conductivity. W. Kosinski and W. Kelly (1981) presented data showing a direct relation between permeability and apparent formation factor and another direct relation between transmissivity and normalized aquifer resistance. Frohlich R. and Kelly, W.E (1985), showed a direct empirical relation between hydraulic conductivity and transverse resistivity, and empirical relation between hydraulic conductivity and transverse resistivity. Mazac, et. al., (1985), studied the Factors influencing relations between electrical and hydraulic prosperities of aquifers and aquifer materials. A general hydrogeophysical model was used to demonstrate that at the aquifer scale a variety of relations might be expected. R.K. Frohlich, et. al., (1996), studied the relationship between hydraulic conductivity and aquifer resistivity in fractured crystalline bedrock, Rhode Island. Reverse relation between hydraulic conductivity and aquifer resistivity has been found. This result agree with theoretical calculations by Brown (1989), laboratory sample measurements by Mazac et al, (1990), and field data relationship by Heigold et al (1979).
B) Laboratory scale: David Huntley (1987) performed laboratory experiments to show the importance of matrix conduction. He showed that the ratio between the measured bulk resistivity and the measured fluid resistivity, the apparent formation factor varies significantly with varying fluid resistivity for the range of normal ground water salinities.

3.3 Theoretically petrophysical based models:

The accuracy of determining the porosity, the filtration coefficient and transmissivity of an aquiferous reservoir rock, the mineralization and actual flow velocity of underground water in a percolation medium by means of surface geoelectric methods is discussed via synthetic data (O. Mazac, et. al. 1978). The results of theoretical analysis enable the accuracy in determining the fundamental hydrogeological parameters by the VES method. R.K.Frohlich, (1994), the relationship between resistivity and hydraulic conductivity is discussed on the Kozeny- Carmen equation. The uses and abuses of the Archie equations are modelled by Worthington, (1993) using Waxman and Smits equation (1968).
Generally, geophysics assisted groundwater exploration is based on empirical relationships between electric and hydraulic units. Empirical laws are unsatisfactory, as they do not provide an understanding of any potential physical law. However, similar relationships must be established in new areas. The dependence between (K) and (R) remains nonunique; a simple predictable K-R relationship can not be expected.

Some previous studies combine two or more regimes such that, D.W. Urish (1981), where, theoretically three-phase parallel resistor model, supported by data from laboratory tests assumed inverse correlation between porosity and hydraulic conductivity. From empirical and theoretical model a positive correlation between apparent formation factor and hydraulic conductivity is shown. The model demonstrates that intergranular surface conductance is an important factor at small grain size and high pore water resistivities, operating to lower the apparent formation factor. W. E. Kelly and P. F. Reiter (1984), where the influence of aquifer anisotropy caused by layering on the relation between resistivity and hydraulic conductivity was studied with idealized analytic and numerical models.
It is worthily mentioned that all these relations are site restricted and have no potentially physical law; in addition, the physical relation between hydraulic conductivity and aquifer resistivity is not completely understood. It has a direct correlation in some studies and reverses in others. The main target of this paper is to study the effect of water saturation in such relation.

4. Influence of water saturation in the electric resistivity-hydraulic conductivity relationship.

Archie's first and second laws show the relation between bulk resistivity and formation factor. Formation factor could be linked to hydraulic conductivity by Kozeny-Carman equation. One of the most recent modifications of this equation is made by Börner and Shön (1991). They obtained the following expression for the estimation of hydraulic conductivity of unconsolidated sediments (sand, gravel, silt) (Lesmes and Friedman, 2005):

$$K_s = \frac{a}{F S_{p[el]}^c} = \frac{a}{F(10^5 \sigma''_{1Hz})^c} \tag{17}$$

Where K_s is the hydraulic conductivity in m/s, F is the apparent formation factor, $S_{p[el]}$ is the electrically estimated specific surface area per unit volume (μm^{-1}), σ'' is the imaginary conductivity component measured at 1 Hz (S/m), a is a constant equals 10^{-5}, C is a constant ranges between 2.8 and 4.6 depending on the material type and the method used to measure Ks.
Accordingly, the modified Kozeny-Carman equation (Eq. 17) and Archie's first and second laws (Eqs. (2) and (3)) should control the relationship between hydraulic conductivity (K) and formation resistivity (R_o) in both saturated and non-saturated sediments.
Khalil and Fernando (2009) numerically analyzed two important equations: (1) Archie's second (eq.3), which controls the relation between porosity, water saturation, and formation factor, (2) Kozeny-Carman model (eq.17), which controls the relation between formation factor and hydraulic conductivity. Beginning with the generalized Archie's second law, using a =1, m=n=2, and proposed values of porosity and water saturation ranging from 0.2 to 1 with an increment of 0.2. They calculated the net product of porosity (φ) and water saturation (S_w), which is the volumetric water content (θ).

$$\theta = \varphi . S_w \tag{18}$$

Figure (3, a) shows the relation between intrinsic formation factor and porosity when water saturation equals one. Figure (3, b) shows the same relation when porosity equals water saturation. The two cases (Fig 3, a, b) resulted in an inverse power relationship with a correlation coefficient equals one.

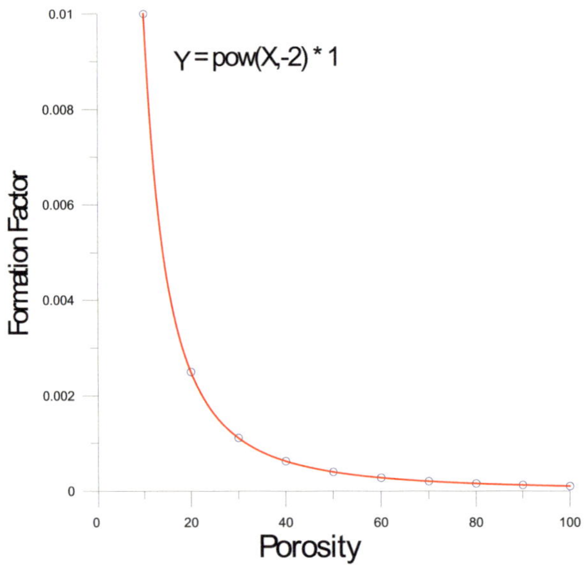

(a)

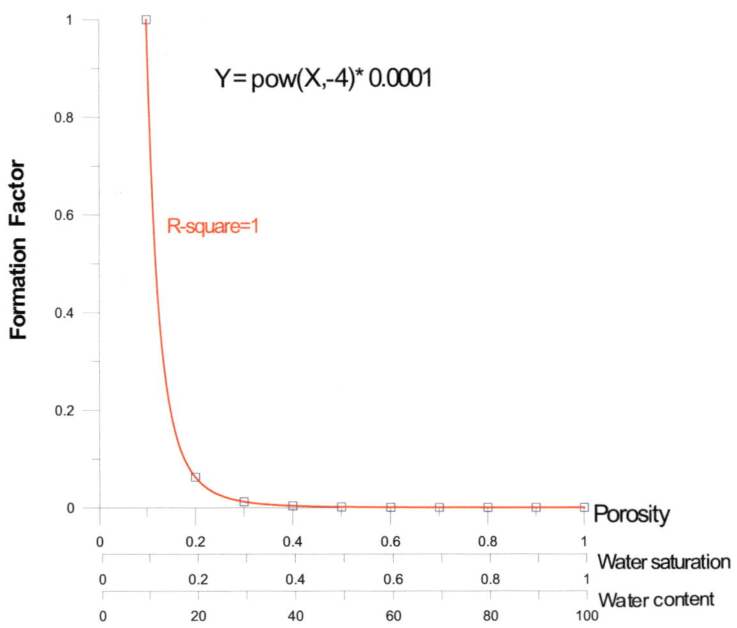

(b)

Fig. 3. Analytical relation between formation factor, porosity, water saturation, and water content when A)-water saturation =1, and B) - porosity = water saturation.

In the case where water saturation and porosity changes inversely to each other, they got the following relation (Fig.4)

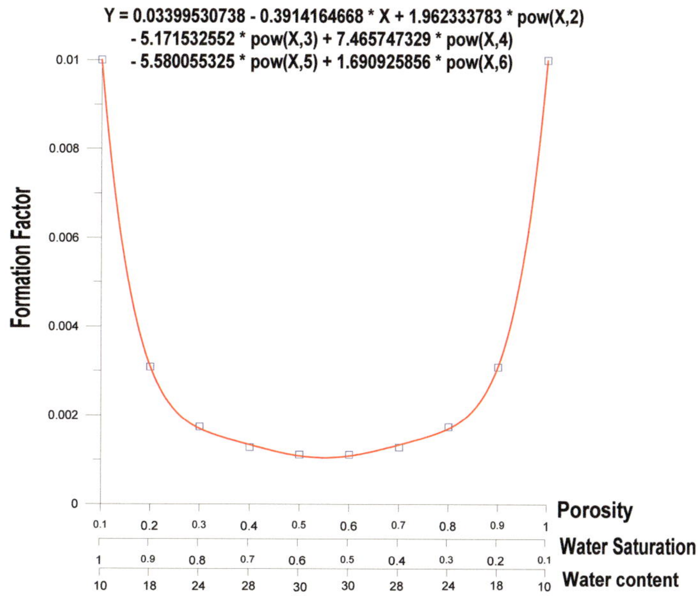

Fig. 4. Analytical relationship between Formation factor, porosity, water saturation and water content when porosity ≠ water saturation.

Archie's law in this case has deviated from it is traditional power low to a polynomial correlation of sixth order. In the right half of the curve, where porosity is lower than water saturation, and lower than 50%, a considerable inverse polynomial relation has achieved. In the left half of the curve, where porosity is higher than water saturation, and higher than 50% (poorly saturated sediments) a direct polynomial relation exist. In this part of the curve Archie's second low does not deviated from its power low to a polynomial correlation only but it breaks down also, where formation factor has a direct correlation with porosity and water saturation. However, for practical purposes, a direct correlation between (F) and (φ) is in common usage (Börner, et. al., 1996). Martys, (1999) used Lattice Boltzmann method to numerically simulate the diffusive transport of ions in two classes of partially-saturated porous media as a function of saturation and wetting properties. At high saturations, good agreement is found between his estimates of diffusivity and that predicted by the semi-empirical Archie's second law. At lower saturations, it is found that Archie's second law breaks down as percolation effects become important. His study resulted in an empirical polynomial function between relative diffusivity ($\frac{\sigma_{bi}}{\sigma_b}$) and water saturation (Sw), where σ_{bi} is the electrical conductivity of fluid and σ_b the electrical conductivity of wetted (partially saturated) porous material.

Since, figure (4) describes two different hydrogeological media; they are separated and presented in figures (5a, and b).

Figure (5) describes the relation when water saturation > 50%> porosity (5.a) and when water saturation < 50% < porosity (5.b). The best fit to the analytical data (correlation coefficient = 1) is the polynomial regression fourth order (blue line), where power correlation shows a lower fitting (red line) in the two cases. Figure (5, a) still reflect the inverse relation between intrinsic formation factor and both porosity and water saturation.

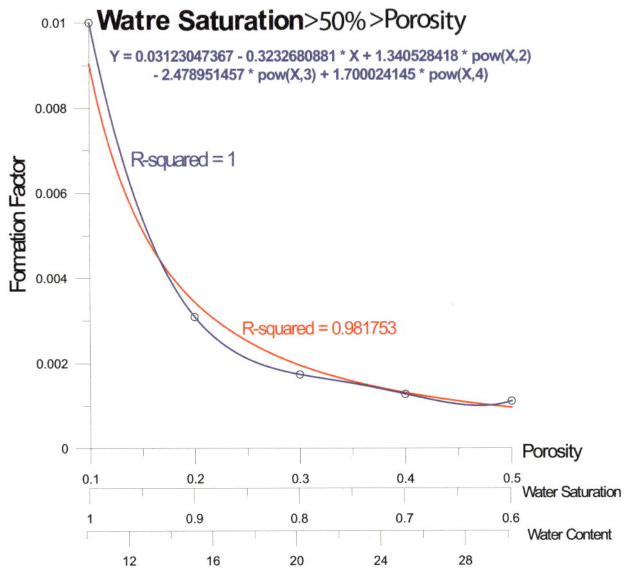

(a)

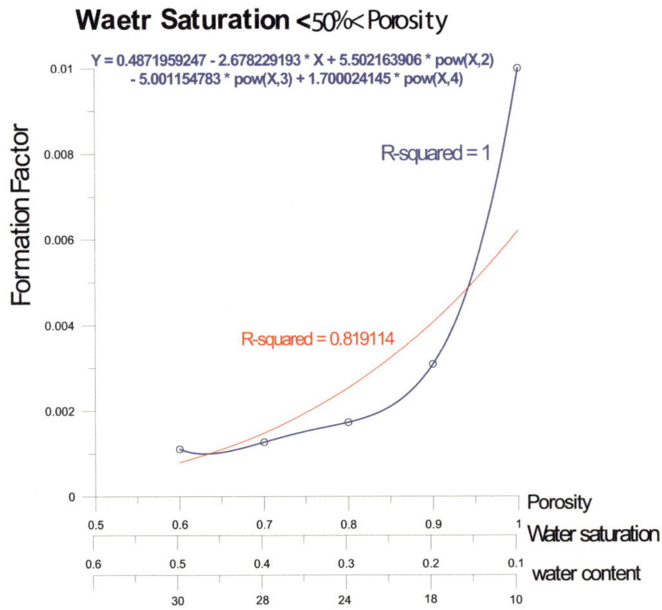

(b)

Fig. 5. Analytical relationship between Formation factor, porosity, water saturation and water content in the two different cases.

Whereas figure (5, b) reflects a direct correlation between intrinsic formation factor and both porosity and water saturation, which is in agreement with Martys, (1999) and (Börner, 1996). Applying these direct and inverse relations in the modified Kozeny-Carman model (Eq. 4), resulted in an inverse correlation between hydraulic conductivity and formation factor in the first case (fig.5, a) and a direct correlation in the second case (fig 5, b). Comparing these

results with some published empirical relations concluded between aquifer hydraulic conductivity from pumping test and formation factor, shows an agreement (Fig.1).
In the view of present analysis (Figs, 3, 4, and 5), we can expect a group of relations between hydraulic conductivity (K) and formation resistivity (Ro), differ in mathematical expressions and hence in curve form. These relations could be classified into 3 characteristic cases:

1. An inverse power relation in fully saturated aquifers and when porosity equals water saturation.
2. An inverse polynomial relation in unsaturated aquifers, when water saturation higher than 50%, higher than porosity.
3. A direct polynomial relation in poorly saturated aquifers, when water saturation lower than 50%, lower than porosity.

In the next section, the present results are compared with some previously published empirical relations between aquifer resistivity and hydraulic conductivity in different geographic locations and hydrogeologic conditions with a comparison between expected porosity and saturation from our models with that measured, as possible as the data is available.

4.1 First category :(fully saturated aquifer or water saturation equals porosity)

Two case studies (Fig.6, a and b) are collected: (A) fractured crystalline bedrock, central landfill, Rhode Island, USA (Frohlich, et. al., 1996), (B) Granitic host rock, (OUC), Hyderabad, A.P., India (K.P.Singh, 2005).

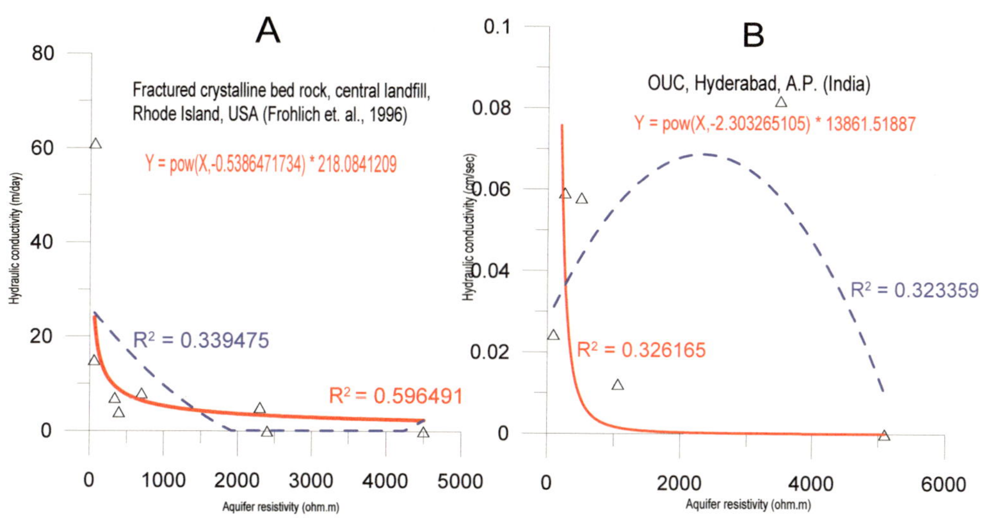

Fig. 6. Empirical relationship between hydraulic conductivity and aquifer resistivity in different locations. (Red sold line is the power relation; blue dashed line is the polynomial relation)

A considerable inverse power correlation between hydraulic conductivity and aquifer resistivity exist in the two case studies. The correlation coefficient of the power relation is higher than that of polynomial in the two cases. Geologically, all cases are from fractured hard rock aquifers. The fractured crystalline bedrock, central landfill, Rhode Island, USA (fig.6-A) is characterized by high fractured granite, high hydraulic conductivity, no primary permeability and hydraulic flow is restricted to fractures, and no clay, where weathering product of granite decomposition, have been washed out by glacial melt waters (Frohlich, et. al., 1996). Water resistivity ranges from 41 to 125 Ohm.m (Frohlich, et. al., 1996). Estimated porosity from the published data of formation resistivity (Ro), water resistivity (Rw), and formation factor (F) ranges from 19 to 82 %, assuming that a=1 and m=2, in Archie's first law.

Data published by K.P.Singh, (2005) were measured in Osmania University Campus (OUC), Hyderabad A.P. (India) for the fractured Granitic aquifer of Archaean age.

The available information of the two fractured hard rock aquifers and the empirical inverse power correlation indicate that they are in a good agreement with the first analytical model (fig.3).

4.2 Second category (water saturation > 50% > porosity)

Three case studies have non-linear inverse correlation between hydraulic conductivity and formation resistivity: Glacial outwash aquifer in central Illinois, USA, (Heigold, et. al., 1979), Banda area U.P., India, (Niwas and Singhal, 1985), and Mount Tsukuba, Central Japan, intact rock aquifer (Sudo et. al., 2004).

From Figure (7), the data are correlated as inverse polynomial with more correlation coefficient than that of power correlation, which in agreement with figure (5, A).

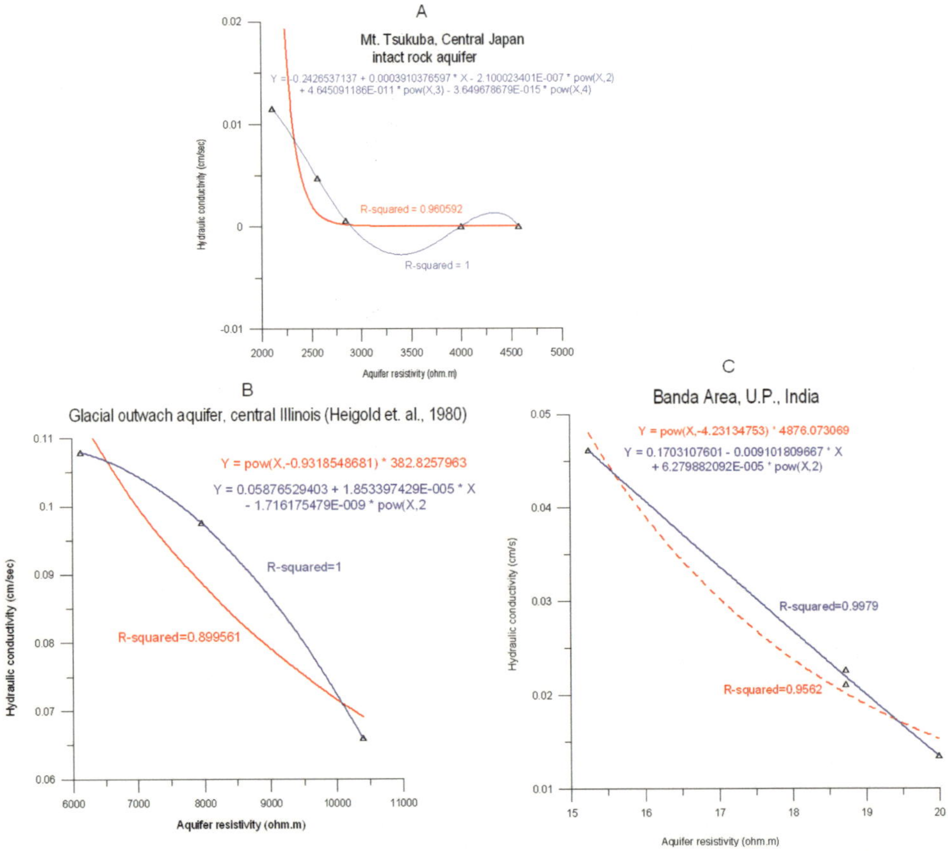

Fig. 7. Empirical relationship between hydraulic conductivity and aquifer resistivity in different locations. (Red dashed line is the power relation; blue solid line is the polynomial relation).

As for Glacial outwash aquifer in central Illinois, USA, (Heigold, et. al., 1979), the explanation of this inverse relation was a problematic. Because Kelly (1977) found a direct linear relationship between hydraulic conductivity and resistivity of the water bearing deposits in two New England aquifers composed of the same glacial deposits of sand and gravel and one case study of direct relation in glacial deposits is discovered later by Frohlich and Kelly (1985). The inverse correlation was reasoned due to more poorly sorted sediments near the head of the Niantic-Illiopolis aquifer, which are responsible not only for reduced porosity and thus less hydraulic conductivity, but also for an increase in the volume of low conductivity solids which increase the resistivity of the aquifer (Heigold et. al., 1979). Kelly et al, (1984) explained the inverse relation due to the presence of clay, although the clay fraction of the aquifer was quite small (less than 4%) (Heigold et. al., 1979). Frohich, (1994) explained this inverse relation due to; they have only three data points.
Heigold et. al., (1979) in his paper, measured the porosity of these three samples; they are 26%, 32%, and 39%. Sieve analysis made on each sample indicates that the clay fraction of the aquifer was quite small (less than 4%). Water resistivity is 1818 ohm.cm, and total dissolved solids are 490 ppm. All mentioned parameters of this aquifer are in agreement with our approach in particular the porosity values. The Mount Tsukuba, Central Japan, intact rock aquifer is covered by homogeneous and fine-grained granite of late Cretaceous to early Palaeogene age. The relation between resistivity and hydraulic conductivity is based on electrical logging and in-situ permeability data from boreholes (Sudo et. al., 2004).
Concerning Banda area U.P., India, the presence of hard rock lithologies in the area may be the cause the negative correlation of the variation in permeability with resistivity (Singh, 2005). This type of inverse correlation typically is found in saturated fractured hard rock aquifer, as previously discussed, but the polynomial correlation is attributed to dissimilarity between porosity and water saturation.

4.3 Third category (water saturation < 50% < porosity)

This category contains one case study for weathered hard rock aquifer is in Mt.Tsukuba, Central Japan, (Sudo, et. el. 2004).
Hydraulic conductivity of this aquifer has an ideal fourth order polynomial direct correlation with the aquifer resistivity (Fig.8). The correlation coefficient of polynomial relation (in blue) is higher than power relation (in red). The mathematical characteristics of this sample classify it in the third category of our analytical models (Fig.5-B), where porosity is higher than 50%, higher than water saturation. The category highlights on the effect of low saturation on the relation between hydraulic conductivity and resistivity of porous media, where in low saturation conductivity of the electrical double layer increases, and surface conductance becomes the main transport mechanism (Pfannkuch, 1969, Urish, 1981, Brovelli, et al., 2005).
The data of Mt. Tsukuba, central Japan weathered rock aquifer are sampled from fine-grained Granitic rocks with cracks. The cracks have approximately 2-mm-thick fillings (Sudo, et. el. 2004). It is worthily to mention that, the resistivity and hydraulic conductivity data of Mount Tsukuba, Central Japan, in both intact and weathered rock aquifer reflect perfectly the analytical relation in the form of polynomial forth orders. This is may reasoned to the nature of the data, where resistivity data are extracted from resistivity log, and permeability data are from in-situ permeability measurements (Sudo, et. el. 2004). It is important to mention that such direct relation between hydraulic conductivity and aquifer

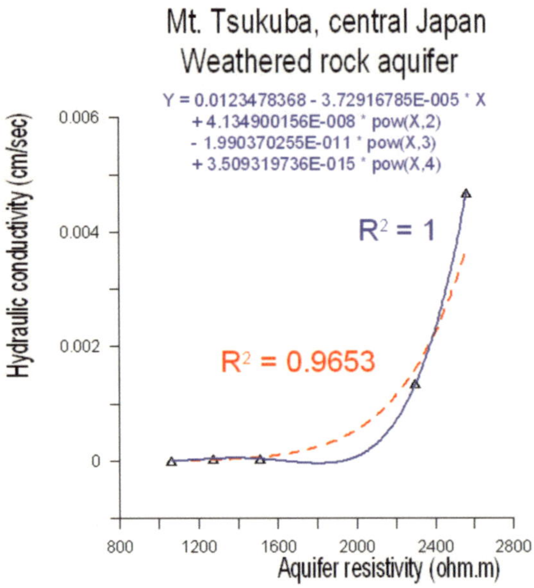

Fig. 8. Empirical relationship between hydraulic conductivity and aquifer resistivity in Mt. Tsukuba, central japan (Red dashed line is the power relation; blue solid line is the polynomial relation).

resistivity could be resulted also in case of high clay content and/or high groundwater resistivity aquifers, where surface conductance effect resulted on the surface of clay mineral or sand imbedded in fresh water became the main transport mechanism, and Archie's low in these cases breaks down (Huntley, 1987, Alger, 1966, Worthington, 1993, Vinegar and Waxman, 1984, Pfannkuch, 1969).

5. Conclusion

Studying the relationship between hydraulic conductivity and electrical resistivity is one of the most interesting objects in both geophysics and hydrogeology. Hydrogeological methods still have some disadvantages, and the measured hydraulic conductivity value is a function of the measuring method. Geophysical methods as well still empirical and site restricted. Hydraulic conductivity and electric resistivity relationship showed a positive non-linear relation in some studies and a negative non-linear relation in others. This is attributed to the relation between apparent formation factor, which is directly proportional to the bulk resistivity, and the intrinsic formation factor, which is inversely proportional to porosity. When the hydrogeological system is in agreement with Archie's conditions (fully saturated, free of clay, high salinity water), the apparent formation factor is equivalent to intrinsic formation factor, and any relation between hydraulic conductivity and bulk resistivity should depends on the porosity. In the other side, when the hydrogeological system disagrees with Archie's conditions, the apparent formation factor is not longer equivalent to intrinsic formation factor. Any relation between hydraulic conductivity and bulk resistivity in this case should depend on surface conductance effect, which is a function of grain size and the specific surface area of the grains. An analytical approach has been illustrated here to show the effect of water saturation on the electric resistivity-hydraulic conductivity relationship. The results are categorized into three non-linear relations. Some field scale relationships empirically support the present analytical approach.

6. References

Alger RP (1966) Interpretation of electric logs in fresh water wells in unconsolidated formations, Soc of Prof Well Log Analyst Trans, Art CC, 1–25.

Archie GE (1942) The electrical resistivity log as an aid in determining some reservoir characteristics. American Institute of Mineral and Metal Engineering. Technical publication, 1442, Petroleum Technology, pp. 8–13.

Batu, V., 1998. *Aquifer Hydraulics; A Comprehensive Guide to Hydrogeologic Data Analysis.* John Wiley & Sons, Inc. New York, NY.

Bear, J., (1972). Dynamics of Fluids in Porous Media. American Elsevier. New York, NY.

Börner FD, (2006) Complex conductivity measurements. In Kirsch R (ed) Groundwater geophysics: a tool for hydrogeology. ISBN 13 978-3-540-29383-5 Springer Berlin Heidelberg New York.

Börner FD, Schopper JR, Weller A (1996) Evaluation of transport and storage properties in the soil and groundwater zone from induced polarization measurements. Geophys Prospect 44:583–601. doi: 10.1111/j.1365-2478.1996.tb00167.x

Börner, F. D., and J. H. Schön, (1991) A relation between the quadrature component of electrical conductivity and the specific surface area of sedimentary rocks, Log Anal., 32: 612–613.

Brown, S. R., (1989) Transport of fluid and electric current through a single fracture. Journal of Geophysical research, 94 (37): 9429-9438.

Butler, D. B., and Knight, R. J. (1998) Electrical conductivity of steam-flooded, clay-bearing geologic materials. GEOPHYSICS, V. 63, No. 4, P. 1137–1149.

Butler, J. J. (2005) Hydrogeological methods for estimation of spatial variations in hydraulic conductivity. (In Hydrogeophysics, p.23-58.).

Carman, P. C., 1937. Fluid flow through granular beds. Transactions, Institution of chemical engineers, London, 15: 150466.

Carman, P. C., 1956. Flow *of gases through porous media.* Butter worth, London.

Carothers JE (1968) A statistical study of the formation factor relation. Log Anal 9(5):13–20

Chandra, S., Ahmed, S., Ram, A., and Dewandel B. (2008) Estimation of hard rock aquifers hydraulic conductivity from geoelectrical measurements: a theoretical development with field application. Journal of Hydrology, V. 357, I. 3-4, P.218-227, doi:10.1016/j.jhydrol.2008.05.023

Chappelier, D., 1992. *Well logging in hydrogeology.* A.A.Balkema Publishers, 175p.

Chapuis, R. P. and Aubertin, M., 2003. Predicting the coefficient of permeability of soils using the Koneny-Carman equation. EPM-RT2003-03. department CGM, Ecole polytechnique de Montrial. Canada.

Deppermann, K. (1954) Die Abhangikeit des scheinbaren Widerstandes vom Sonden abstand bei der vierpunkt-Methode, Geophysical prospecting, II : 262-273.

Ekwe AC, Onu NN, Onuoha KM (2006) Estimation of aquifer hydraulic characteristics from electrical sounding data: the case of middle Imo River basin aquifers, south-eastern Nigeria. J Spatial Hydrol 6(2):121–132

Evans, K., Beavan, J., Simpson, D., 1991. Estimating aquifer parameters from analysis of forced fluctuation in well level: An example from the Nubian Formation near Aswan, Egypt, 1. Hydrogeological background and large-scale permeability estimates. *Journal of Geophysical Research.* 96, B7, 12.127-12.137.

Frohlich R. K and Kelly W. E. (1985) the relation between hydraulic transmissivity and transverse resistance in a complicated aquifer of glacial outwash deposits. Journal of hydrology. 79 (3-4): 215-229.

Frohlich, R. K., (1994) The electric- hydraulic relationship. A geophysical model. Trends in hydrogeology. 1 : 347-358.

Frohlich, R. K., Fisher, J. J. and E. Summerly (1996) Electric-hydraulic conductivity correlation in fractured crystalline bedrock: Central Landfill, Rhode Island, USA. Journal of Applied Geophysics 35: 249-259.

Gerasimos A. T. (1985) A study of the hydrogeophysical properties of fissured aquifers using a double porosity model. *Journal of Hydrology,* 78 (1985) 331—344, 0022-1694/85/$03.30

Gomez-Rivero O (1977) some considerations about the possible use of the parameters a and m as a formation evaluation tool through well logs. Trans SPWLA 18th Ann Logging Symp: J 1–24

Heigold, P. C., Gilkeson, R. H., Cartwright, K. and Reed, P. C. (1979) Aquifer transmissivity from surficial electrical methods. Ground water,. 17 (4) :338-345.

Hill, H.J. and Milburn, J.D., (1956) Effect of clay and water salinity on electrochemical behaviour of reservoir rocks. Trans. AIME, 207: 65-72.

Hubbert, M. King (1940) The theory of groundwater motion. Journal of Geology. 48 (8): 785-944.

Huntley, D. (1987) Relations between permeability and electrical resistivity in granular aquifers. Ground water. 24 (4): 466-474.

Kallergis, G., 1999. Applied-Environmental Hydrogeology. Technical Chamber of Greece, Athens, p. 330.

Kelly, W. E, and Reiter, P., (1984) Influence of Anisotropy on relations between electrical and hydraulic properties of aquifers. Journal of Hydrology 74: 311-321.

Kelly, W.E. (1977) Geoelectric sounding for estimating aquifer hydraulic conductivity. Groundwater. 15 (6) : 420-425.

Khalil MA, Abd-Alla MA (2005) An approach to estimate hydraulic parameters and water quality from surface resistivity measurements at wadi El-Assuity area, Egypt. NRIAG J Geophys, Special issue: 267–281

Khalil, MA and Santos, FA (2009) Influence of degree of saturation in the electric resistivity–hydraulic conductivity relationship. Surveys in Geophysics. 30:601-615. DOI 10.1007/s10712-009-9072-4

Khalil, M. and Fernando M. (2010): Real surface conductivity component: an approach to estimating hydraulic conductivity for fresh water aquifers characterised with clean sand, (submitted to Journal of hydrology).

Kim, J., Sultan, M., 2002. Assessment of the long-term hydrologic impacts of Lake Nasser and related irrigation projects in southwestern Egypt, *Journal of Hydrology*, 262, 68-83.

Kim, J., Sultan, M., 2002. Assessment of the long-term hydrologic impacts of Lake Nasser and related irrigation projects in southwestern Egypt, *Journal of Hydrology*, 262, 68-83.

Kirsch, R., (ed.) (2006) Groundwater Geophysics, Springer.

Kosiniski, W.K and Kelly, E.W. (1981) Geoelectrical sounding for predicting aquifer properties. Groundwater. 19: 163-171.

Kozeny, J. 1927. Ueber Kapillare leitung des wassers in Boden. Sitzungsber Akad. Wiss., Wein, 136 (2a): 271-306.

Krahl, J., Kauffmann, G., Kozur, H., Richter, D., Forster, O., Heinritzi, F., (1983). Neue Daten zur Biostratigraphie und zur tektonischen Lagerung der Phyllit-Gruppe und der Trypali- Gruppe aufder Insel Kreta (Griechenland). Geol. Rundsch. 72, 1147–1166.

Kurniawan (2002) Evaluation of the hydrocarbon potential in low-salinity shaly sand MSc thesis, Faculty of the Louisiana State University and Agricultural and Mechanical College.pp. 86.

Lee Slater (2007) Near Surface Electrical Characterization of Hydraulic Conductivity: From Petrophysical Properties to Aquifer Geometries—A Review. Surveys in Geophysics. 28:169–197.

Lesmes, D. P. and Friedman, S. P. (2005) Relationships between the electrical and hydrogeological properties of rocks and soils, (in Hydrogeophysics :87-128.)

Lorne, B., Perrier, F., and Avouac, J. (1999) Streaming potential measurements,2. Relationship between electrical and hydraulic flow patterns from rock samples during deformation. JOURNAL OF GEOPHYSICAL RESEARCH, V. 104, NO. B8, P. 17,879-17,896. 0148-0227/99/1999JB900155509.00

Martys NS (1999) Diffusion in partially-saturated porous materials. Mater Struct (Materiaux et Constructions) 32:555–562

Mazac , O., Landa, I., and Skuthan, B. (1978) Information capacity of some geoelectrical methods applied to hydrogeological survey. Proc. 23rd geophysical symposium, Verna, Bulgaria: 460-472.

Mazac, O., Cislerova, M., Kelly, W.E., Landa, I, and Venhodova, D., (1990) Determination of hydraulic conductivities by surface geoelectrical methods. In: ed. S. Ward, Geotechnical and Environmental Geophysics, Vol. II. Soc. Explor. Geophys: 125-131.

Mazac, O., Kelly, W., and Landa, I. (1985) A hydrogeophysical model for relations between electrical and hydraulic properties of aquifers. Journal of Hydrogeology 79: 1-19.

Mbonu PC, Ebeniro JO, Ofoegbu CO, Ekine AS (1991) Geoelectric sounding for the determination of aquifer characteristics in parts of the Umuahia area of Nigeria. Geophys 56(2):284–291

Niwas S, Singhal DC (1981) Estimation of aquifer transmissivity from Dar Zarrouk parameters in porous media. J Hydrol (Amst) 50:393–399.doi:10.1016/0022-1694(81)90082-2

Niwas S, Singhal DC (1985) Aquifer transmissivity of porous media from resistivity data. J Hydrol (Amst) 82:143–153. doi:10.1016/0022-1694(85)90050-2

Patnode, H. W., and Wyllie, M. R. (1950) the presence of condutive solids in reservoir rocks as a factor in electric log interpretation, J. Pet. Technol. , 189, 47-52.

Paul F. Worthington (1976) Hydrogeophysical equivalence of water salinity, porosity and matrix conduction in arenaceous aquifers. Groundwater 14 (4) : 224-232.

Pflannkuch HO (1969) On the correlation of electrical conductivity properties of porous system with viscous flow transport coefficients. Proceedings of the IAHR First International Symposium on Fundamentals of Transport Phenomena in Porous Media, Haifa, pp. 42–54

Pflannkuch. H.O. (1969) On the correlation of electrical conductivity properties of porous system with viscous flow transport coefficients. Proceedings of the IAHR First International symposium on fundamentals of transport phenomena in porous media, Haifa : 42-54.

Porter, C.R. and Carothers, J.E, (1970) Formation factor-porosity relation derived from well log data. Trans. SPWLA11th Ann. Logging Symp: 1-19.

REGWA, 2003. Report on RML Well. The General Company for Research and Groundwater.

Rubin Y, Hubbard S (2005) Hydrogeophysics, water science and technology library, vol 50. Springer, Berlin, p 523 Schon J (1983) Petrophysik. Akademie-Verlag, Berlin

Schon, J., (1983) Petrophysik. Akademie-Verlag, Berlin, 405 pp.

Scott, J.B. (2006) The origin of the observed low-frequency electrical polarization in sandstones. GEOPHYSICS,V. 71, NO. 5; P. G235–G238,10.1190/1.2258092

Shah, P. H. and Singh,D. N. (2005) Generalized Archie's Law for Estimation of Soil Electrical Conductivity. Journal of ASTM International, Vol.2, I.5, DOI: 10.1520/JAI13087

Singh, K. P (2003) Geo-electrical exploration for groundwater in a Hard Rock Region of Hyderabad, India, First Break, 21: 29–34.

Singh, K. P (2005) Nonlinear estimation of aquifer parameters from surfficial resistivity measurements. Hydrol.Earth Sys. Sci. Discuss., 2: 917-938.

Soupios P, Kouli M, Vallianatos F, Vafidis A, Stavroulakis G (2007) Estimation of aquifer hydraulic parameters from surficial geophysical methods: a case study of Keritis basin in Chania (Crete-Greece). J Hydrol (Amst) 338:122–131. doi:10.1016/j.jhydrol.2007.02.028

Sudo, H., Tanaka, T., Kobayashi, T., Kondo, T., Takahashi, T., Miyamoto, M., and M. Amagai, (2004) Permeability imaging in granitic rocks based on surface resistivity profiling. Exploration Geophysics 35: 56–61

Sundberg, K., (1932) Effect of impregnating waters on electrical conductivity of solids and rocks. Trans. AIME, 97: 367-391.

Telfold, W. M., Geldert. L.P. and Sheriff, R. E. (1991) Applied Geophysics. Cambridge Univ. New York.

Timur, A., Hemkins, W.B. and Worthington, A.E. (1972) Porosity and pressure dependence of formation resistivity factor for sandstones. Trans. CWLS 4th Formation Evaluation Symp., 30pp.

Urish DW (1981) Electrical resistivity–hydraulic conductivity relationships in glacial outwash aquifers. Water Resour Res 17(5):1401–1408. doi:10.1029/WR017i005p01401

VinegarHJ, WaxmanMH(1984) Induced polarization of shaly sands. Geophysics 49(8):1267–1287. doi:10.1190/ 1.1441755

Waxman MH, Smits LJM (1968) Electrical conductivities in oil bearing sands. Journal of the society of Petroleum Engineerins 8:107–122

Worthington PF (1993) The uses and abuses of the Archie equations. 1. The formation factor–porosity relationship. J Appl Geophys 30:215–228. doi:10.1016/0926-9851(93)90028-W

Yadav GS (1995) Relating hydraulic and geoelectric parameters of the Jayant aquifer, India. J Hydrol (Amst) 167:23–38. doi:10.1016/0022-1694(94)02637-Q

Yadav GS, Kumar R, Singh PN, Singh SC (1993) Geoelectrical soundings for aquifer characterization around Jayant colony-Singrauli, Sidhi District, MP. J Assoc Expl Geophysists XIV(3):123–131.

Yadav, G.S., and Abolfazli (1998) Geoelectrical sounding and their relationship to hydraulic parameters in semi-arid regions of Jalore, north-western India. J. Applied Geophysics, 39: 35-51.

3

Hydraulic Conductivity and Water Retention Curve of Highly Compressible Materials- From a Mechanistic Approach through Phenomenological Models

Serge-Étienne Parent[1], Amir M. Abdolahzadeh[2], Mathieu Nuth[3] and Alexandre R. Cabral[4]
[1]*Department of Soils and Agrifood Engineering, Université Laval, (Formerly, Ph.D. student, Université de Sherbrooke),*
[2,3,4]*Department of Civil Engineering, Université de Sherbrooke, Canada*

1. Introduction

Highly compressible material (HCM), such as clay and materials containing high organic content, are increasingly used for applications in geoenvironmental engineering and agriculture (Nemati, et al., 2002; Paquet & Caron, 1993). In particular, deinking by-products (DBP) - a clay and fibers spongy material produced in the early stage of the paper recycling process - has been suggested by numerous authors as a cover material over landfills (Kraus, et al., 1997; Moo-Young & Zimmie, 1996; Cabral, et al., 1999; Burnotte, et al., 2000; Kamon, et al., 2002; Parent, 2006; Robart, 1998; Bédard, 2005) and mine waste top covers (Cabral, et al., 1999) as well as structural enhancement material in agricultural soils (Nemati, et al., 2000). The challenge with using DBP as part of a landfill cover was reviewed by Cabral et al. (2007) and Abdolahzadeh et al. (2010; 2008).

The unsaturated hydraulic properties of DBP are fundamental for the design of top covers. Those properties are namely water retention curve (WRC), also commonly called soil-water characteristic curve (SWCC[1]) and hydraulic conductivity function (*k*-function). In the early stage of our projects concerning DBP, we needed to determine their hydraulic properties. DBP revealed to be highly compressible under the effect of suction, and classical models to represent the WRC of porous materials (Brooks & Corey, 1964; van Genuchten, 1980; Fredlund & Xing, 1994) usually do not take into account volume change during drying. Indeed, when a porous material dries, the capillarity phenomenon induces a tensile stress into the pores. As a result, the pores may shrink upon drying. Variations in pore structure

[1] The expression "water retention curve" is preferred over the widely used "soil-water characteristic curve", because many porous materials, like DBP, are not soils. Moreover, to state that the water retention curve is characteristic may only be appropriate if it is use to distinct a porous material from others.

resulting from variations of suction would influence water retention characteristics, i.e. air-entry value (AEV), desaturation slope and residual trend.

Moreover, because of experimental difficulties involved in direct determination of the hydraulic conductivity function (*k*-function), the conventional practice is to indirectly estimate the *k*-function based on WRC models (Fredlund, et al., 2002; Huang, et al., 1998; Leong & Rahardjo, 1997; Fredlund, et al., 1994; Mualem, 1976). Huang et al. (1998) proposed a minor adaptation of the Fredlund et al. (1994) *k*-function model, derived from the Childs & Collis-George (1950) model. Reliability and validity of *k*-functions depends not only on the *k*-function model selected, but also on accuracy of the WRC model used to derive the *k*-function. As a result, there is a need for more suitable methods to determine the WRC and *k*-functions of HCMs.

This research project provides a mechanistic approach based on a compression energy concept accounting for pore shrinkage under the effect of suction. A clear distinction between non-compressible materials, compressible materials and highly compressible materials is also made. Highly compressible materials (HCMs) are defined as porous materials undergoing considerable strain as an effect of external stresses or internal pressure. For HCMs, tensile stresses will induce shrinkage even for suction levels greater than the AEV. An experimental technique was developed to determine the WRC of HCMs. The testing procedure was based on the axis-translation technique and allowed for simultaneous and continuous determination of the volumetric water content and void ratio of soil samples during desaturation (Cabral, et al., 2004). The results obtained from the developed testing procedure indicate that our material behave like a HCM.

In this chapter, a WRC model is proposed for suction-driven shrinkage of HCMs. The proposed model, based on a model originally proposed by Fredlund et al. (1994), was validated using experimental data published by Huang (1994). The results of representative tests to determine the WRC of DBP are discussed, as well as their associated computed k-functions.

This study also provides a procedure to determine the *k*-function of HCM, based on hydraulic conductivity tests performed on saturated samples and on suction test data performed in order to obtain the AEV. Suction tests must be performed so that volume changes can be monitored using methodologies such as proposed by Cabral et al. (2004). The procedure was validated by comparing results of actual unsaturated hydraulic conductivity tests (experimental data for a compressible soil published by Huang (1994) with predictions made by applying the procedure.

2. Definitions and concepts

Volume change is a change of porosity in response to a change in stress state (Fredlund & Morgenstern, 1976). A porous material can be compressed by either mechanical or suction-induced tensile stresses. A mechanical stress is defined as an external stress (i.e. a pressure exerted at the boundaries of a sample) or as a gravity-induced stress. A mechanical stress can thus be generated by a weight or an impact, and accentuated by vibration. Suction-induced tensile stresses rise from water menisci that attract particles to each other. Water suction (ψ) is a reactive pressure that equilibrates with pressure caused by gravitational forces, osmosis or pneumatic pressures. Osmotic suction will be dominant only at high suction ranges and are not considered in this report.

Pore shrinkage may happen during desaturation, resulting in modifications of hydraulic properties. In this section, a mechanistic theory for the compression of porous material is introduced, leading to a phenomenological model describing the water retention curve (WRC) of highly compressible materials (HCM) and a procedure to predict the hydraulic conductivity function (*k*-function) of HCM using saturated samples.

2.1 Porous material compression

Many authors showed that, as long as the pores are saturated, the compression processes induced by mechanical and suction stresses can be considered identical (Khalili & Khabbaz, 1998). However, unsaturated porous materials must behave differently (Toll, 1988).

2.1.1 Saturated porous media

The principle of effective stress is probably the most fundamental in soil mechanics. In its general form, it is defined as the difference between total stress and pore water pressure (Holtz & Kovacs, 1981; Terzaghi, 1936), as follows:

$$\sigma' = \sigma - u \tag{1}$$

where σ' is effective stress (kPa), σ is net or total stress (kPa), u is pore water pressure ($u= -\psi$) and ψ is matric suction (kPa). We assume that stress is isotropic, and that σ' designates the mean value of effective stress in this chapter, unless otherwise indicated. Note that u_a, the air pressure, is supposed to be the same for σ and u.

As shown in the following sections, porosity (or void ratio) decreases with increasing total stress or decreasing pore pressure (increasing suction), but both stresses are acting differently on the pore structure.

2.1.2 Unsaturated porous materials

For porous materials in the unsaturated state, variation in suction does not result in the same change in effective stress as the equivalent variation in applied mechanical stresses. In the case of a pressure state, a water pressure stress is applied on every particles of a porous material. However, when a porous material is under suction, the pores may not be fully saturated and tensile forces, that attract pore walls together, may not be applied over all the pores. Bishop (1959) proposed a factor χ that factors tensile stress under unsaturated conditions. This factor was later transformed to a function by Khalili & Khabbaz (1998), as follows:

$$\sigma' = \sigma - \chi(\psi)u = \sigma + \chi(\psi)\psi \tag{2}$$

where $\chi(\psi)$ is an effective stress function attaining the value of 1 for saturated porous materials and 0 when the porous material is completely dry (no units).

The Bishop's effective stress is often considered as a natural extension of the saturated effective stress to unsaturated soils (explained further in section 2.1.2.1). Equation 2 reveals that the external stress and the suction can both have an influence on the effective stress. The relevancy of the unsaturated effective stress has been questioned in the literature, especially regarding the phenomenon of wetting collapse. Recent works clarified the definition and implications (Jardine, et al., 2004; Nuth & Laloui, 2008) of Bishop's stress. They justified that the latter is an adequate representation of the state of stress within the unsaturated soils, and a convenient constitutive stress variable.

The constitutive behavior of soils is complex due to the intrinsic non-linearities as well as irreversible phenomena (i.e. plasticity). For the sake of simplicity, only monotonic loading paths (drying tests) will be considered in the rest of the paper. This simplification allows us to use phenomenological models based on non linear elasticity. The modeling of cyclic test paths using the theory of elasto-plasticity is out of the scope of the paper. D'Onza et al. (2010) present a review of the recent, advanced models for unsaturated soils coping with elasto-plastic features.

Fredlund & Morgenstern (1976) proposed that Bishop (1959)'s equation describes a constitutive behavior of unsaturated soils. The planes plotted in Fig. 1 presents such behavior for a single material. Porosity (light gray plane in Fig. 1) decreases with both increasing suction and net stress, but the $\chi(\psi)$ function reduce the contribution of suction to shrinkage while the material desaturates. Compared to the WRC at low porosity, a WRC at high porosity is characterized by a lower air-entry value (AEV) and a steeper desaturation slope. Salager et al. (2010) considered the constitutive desaturation plane as a hydromechanical phenomenon and adopted the term "soil-water retention surface", contrasting with the usual expression "soil-water retention curve", the equivalent of the term "water retention curve" used in this report. Consequently, the WRC would be the projection of the "water retention plane" on the suction axis (dark plane in Fig. 1).

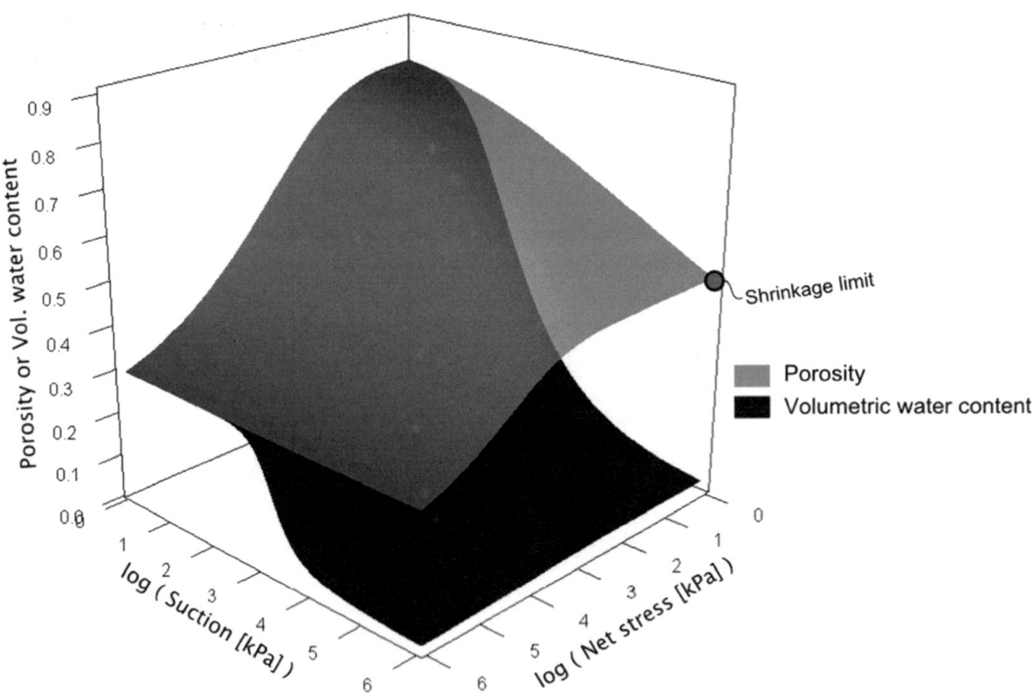

Fig. 1. Fredlund and Morgenstern constitutive model

2.1.2.1 The capillary-induced compression energy

Parent (2006) related the capillary induced volume change behavior with the concept of compression energy. Indeed, stress is by definition a volumetric concentration of energy:

$$Stress = \frac{Force}{Surface} = \frac{Force \times Distance}{Surface \times Distance} = \frac{Energy}{Volume} \tag{3}$$

or

$$kPa = \frac{kN}{m^2} = \frac{kN \cdot m}{m^2 \cdot m} = \frac{kJ}{m^3} \tag{4}$$

In order to collapse, a porous material needs an energy source that is able to contract the pores. This energy can be of two major sources:

- **Mechanical energy**: The sample is compressed under the effect of gravity. The effect can be accentuated thanks to a vibration energy, weight or impact. For example: the mass of an overlying soil, a vibrating compactor, a falling proctor hammer.
- **Capillary energy**: Drying a sample induces tensile internal stresses caused by water menisci, which forces particles to reorient and attract to each other, hence leading to shrinkage (Baumgartl & Kock, 2004). For example: (1) soil surfaces exposed to the sun may dry, shrink and crack, creating shriveled pattern, and (2) a strain may be observed on the slope of a drained, recently dug ditch.

Whereas the mechanically induced compression energy is distributed into the solid matrix of the porous material, the energy induced by capillary forces is distributed into the water phase contained in the pores of a porous material, as follows:

$$[E]_{mec} = \sigma \times \frac{V_s}{V_s} = \sigma \tag{5}$$

$$[E]_{cap} = \psi \times \frac{V_w(\psi)}{V_p(\psi)} = \psi \times S(\psi) \tag{6}$$

where$[E]_{mec}$ is the concentration of mechanically induced compression energy, σ is the total (mechanical) stress, V_s is the solid volume of a porous material sample, $[E]_{cap}$ is the concentration of capillary induced compression energy, ψ is the suction, V_w is the volume of water into a porous material sample, V_p is the volume of the pores and $S(\psi)$ is the degree of saturation as a function of suction, or the water retention curve (WRC).
If the porous material is fully saturated, then Equation 6 yields that $[E]_{cap} = \psi$. This shows that the concentrations of mechanically and capillary induced energy are equivalent: for $\psi = \sigma$, $[E]_{cap} = [E]_{mec}$ as long as the porous material is saturated. The AEV is defined by the suction value where significant loss of water is observed on the S versus ψ curve, i.e. in the region of the inflection point determined using the procedure proposed by Fredlund & Xing (1994)[2]. Accordingly, when the largest pores of a material are drained, close to the AEV, $S(\psi)<1$. Thus, $[E]_{cap} < [E]_{mec}$. As Toll (1995) stated, for suction values higher than the AEV, the largest pores have been already compressed (shrinkage happens in the pores containing water). We may add that the finer pores need more energy to be compressed and are less susceptible to volume loss. For this reason, the shrinkage limit, which is defined as the stabilized, minimal volume of soil (see Fig. 1 and Fig. 2), is often assumed being reached at

[2] The tangent method (Fredlund & Xing, 1994) defines the AEV as the suction value in the inflection zone of the WRC on the degree of saturation scale where the projection of the desaturation slope crosses the degree of saturation value of 100%, generally determined on a log scale along the suction axis.

the AEV. Nevertheless, we infer in this chapter that for materials that are highly compressible, $[E]_{cap}$ may be sufficient to keep on inducing compression for suction values greater than the AEV. Hence, the shrinkage limit may be observed for suction values higher than the AEV.

From the compression energy concept, it can be inferred that $\chi(\psi)$ may be somehow related to $S(\psi)$. Indeed, the pore water pressure component of the effective stress ($\chi(\psi)\cdot\psi$) should null when the porous media is saturated ($\psi=0$ and $\chi=1$), and also be null when it is dry ($\psi = 10^6$ kPa - the theoretical suction value that corresponds to a null water content (Fredlund & Xing, 1994) - and $\chi=0$).Hence, the pore water pressure component of the effective stress in unsaturated state - i.e. $\chi(\psi)\cdot\psi$ - reaches a maximal value at a certain suction value between complete and null saturation. This behavior can be easily observed when wet and dry beach sands flows through our fingers, but when the sand is partially saturated, particles stick together, making possible the construction of a sand castle. However not supported by a mechanistic model, Bishop (1959)'s approach was used by Khalili & Khabbaz (1998), who proposed an exponential empirical relationship between χ and ratio ψ/ψ_{aev} (where ψ_{aev} is the AEV), allowing the determination of $\chi(\psi)$ for most soils with an equation similar to the one proposed by Brooks & Corey (1964) for WRC curve fitting:

$$\chi(\psi) = \begin{cases} \left(\dfrac{\psi}{\psi_{aev}}\right)^{\iota} & if \ \psi \geq \psi_{aev} \\ 1 & if \ \psi \leq \psi_{aev} \end{cases} \tag{7}$$

where ψ_{aev} is the suction at the air-entry value (AEV) and ι is an empirical parameter estimated to be equal to -0.55 by Khalili & Khabbaz (1998).

It is possible to force parameter χ to reach a null value at 10^6 *kPa* using the function $C(\psi)$ in Equation 10, presented after.

$$\chi = \begin{cases} \left(1 - \dfrac{ln\left(1+\frac{\psi}{C_r}\right)}{ln\left(1+\frac{\psi}{10^6}\right)}\right) \times \left(\dfrac{\psi}{\psi_{aev}}\right)^{\iota} & if \ \psi \geq \psi_{aev} \\ 1 & if \ \psi \leq \psi_{aev} \end{cases} \tag{8}$$

The optimum of compression capability by means of suction using χ is coherent with Fredlund (1967)'s conceptual behavior, that was treated later on by Toll (1995). The latter suggested that void ratio of a normally consolidated soil decreases as suction increases and levels off slightly after the AEV, *i.e.* where "[...] *the suction reaches the desaturation level of the largest pore (either due to air entry of cavitation) and air starts to enter the soil. The finer pores remain saturated and will continue to decrease in volume as the suction increases. However, the desaturated pores will be much less affected by further changes in suction and will not change significantly in volume. The overall change will therefore be less than in a mechanically compressed saturated soil, and the void ratio - suction line will become less steep than the virgin compression line*[3] ".

A schematic representation of Fredlund (1967)'s conceptual behavior is shown in. Fig. 2. As pores lose water under the effect of suction, porosity follows the virgin compression line and the water retention curve (WRC). Porosity stabilizes at suction values slightly higher than the AEV. The asymptote toward which the curve converges is the shrinkage limit.

[3] Toll (1995), page 807.

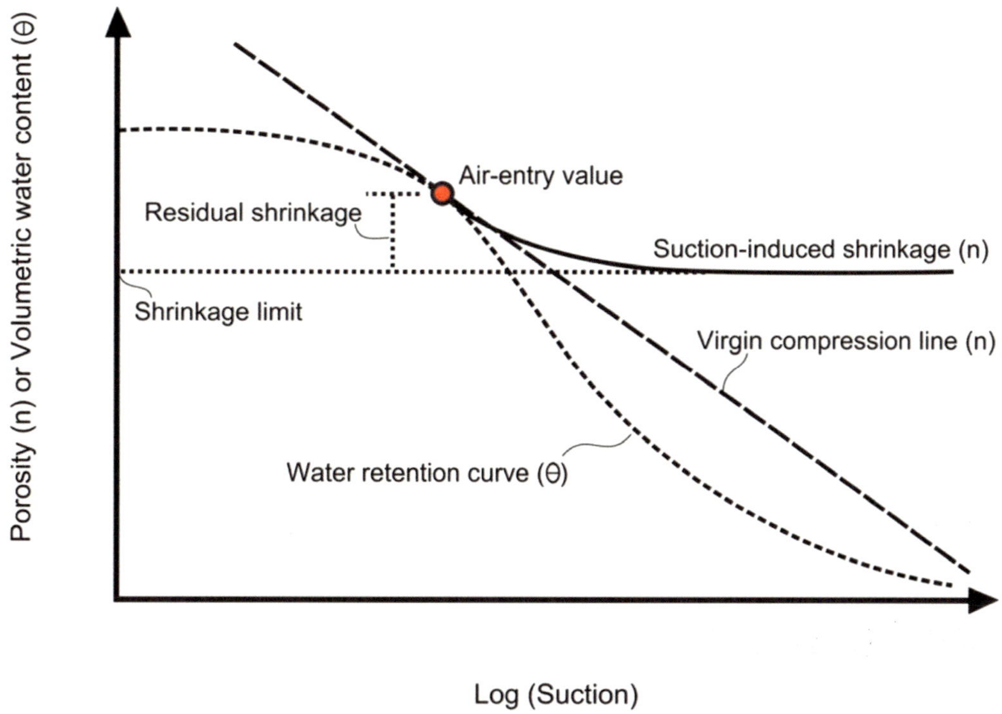

Fig. 2. Conceptual scheme representing shrinkage

Most data from the literature come from soils and show a shrinking behavior similar to the one schematically presented in Fig. 2, where the shrinkage limit is reached in the area of the AEV. However, it is shown by the compression energy concept (Equation 6) that capillary stresses are still active for suction values beyond the AEV. Fig. 3 shows a hypothetical desaturation curve and porosity function of a highly compressible material. The desaturation curve is expressed both in terms of volumetric water content and degree of saturation, the later being printed for sake of comparison with the $\chi(\psi)$ function (Equation 8). The concentration of capillary energy - $S(\psi)\cdot\psi$ - is plotted asides the suction component of the effective stress - i.e. $\chi(\psi)\cdot\psi$.

It can be observed that $S(\psi)$ is similar to the more generic $\chi(\psi)$ function (using ι=-0.55), leading to similar $[E]_{cap}$ and $\chi(\psi)\cdot\psi$ energy curves (which may not be the case for all porous materials). These curves increase linearly with suction from 0 to the AEV. As the hypothetical material presented here is qualified as "highly compressible", its porosity can decrease with increasing suction far beyond the AEV. However, it is worth mentioning that increasing $[E]_{cap}$ or $\chi(\psi)\cdot\psi$ does not necessarily mean that porosity decreases, because the energy may not be sufficient or adequate to cause shrinkage, particularly if the capillary stress is applied to the smallest pores.

It may be added that as the suction component of compression energy is null at complete desaturation, a rebound may be observed (although it was not yet observed in laboratory), similar to the one observed when mechanical stress is released from a soil sample submitted to an oedometer test.

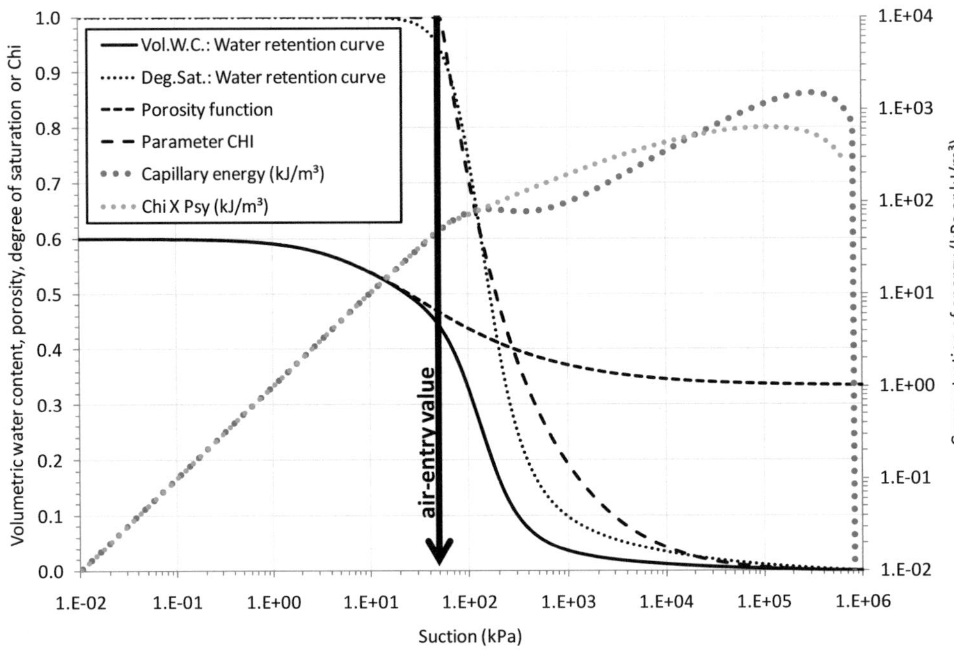

Fig. 3. The energy of compression conceptual approach and the variation of parameter χ

2.1.2.2 Defining material compressibility with suction

Any porous material is virtually compressible if it undergoes a sufficiently high level of stress. In the particular case of soils, the coefficient of compressibility is determined by consolidation tests and used for constitutive modeling (Roscoe & Burland, 1968). The compressibility is thus commonly regarded as a mechanical property characterizing the response of a material to an external, mechanical, stress. Yet, when describing the material response to suction changes, the term compressible material is not clearly defined in the literature. A clear definition is needed to proceed further. Using sensitivity of materials to suction, three categories were thus created:

- non compressible materials (NCM), e.g. ceramic, concrete;
- compressible materials (CM), e.g. sand, silt;
- highly compressible materials (HCM), e.g. fine-grained clays, peat, deinking by-products.

The definition considers a relationship between void ratio and gravimetric water content (w), commonly called the *soil shrinkage characteristic curve* (Tripathy, et al., 2002). However, because compressible porous materials are not necessary soils, this curve will be called *pore shrinkage characteristic curve* (PSCC) in this chapter.

Fig. 4 shows a schematic representation of three PSCCs. The NCM (coarse dashed line) does not shrink under the effect of suction. The CM (fine dashed line) shrinks only when it is saturated. Finally, the highly compressible material (solid line) shrinks over a range of suction that goes beyond the AEV (e.g. as shown by Kenedy & Price (2005) for peat). In other words, the capillary energy is not high enough to produce significant shrinkage to the NCM. The CM shrinks under suction, but the capillary energy is not high enough to produce significant shrinkage at suction values higher than the AEV. As for HCM, the compression energy induced by capillary forces makes the pores shrink even for suction values beyond the AEV. The sigmoïdal effect represented on the HCM curve is due to an asymptotical tendency to reach the shrinking limit at high suction values, near complete desaturation (theoretically at a suction of 10^6 kPa). This behavior is treated in the "results and discussion" section, hereafter.

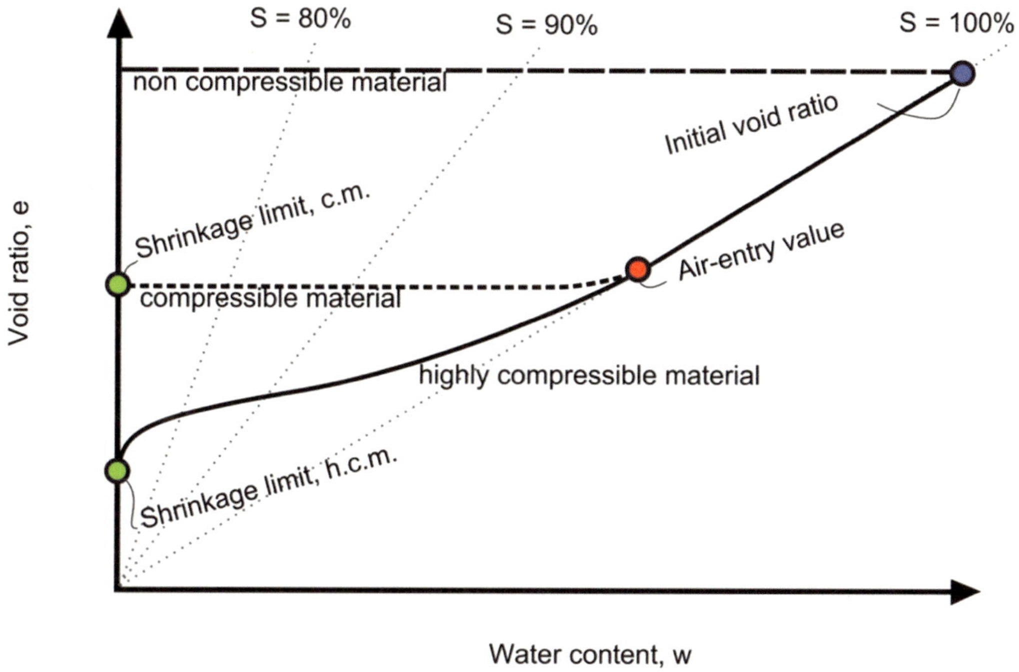

Fig. 4. Schematic representation for the definition of non compressible materials, compressible materials and highly compressible materials (S is degree of saturation)

2.2 The Water Retention Curve

The relationship between water content and suction in a porous material is commonly called the Water Retention Curve (WRC) and constitutes a basic relationship used in the prediction of the mechanical and hydraulic behaviors of unsaturated porous materials used in geotechnical and soil sciences. The theory associated with the prediction of the engineering behavior of unsaturated soils using the WRC is presented by Barbour (1998). Leong & Rahardjo (1997) summarize the equations to model the WRCs, mainly of the non-linear, fully reversible type. A review of recent models for WRC including capillary hysteresis, drying-wetting cycles, irreversibilities and material deformations is proposed by Nuth & Laloui (2008). Again, only the drying (desaturation) branch of the WRC will be studied here. Fig. 5 shows a schematic representation of a set of WRCs for the same material consolidated to different initial void ratios. It has been explained before that a porous material may shrinks while it dries. The various WRCs presented in terms of volumetric water content θ versus suction ψ in Fig. 5 superimpose onto a single desaturation branch (solid thick line in

Fig. 5) for suction levels that are higher than the AEV (ψ_{aev}) of each single curve (Fredlund, 1967; Toll, 1988). This common desaturation branch is equivalent to the virgin consolidation of clayey soils. The AEV is a value of suction where significant water loss is observed in the largest pores of a specimen. As shown later, the AEV depends on the initial void ratio and on how the void ratio changes with suction. It is important to note that for HCMs, the AEV should be determined on a degree of saturation versus suction plot, rather than on the volumetric water content versus suction curve, because the volumetric water content of a sample can start to drop without emptying its pores. Indeed, if it is assumed that the volume of water expelled is equal to the decrease in void ratio, the volumetric water content decreases whereas the degree of saturation remains the same (Fig. 5).

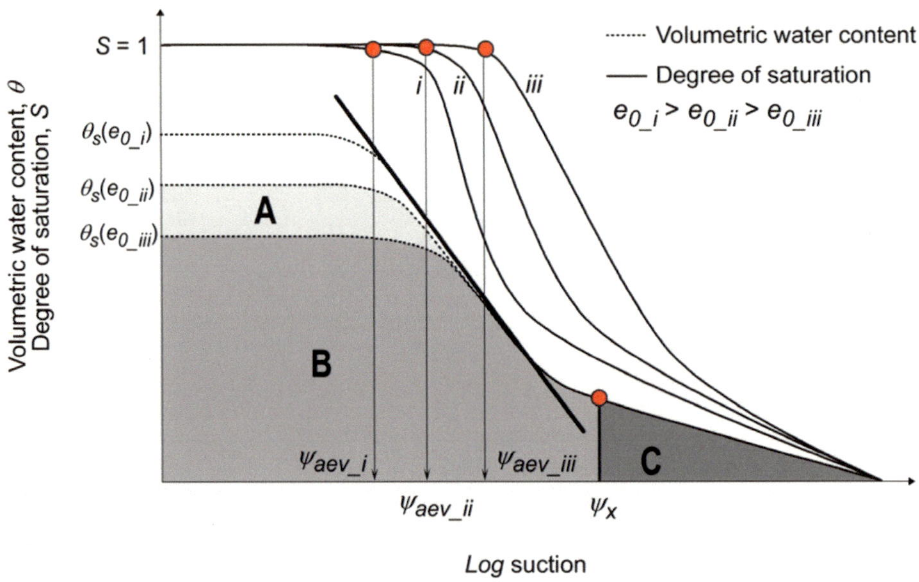

Fig. 5. Water retention curves for a material initially consolidated to different void ratios

The shape of the WRC is mainly influenced by the soil pore size distribution and by the compressibility of the material (Smith & Mullins, 2001). Pore size distribution and compressibility depend on initial water content, soil structure, mineralogy and stress history (Simms & Yanful, 2002; Vanapalli, et al., 1999; Lapierre, et al., 1990). Volume change (shrinkage) during desaturation can markedly influence the shape of the WRC. Emptying voids as suction increases may lead to a reduction in pore size, which in turn affects the estimated volumetric water content (θ) and degree of saturation (S). Accordingly, taking into account volume change during suction testing is of great importance, be it in the laboratory or in the field, in order to avoid eventual flaws in the design of geoenvironmental and agricultural applications, be it a misinterpretation of strain, hydraulic conductivity or water retention (Price & Schlotzhauer, 1999). Cabral et al. (2004) proposed a testing apparatus based on the axis translation technique to measure volume change continuously during determination of the WRC of HCMs. This apparatus is presented in the "Materials and methods" section.

An extensive body of literature exists regarding the experimental determination of the WRC (Smith & Mullins, 2001). Although there are multiple procedures to determine WRCs, the volumetric water content of a HCM specimen cannot be accurately obtained from a single test. Indirect methods based on grain size distribution (i.e. a measure of the pore size distribution) are also widely used to obtain WRCs (Aubertin, et al., 2003; Zhuang, et al., 2001; Arya & Paris, 1981). However, these methods are not suitable to fibrous materials, such as deinking by-products, and do not consider the reduction in pore size when suction increases (nor the distribution of this reduction among the pores).

In fact, most precursor models employed to fit WRC data have been developed assuming that the material would not be submitted to significant volume changes (Brooks & Corey, 1964; van Genuchten M. T., 1980; Fredlund & Xing, 1994). In particular, the WRC model proposed by Fredlund & Xing (1994) was elaborated based on the assumption that the shape of the WRC depends upon the pore size distribution of the porous material. The Fredlund & Xing (1994) model is expressed as follows:

$$\theta(\psi) = \frac{C(\psi)\theta_s}{ln\left(\exp(1) + \left(\frac{\psi}{a_{FX}}\right)^{n_{FX}}\right)^{m_{FX}}}$$

or (9)

$$S(\psi) = \frac{C(\psi)}{ln\left(\exp(1) + \left(\frac{\psi}{a_{FX}}\right)^{n_{FX}}\right)^{m_{FX}}}$$

where θ is the volumetric water content, θ_s is the saturated volumetric water content, S is the degree of saturation, ψ is the matric suction, a_{FX} is a parameter whose value is directly proportional to the AEV, n_{FX} is a parameter related to the desaturation slope of the WRC curve, m_{FX} is a parameter related to the residual portion (tail end) of the curve, $C(\psi)$ is a correcting function used to force the WRC model to converge to a null water content at 10^6 kPa (Equation 10).

$$C(\psi) = 1 - \frac{ln\left(1 + \frac{\psi}{C_r}\right)}{ln\left(1 + \frac{10^6}{C_r}\right)} \tag{10}$$

where C_r is a constant derived from the residual suction, i.e. the tendency to the null water content.

Huang et al. (1998) developed a WRC model that takes into account volume change in the mathematical definition of the WRC. Using experimental data reported in the literature, Huang et al. (1998) assumed, based on experimental evidence, that the logarithm of the AEV was directly proportional to the void ratio obtained at the AEV, as expressed as follows:

$$C(\psi) = 1 - \frac{ln\left(1 + \frac{\psi}{C_r}\right)}{ln\left(1 + \frac{10^6}{C_r}\right)} \tag{11}$$

$$\psi_{aev} = \psi_{aev_{e\prime}} 10^{\varepsilon_\psi(e-e\prime)}$$

where e is the void ratio, e' is a reference void ratio, $\psi_{aev_{e'}}$ is the AEV at the reference void ratio e', ε_ψ is the slope of the $log(\psi_{aev})$ vs. e curve, and ψ_{aev} is the AEV at the void ratio e.
Later, Kawai et al. (2000) validated Huang et al. (1998)'s results. They also proposed that the void ratio at AEV would follow a curve that could be predicted from the initial void ratio defined by Equation 12. This equation was recovered in later studies, namely Salager et al. (2010) and Zhou & Yu (2005).

$$\psi_{aev} = Ae_0^{-B} \tag{12}$$

where e_0 is the void ratio at the beginning of the test, and A and B are fitting parameters.
Nuth & Laloui (2008) proposed a review of the published evidence of the dependency of the AEV with the void ratio and external stress for several materials, which also supports Equations 11 and 12.
An adaptation of the Brooks & Corey (1964) model was used by Huang et al. (1998) to describe the WRC of deformable unsaturated porous media, as follows:

$$S_e = \begin{cases} 1 & if\ \psi \leq \psi_{aev_{e'}} 10^{\varepsilon_\psi (e-e')} \\ \left(\dfrac{\psi_{aev_{e'}} 10^{\varepsilon_\psi (e-e')}}{\psi} \right)^{\lambda} & if\ \psi \geq \psi_{aev_{e'}} 10^{\varepsilon_\psi (e-e')} \end{cases} \tag{13}$$

where S_e is the normalized volumetric water content $[S_e=(S-S_r)/(1-S_r)]$, S_r is the residual degree of saturation, λ is the pore size distribution index for a void ratio e, representing the slope of the desaturation part. Typical values for λ range from 0.1 for clays to 0.6 for sands (van Genuchten, et al., 1991).
Shrinkage reduces the slope of the desaturation part of the WRC. Huang et al. (1998) assumed and provided evidence that, for HCMs, the relationship between λ and void ratio can be represented by:

$$\lambda = \lambda_{e'} + d(e - e') \tag{14}$$

where d is an experimental parameter and $\lambda_{e'}$ is the pore-size distribution index for the reference void ratio e'.
In other modeling frameworks published recently (Ng & Pang, 2000; Gallipoli, et al., 2003; Nuth & Laloui, 2008), the WRC model is coupled with a mechanical stress-strain model. Yet the calibration of these models requires an exhaustive characterization of the mechanical behavior which is not always available in the case of landfills, and out of the scope of this chapter.
It is relevant to note that the Huang et al. (1998) model does not model shrinkage as a function of suction and the partial desaturation for suctions lower than the AEV. A model designed to fit water retention data of a highly compressible material, presented in the results and discussion section, fulfill these gaps.

2.3 The hydraulic conductivity function

The hydraulic conductivity function (k-function) of unsaturated soils can be determined directly, by means of laboratory (McCartney & Zornberg, 2005; DelAvanzi, 2004) or field testing, or indirectly, by empirical, macroscopic or statistical models. Leong & Rahardjo (1997) summarized current models used to determine the k-functions from WRCs. Huang et al. (1998) proposed to take into account the variation in k_{sat} with e in the k-function model, as well as a linear variation in $log(k_{sat})$ with e.

$$k(e) = k_{sat}(e) \cdot k_r(\psi)$$
$$k_{sat}(e) = k_{sat_{e'}} 10^{b(e-e')} \tag{15}$$

where $k(e)$ is the hydraulic conductivity, $k_{sat}(e)$ is the saturated hydraulic conductivity at void ratio e , $k_r(\psi)$ is the relative k-function that can be described using a model such as Fredlund et al. (1994) (Equation 16 below), $k_{sat_{e'}}$ is the saturated hydraulic conductivity at the reference void ratio e' and b is the slope of the $log(k_{sat})$ versus e relationship. The relative k-function, k_r, and the void ratio, e, can be a function of either θ or ψ.

Since, for HCM, void ratio is a direct function of suction (Khalili, et al., 2004), it is convenient to use a k-function model integrated along the suction axis, i.e. $k_r(\psi)$. The relative k-function statistical model proposed by Fredlund et al. (1994), adapted from Child & Collis-George (1950)'s model, is expressed as follows:

$$k_r(\psi) = \frac{\int_{ln(\psi)}^{ln(10^6)} \frac{\theta(exp(y)) - \theta(\psi)}{exp(y)} \theta'(exp(y))dy}{\int_{ln(1)}^{ln(10^6)} \frac{\theta(exp(y)) - \theta_s}{exp(y)} \theta'(exp(y))dy} \tag{16}$$

where θ is the first derivative of the WRC model and y is a dummy integration variable representing suction.

It is important to note that, as mentioned by Fredlund & Rahardjo (1993), the Child & Collis-George (1950)'s k-function model, from which Equation 15 and Equation 16 were derived, assumed incompressible soil structure. In fact, the function on the numerator in Equation 16 was integrated from suction value $ln(\psi)$ to the maximum suction value, $ln(10^6)$, while the denominator was computed over the entire suction range, *i.e.* from $ln(0)$ (where $exp(ln(0)) \rightarrow 0$) to $ln(10^6)$. However, the function on the denominator is not the same for two porous materials with different initial void ratios, with different initial θ_s. Consider samples *ii* and *iii* in Fig. 5, the schematic representation of three water retention tests performed with different initial void ratios. It was expected that at suction ψ_x, samples *ii* and *iii* would reach the same void ratio, the same volumetric water content and, as a result, the same hydraulic conductivity. However, considering that the function to integrate is a function of WRC, the denominator of Equation 16 must be larger if calculated over the function derived from the WRC of sample *ii* (areas A+B+C, Fig. 5) then compared to sample *iii* (areas B+C, Fig. 5), leading to different k-functions.

Theoretical explorations can be derived for from better understandings of the mechanism of capillary-induced shrinkage. Such exploration was performed by Parent & Cabral (2004), who proposed means to estimate the k-function of an HCM from water retention tests over the saturated range. This method is presented in the "Results and interpretation" section.

2.4 Synthesis of the theory section

The mechanistic model presented herein is coherent with Bishop (1959)'s empirical model (Equation 2): $\psi \cdot \chi$ is null at 0 and 10^6 kPa and a maximum is observed. The compression energy concept offers a mechanistic perspective that leads to a better understanding. This new paradigm led the authors to three arguments:

1. regarding to suction, definitions can be formulated for non compressible, compressible and highly compressible materials;
2. parameter χ can be used in several manners to deduce the compression behavior of a porous material;
3. water retention curve and k-function models that takes into account volume compression of a porous material when drying needs may be needed.

3. Materials and methods

The materials used in this study, as well as the methods used to determine their properties, are presented in this section. An experimental protocol for the measurement of the water retention curve (WRC) of highly compressible materials (HCMs) is detailed.

3.1 Determination of the water retention curve of deinking by-products

3.1.1 Deinking by-products

Deinking by-products (DBP), also known as fiber-clay, are a fibrous and highly compressible paper recycling by-products composed mainly of cellulose fibers, clay and calcite (Panarotto, et al., 2005) (Fig. 6). The composition of DBP varies significantly with the type of paper recycled and the efficiency of the deinking process employed (Latva-Somppi, et al., 1994). DBP was characterized in the scope of many works (Panarotto, et al., 2005; Cabral, et al., 1999; Panarotto C., et al., 1999; Kraus, et al., 1997; Vlyssides & Economides, 1997; Moo-Young & Zimmie, 1996; Latva-Somppi, et al., 1994; Ettala, 1993). DBP leaves the production plant with gravimetric water content varying from 100% to 190% (Panarotto, et al., 2005). The maximum dry unit weight obtained using the Standard Proctor procedure ranges from 5.0 to 5.6 kN/m^3. The optimum gravimetric water content ranges from 60 to 90%. Fig. 7 presents the consolidation over time of DBP specimens in the laboratory as well as in the field. The field data collected from three sectors of the Clinton mine cover, Quebec, Canada, presented in Figure 7 illustrates the time-dependent nature of the settlements of the DBP and reveals a short primary consolidation phase during the first two months, followed by a long secondary consolidation (creep) phase. Hydraulic conductivity tests were performed in oedometers at the end of each consolidation step in the laboratory. The results are presented in Fig. 8, which shows the saturated hydraulic conductivity obtained for a series of tests performed with samples collected from different sites and prepared at an average initial gravimetric water content of approximately 138% (approximately 60% above the optimum water content). As expected, the saturated hydraulic conductivity increased with increasing void ratio, defining a slope of the mean linear relationship. The parameter b, i.e. the slope of the e versus $log(k)$ linear relationship in Equation 15, equals 0.95. Although the influence of the extreme bottom-left point is minor in the curve-fitting procedure, it may infer that the e versus $log(k)$ relation would be exponential rather than linear. Such relations were obtained by Bloemen (1983) for peat soils. However, in the case presented here, more points would be needed in the 10^{-10} m/s order of magnitude to conclude on the existence of such curved relation.

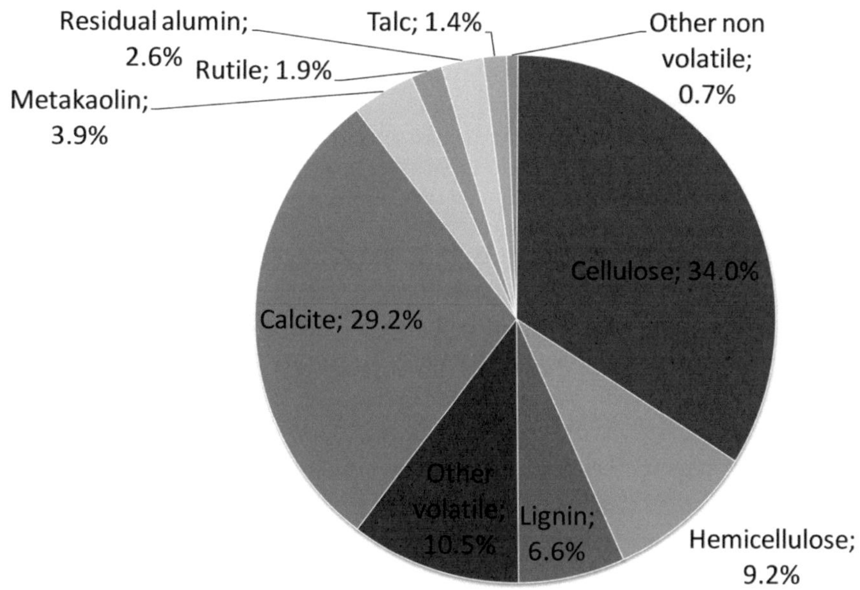

Fig. 6. Average composition of the DBP used in the experimental program (% by weight), adapted from (Panarotto, et al., 2005)

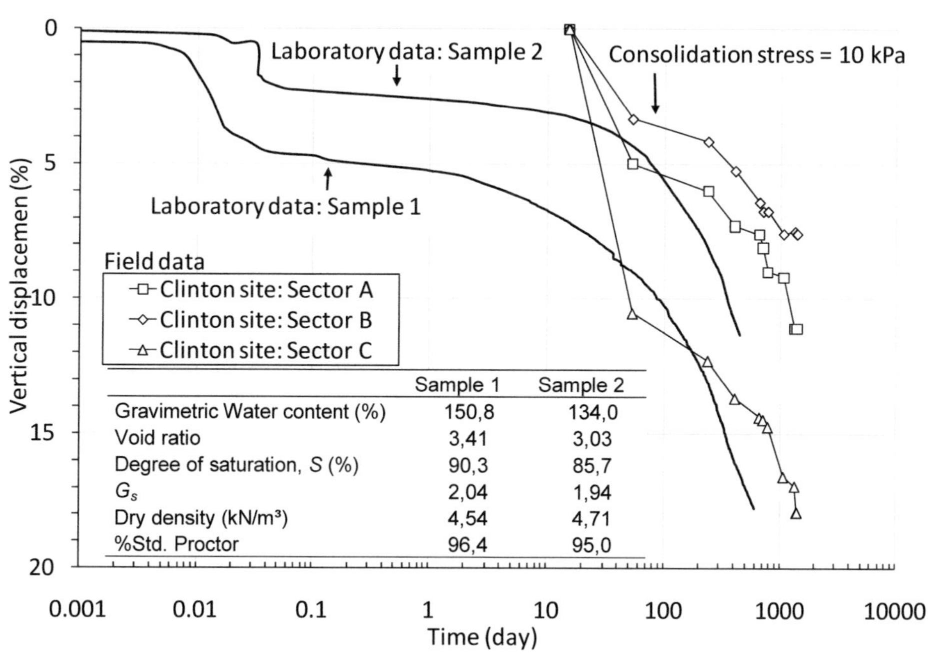

	Sample 1	Sample 2
Gravimetric Water content (%)	150,8	134,0
Void ratio	3,41	3,03
Degree of saturation, S (%)	90,3	85,7
G_s	2,04	1,94
Dry density (kN/m³)	4,54	4,71
%Std. Proctor	96,4	95,0

Fig. 7. Typical consolidation behaviour of deinking by-products from laboratory testing and from field monitoring of three sites (adapted from Burnotte et al. (2000) and Audet et al. (2002))

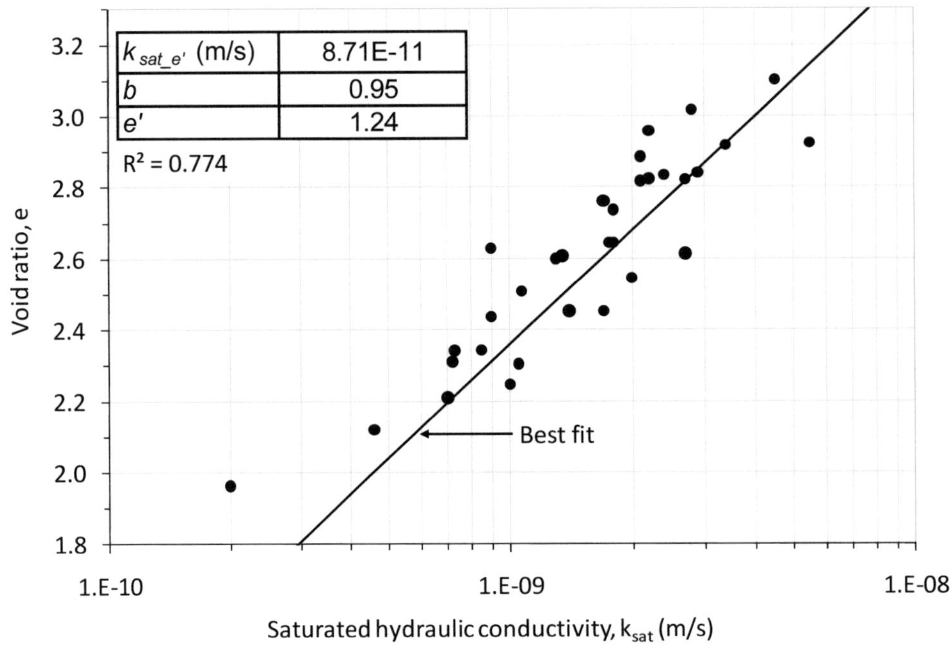

Fig. 8. Void ratio as a function of saturated hydraulic conductivity for deinking by-products

3.1.2 Testing equipment to determine the water retention curve

3.1.2.1 Pressure plate drying test (modified cell test) with continuous measurement of volume changes

Fig. 9 shows a schematic view of the testing apparatus used in this study to obtain the water retention curve (WRC) of DBP. A picture of the apparatus is shown in Fig. 10. The system consists of a 115.4-mm high, 158.5-mm diameter acrylic cell, a pressure regulator to control air pressure applied to the top of the sample and to the burette "CELL", and three valves to control air pressure, water inflow and water outflow. As the air pressure applied on the top of the specimen is increased, water is expelled from the sample and collected in burette "OUT". Any change in volume of the specimen during pressure application results in an equivalent volume of water that enters the cell via the burette CELL. The apparatus thus allows continuous measurement of volume changes, allowing the calculation of volumetric water content at each suction level. Further details of the equipment and testing protocol are described in Cabral et al. (2004).

Cabral et al. (2004) used a porous stone with negligible air-entry value (AEV, 0-*bar* porous stone). However, as suction increased beyond the AEV of DBP, air entering the DBP specimen drained the porous stone. In the present study, testing was performed using porous stones with AEVs of 1 *bar* or 5 *bar* (1 *bar*=101.3 kPa). The use of a 1-*bar* or 5-*bar* porous stone allowed WRC data to be obtained up to suction values of 100 kPa or 500 kPa, respectively. The time needed to reach equilibrium in burettes OUT and CELL after each pressure increment was carefully evaluated. Consistent readings could be made every 24 hours with the 0-*bar* porous stone and after 2 to 5 days for 1-*bar* and 5-*bar* porous stones, depending on the level of pressure applied.

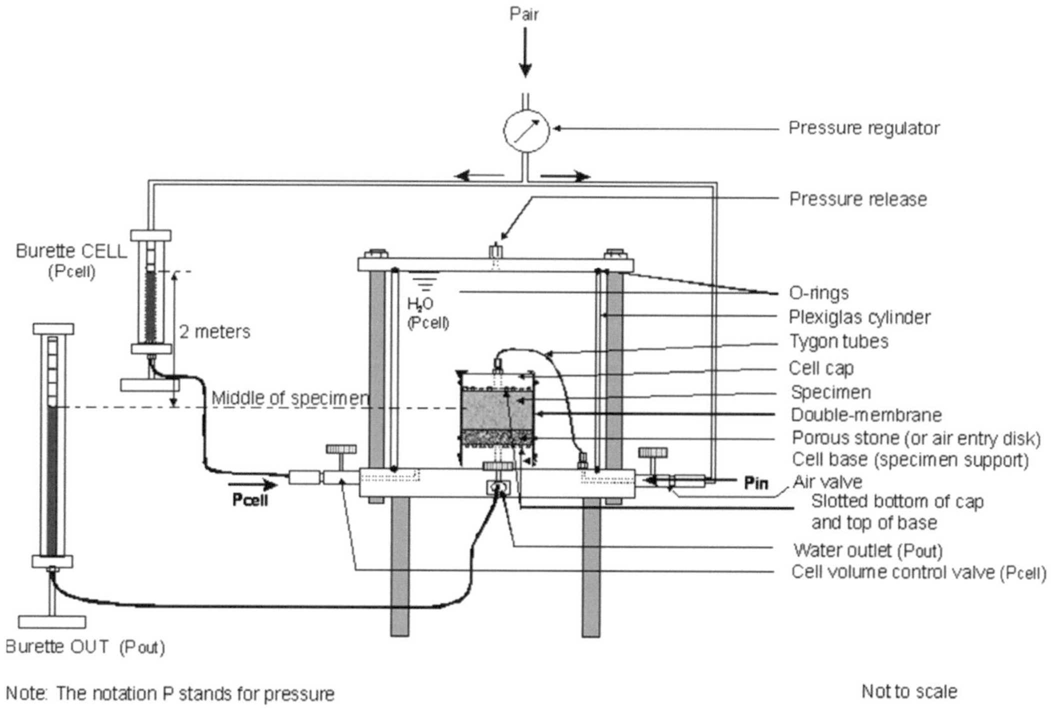

Fig. 9. Scheme of the testing system developed at the Université de Sherbrooke

Fig. 10. Picture of the testing system developed at the Université de Sherbrooke

3.1.2.1.1 Sample preparation

A mass of about 20 kg of DBP was sampled from a pile. From this sample, about 1 kg per test was sampled for the four tests presented in this chapter. Rare gravel particles were

removed. Initial autoclaving of the materials at 110°C and 0.5 *bars* is required to prevent biological activity during testing. Planchet (2001) observed that the use of microbiocide changed the pore structure of DBP by alterating the fibers. Consequently, only autoclaving was performed to prevent microbes to grow into the DBP specimens.
Preliminary tests with DBP showed that the procedure leading to the best reproducibility required compacting three 10 *mm*-thick layers of material by tamping DBP material directly into the cell. For that purpose, a mould and small mortar were designed and constructed (Fig. 11a). The thickness of the layers was controlled using a specially designed piston (Fig. 11b). The initial void ratio of a test was controlled by determining the mass of sample needed to be compacted in each layer. Cabral et al. (2004) provide further details of the procedure for sample preparation and compaction. The characteristics of the samples of DBP used in this study, modified cell tests (MCT) 1 to 4, are presented in Table 2. The data were calculated from the mass of humid material constituting the sample, water content test in a non ventilated oven at 110°C. The relative density was determined thanks to a volumetric method grain density test.

(a) (b)

Fig. 11. Tools used for the modified cell test sample preparation: (a) mould and mortar (b) mould and piston

	MCT1	MCT2	MCT3	MCT4
Gravimetric water content (%)	196.9	153.6	210.3	188.5
Unit weight (*kN/m*³)	11.76	11.58	11.25	11.72
Degree of saturation (%)	99.4	93.3	99.9	99.1
Relative density	1.99	1.99	1.99	1.99
Initial void ratio	3.89	3.28	3.67	2.70
Porous stone used for the test	1-*bar*	1-*bar*	1-*bar*	5-*bar*

Table 2. Characteristics of four specimens tested into the modified cell (MCT) in this study (evaluated after compaction directly into the mold and before the test assembly)

3.1.2.1.2 Testing and calibration

Following compaction, the apparatus was assembled and the consolidation phase initiated. Consolidation was conducted during 120 minutes under a cell confining pressure of 5 kPa.

The valve allowing air into the sample remained closed during this adjustment phase. The pressure was then raised to 20 kPa for a second consolidation phase that lasted 24 hours. A pressure of 20 kPa corresponded approximately to the overburden pressure applied by the protection cover layers to a barrier layer of DBP.
The first point of the WRC was taken at the end of the consolidation phase under 20 kPa, which occurred when the water levels in the burettes CELL and OUT reached equilibrium. At this point, readings were initialized and air pressure increments of 2.5 kPa (irregular increments for MCT4) were applied to the specimen until reaching the suction corresponding to the AEV as clearly identified on the ψ vs. θ plot. Pressure increments of 10 kPa were then applied at suction levels higher than the AEV (irregular increments for MCT4).
The volume of water entering the cell (from burette CELL), corresponding to changes in specimen volume, was recorded during each pressure increment. The water volume reaching the burette OUT was also recorded to indicate the volume of water lost from the MCT specimen. Calibration of the system was conducted to account for the expansion of the cell and the lines. Details of the calibration procedure are provided by Cabral et al. (2004). After appropriate corrections were applied to the recorded values, the water content and degree of saturation of the specimen is determined for the applied air pressure level. Since the axis translation technique was employed, the air pressure corresponded to the suction in the specimen. Stabilization of volumetric water content was reached when two consecutive measurements, taken 24 h apart, show a difference of less than 0.25% in water content for MCT1 to MCT3 and 0.5% for MCT4.
Tests were ended when suction reached the AEV of the porous stone. The cell was then disassembled and the final dimensions and weight of the sample were recorded.
HCMs have usually highly hysteretic behavior, which would affect water retention and flow. In this study, only desaturation was tested. The reader looking for more information about the hysteresis phenomenon on HCMs may refer to Nuth & Laloui (2008).

4. Results and interpretation

This chapter contains mathematical models developed to estimate the hydraulic properties of highly compressible materials (HCM).

4.1 Hydraulic properties of highly compressible materials

This section presents a water retention curve (WRC) model developed from the results of an experimental program performed to determine the WRC of deinking by-products (DBP). This model is able to fit several water retention curves of highly compressible materials using a single set of parameters and is validated using published data. The hydraulic conductivity function (*k*-function) was derived from the WRC proposed model using Fredlund et al. (1994)'s model (Equation 16). Moreover, a model to predict the *k*-function of HCM based on tests with saturated sample is presented and compared to results using Fredlund et al. (1994)'s model.

4.1.1 Results of the experimental program

The results obtained in the experimentation phase of this research program are interpreted in this section, leading to two models:

- a model to fit WRC data of a HCM;
- a model to predict the *k*-function of a highly compressible material (HCM) based on tests with saturated samples.

The first is an adaptation of a common WRC model (Fredlund & Xing, 1994) considering suction-induce consolidation curve. The second is an alternative procedure based on two relationships: void ratio versus saturated hydraulic conductivity and void ratio at the air-entry value (AEV) versus AEV.

4.1.1.1 Model to fit water retention data of a highly compressible material

If porous materials whose void ratios converge toward the same value under increasing suction application, irrespective of their initial void ratio (Fig. 5), it can be expected that, at a certain suction value, the parameters governing the shape of the WRC should reach the same values. Accordingly, the model proposed herein, which is based on the Fredlund & Xing (1994) WRC model, is able to describe multiple WRC test results for the same HCM with different initial void ratios using a single set of parameters. The adaptation consists in the variation of the four parameters of the Fredlund & Xing (1994) model (a_{FX}, n_{FX}, m_{FX} and θ_s) with void ratio. In this section, a void ratio function model, a WRC model and a hydraulic conductivity function (*k*-function) model are presented.
Rode (1990) approximated the effect of hysteresis and found that the effect on the calculated water content was not large. However, investigations performed by Price & Schlotzhauer (1999) showed that the shrinkage behavior of peat, a material with high organic content like DBP, was highly hysteretic. Nevertheless, this aspect is not covered in the present study. Considering hysteresis would lead to the prediction of lower water content and hydraulic conductivity values for a same suction value.

4.1.1.1.1 The water retention curve model

For HCMs, a gradual desaturation takes place before the AEV is reached, as shown for peat by Weiss et al. (1998), Schlotzhauer & Price (1999) and Brandyk et al. (2003), and for DBP by Cabral et al. (2004). Therefore, the two-phase behavior of Huang et al. (1998)'s model (Equation 13) may lead to a model bias. The Fredlund & Xing (1994) model (Equation 9) was adapted in this study to account for volume changes, including the region where suction is lower than the AEV.
Fig. 12 shows the relationship between $log(\psi_{aev})$ and e_{aev} obtained from suction tests performed using DBP (MCT1 to MCT4 and tests 15 and 16 from Cabral et al. (2004)). Parameters of Equation 11 obtained from R^2 maximization over the data are shown in Table 3. In the cases where a 0-*bar* porous stone was employed (tests 15 and 16), AEV was considered to be equal to the suction value where air broke through the sample (which is an approximation). In the cases where 1-*bar* or 5-*bar* porous stones were employed (MCT1 to 4), the AEV was considered to be equal to the suction value where significant loss of water was observed on the S versus ψ curve. The relationship between $log(\psi_{aev})$ and e_{aev} is approximately linear and can be described using Equation 11. According to Fredlund et al. (2002), the parameter a_{FX} defines the lateral position of the WRC and is linearly proportional to AEV. Consequently, for HCM, the variation of a_{FX} with void ratio can be stated to be similar to the variation of ψ_{aev}, as in Equation 17. This model was preferred over Kawai et al. (2000)'s model (Equation 12), being more closely related to experimental.

$$a_{FX} = a_{FXe'} 10^{\varepsilon_a (e-e')} \tag{17}$$

where $a_{FXe'}$ is the value of a_{FX} at a reference void ratio e' and ϵ_a is the slope of the $log(a_{FX})$ and e_{aev} curve. In order to reduce the number of parameters, the value of e can be set to e_c. For $e = e_c$, the more compressible the material is, the closer $a_{FXe'}$ will be from residual suction (C_r). Moreover, since $a_{FX} \propto \psi_{aev}$, $\varepsilon_a = \varepsilon_\psi$.

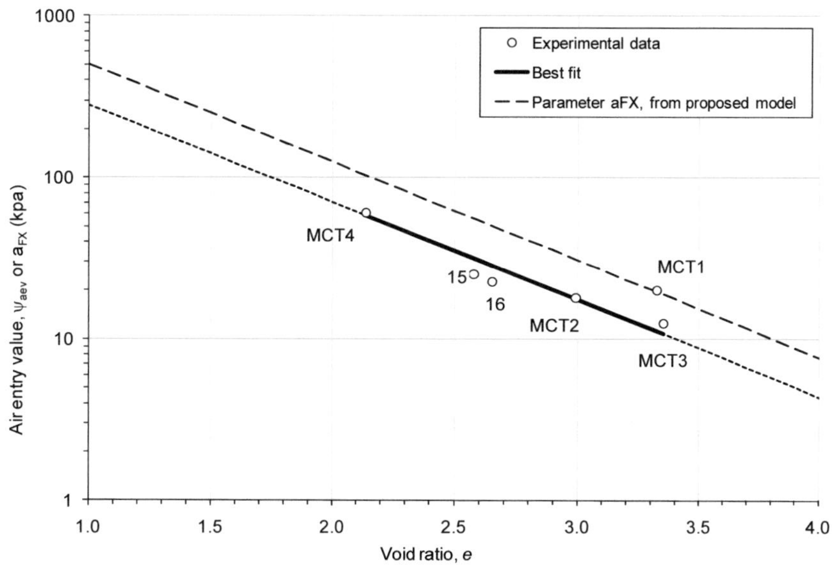

Fig. 12. Air-entry value as a function of void ratio at the air-entry value

$\psi_{aev\_e'}$	731
ε_{ψ}	-0.603
e'	0.31
R^2	0.891

Table 3. Parameters of Equation 11 obtained from R^2 maximization over MCT1 to MCT4 and tests 15 and 16 from Cabral et al. (2004)

In the Huang et al. (1998) model, the parameter λ is assumed to vary linearly with e. In the proposed model, the same assumption is made concerning the parameters n_{FX} and m_{FX} (Equation 9), whose variation with void ration are described as follows:

$$n_{FX} = n_{FXe'} + \varepsilon_n(e - e') \tag{18}$$

$$m_{FX} = m_{FXe'} + \varepsilon_m(e - e') \tag{19}$$

where $n_{FXe'}$ and $m_{FXe'}$ are respectively the Fredlund & Xing (1994) parameters n_{FX} and m_{FX} at the reference void ratio e', ε_n and ε_m are regression parameters obtained by least square minimization of WRC data. Note that the slope of the WRC tends to be null when n_{FX}, tends to unity (and void ratio tends to its suction-induced shrinkage limit).

4.1.1.1.2 The void ratio function

Fig. 13 shows pore-shrinkage characteristic curve (PSCC, i.e. void ratio versus water content) data from suction tests with DBP performed in order to obtain the WRC (tests MCT1 to MCT4), using the procedure described in the Materials and Methods section. The initial void ratio of the four tests ranged between 2.70 and 3.89 (Table 2.). The creep phases occurring for values lower than AEVs for every tests indicate that DBP is a HCM.

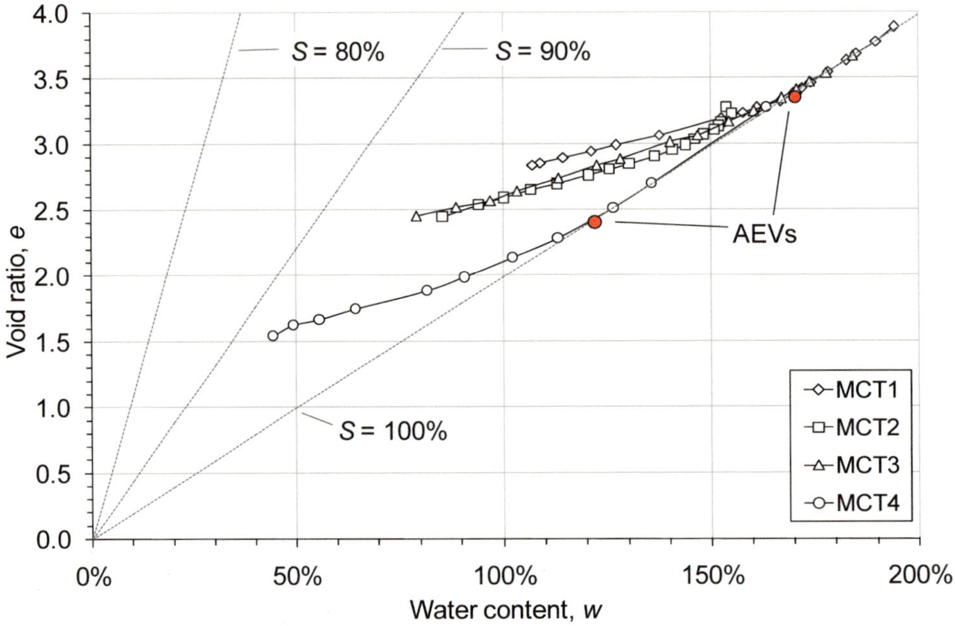

Fig. 13. Void ratio versus water content for deinking by-products under suction plus a total stress of 20 kPa

A void ration function (e-function) must be defined in order to estimate the variation of Fredlund & Xing (1994) model's parameters. Suction induced in a porous material is a stress that may result in pore shrinkage. Fig. 14 presents void ratio versus suction for four representative modified cell tests with DBP (MCT1 to 4). A total stress of 20 *kPa* was applied for all tests and several increments of suction were imposed. These results are compared to oedometer tests (thick line in Fig. 14). In the range of effective stresses applied in these tests, the exponential shape of the consolidation curve follows the same pattern as the suction induced consolidation behavior. However, the comparison has no quantitative value due to the fact that the axial stress measured in oedometer tests cannot be compared to the volumetric (mean) stress measured in desaturation tests.

Based on results for compressible soils, Huang et al. (1998) assumed that changes in void ratio occurred only for values of matric suction less than the AEV. However, the results obtained from tests MCT1 to 4, presented in Fig. 14, indicate that the above-mentioned assumption does not apply to HCMs such as DBP. Indeed, the void ratio continued to decrease significantly as suction increased, even for suction levels greater than the AEV. Salager et al. (2010) proposed and equation to link void ratio and suction. However, their model was not suitable to adequately fit our data, namely because it was not meant to consider the convergence of several tests to a unique shrinkage limit. The alternative exponential model in Equation 20, adapted from Ratkowski (1990), is proposed to describe the variation of void ratio as a function of suction as follows:

$$e_i = e_c + (e_{oi} - e_c)(1 + c_1 e_{oi} \sigma')^{c_2} \tag{20}$$

where e_i is the void ratio at effective stress σ' for test i, e_{0i} is the void ratio in the beginning of the suction test i, and c_1, c_2 and e_c are fitting parameters. The parameter e_c is the void ratio at which the e vs. σ' curves obtained from compression or consolidation tests conducted using

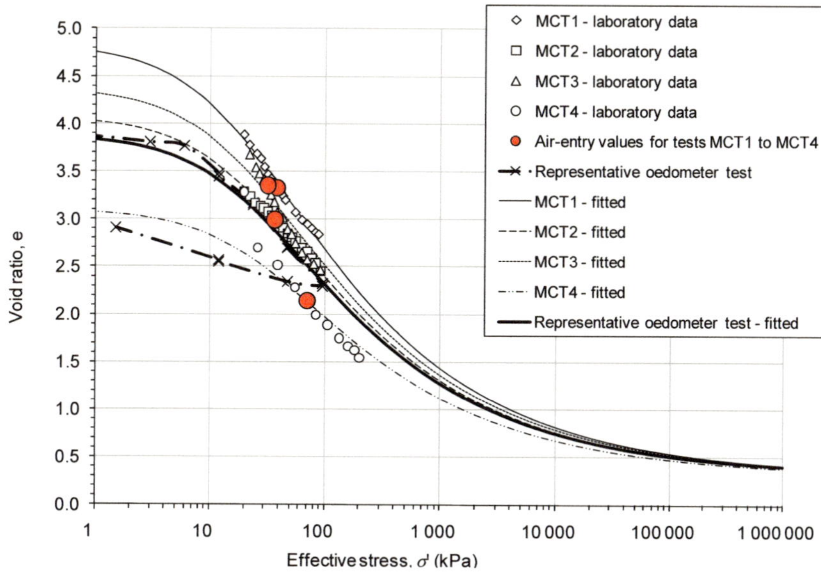

(a)

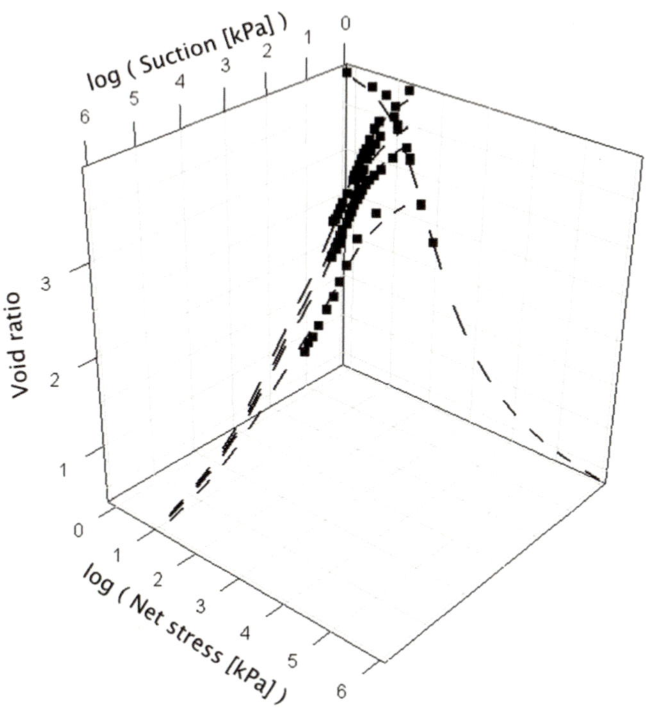

(b)

Fig. 14. (a) Void ratio versus effective stress $\sigma' = \sigma + S(\psi)\psi$ the effective stress is the axial one in the case of the oedometer Test) and (b) decomposed versus suction and net stress for four MCTs and one representative oedometer test specimens prepared at different initial void ratios converge, the parameters c_1, c_2 and e_c for DBP were obtained by a least square optimization technique (their values are presented in Table 4.), σ is total (mechanical) stress and ψ_{e_c} is the suction observed in the vicinity of e_c. The effective stress is defined in Equation 6, i.e. $\sigma' = \sigma + S(\psi)\psi$, where $S(\psi)$ is the WRC.

Fig. 3 shows that the suction component of compression energy may theoretically drop to zero at complete desaturation. Accordingly, although not observed yet in laboratory, a rebound should be observed, similar to the one observed when mechanical stress is released from a soil sample submitted to an oedometer test. This rebound is not described by the void-ratio function of Equation 20. The apparatus did not allow suction values higher than 500 kPa. Nevertheless, the *e*-function is of Equation 20 is an exponential function where two curves with different e_{0i} values converge to a threshold value of e_c. Such convergence observed by Boivin et al. (2006) gives confidence in the assumption that the convergence towards an asymptotic void ratio value is still valid for suction values higher than 500 kPa. The *e*-function regression of Equation 20 was fitted to MCT and oedometer data by maximizing R^2. The parameters are shown in Table 4.

	MCT1	MCT2	MCT3	MCT4	Oedometer
e_0	4.84	4.09	4.38	3.11	3.88
e_c			0.31		
c_1			0.01097		
c_2			-0.349		
R^2			0.980		

Table 4. Parameters of the e-function for tests MCT1 to 4 and PPCT1 to PPCT4

The proposed WRC model (Equation 21) is then obtained by inserting Equation 20 into equations Equation 17 to Equation 19, and inserting Equation 17 to Equation 19 into Equation 9, with $e' = e_c$:

$$S(\sigma,\psi) = \frac{\left(1 - \dfrac{\ln\left(1+\dfrac{\psi}{C_r}\right)}{\ln\left(1+\dfrac{10^6}{C_r}\right)}\right)}{\ln\left(exp(1) - \left(\dfrac{\psi}{A}\right)^B\right)^C} \qquad (21)$$

Where

$$A = a_{FXe'} 10^{(e_{0i}-e_c)(1+c_1 e_{0i}(\sigma+S(\psi)\psi))^{c_2}}$$

$$B = n_{FXe'} + \epsilon_n (e_{0i} - e_c)\left(1 + c_1 e_{0i}(\sigma + S(\sigma,\psi)\psi)\right)^{c_2}$$

$$C = m_{FXe'} + \epsilon_m (e_{0i} - e_c)\left(1 + c_1 e_{0i}(\sigma + S(\sigma,\psi)\psi)\right)^{c_2}$$

The degree of saturation in Equation 21 is present on both sides of the equation, and its isolation is not possible. The proposed strategy is to evaluate a single $S(\sigma,\psi)$ based on a WRC function (Fredlund & Xing, 1994; Brooks & Corey, 1964; van Genuchten M. T., 1980) representing a whole data set and insert that function on the right-wing side of Equation 21.

The proposed WRC model is supported by a documented theoretical framework that supposes that it can be applied to most HCMs. The next section presents a validation procedure on a compressible silty sand.

4.1.1.1.3 Validation of the proposed model

In order to validate the proposed model (Equation 21), the procedure for obtaining the WRC and predicting the *k*-function was applied to experimental data published by Huang (1994) for a series of tests with a compressible silty sand from Saskatchewan, Canada. These tests were performed using pressure-plate cells. Changes in volume during testing were not recorded. As a consequence, the variation of void ratio with suction had to be derived from the results of flexible-wall permeability tests performed by Huang (1994) with specimens with similar initial void ratios.

In order to apply the proposed model, it is supposed that the silty sand behaves like a CM, i.e. void ratio converges toward a single value (the shrinkage limit). The proposed model was applied to fit the results of suction test data for specimens PPCT13, PPCT16 and PPCT22, taken from Huang et al. (1998). The parameter a_{FX} was obtained using a $log(a_{FX})$ and e_{aev} curve, where the AEV is determined on the S versus ψ curve. Fig. 15 shows the results of Huang et al. (1998) fitted with the proposed model. It can be observed that, using a single set of parameters, the proposed model superimposes the experimental results (R^2 = 0.978) rather well. At a value of about 700 *kPa*, the three curves practically merge into a same desaturation curve. The data ranged between 0 and 300 *kPa*. The predicted degrees of saturation corresponding to suction values higher than 300 *kPa* are extrapolated to reach a null value at a suction value of 10^6 *kPa*.

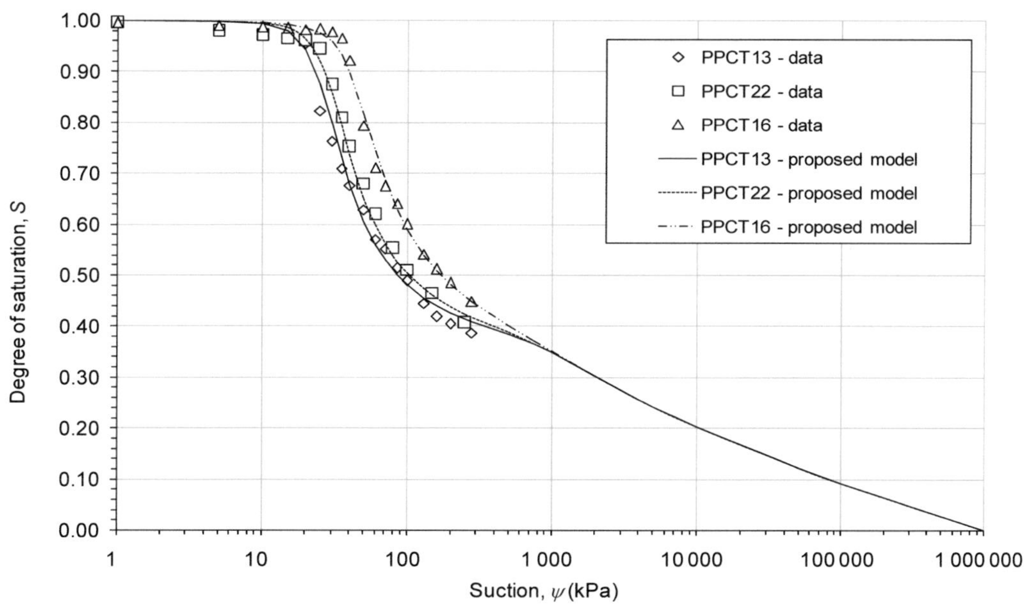

Fig. 15. Proposed water retention curve model to represent desaturation of a silty sand – data from Huang (1994)

Fig. 15 presents the results of three flexible-wall unsaturated hydraulic conductivity tests (FWPT2, FWPT3 and FWPT6) performed by Huang (1994) with the same Saskatchewan silty sand. These tests were chosen because their initial void ratios are similar to those of PPCT13,

PPCT22 and PPCT16, respectively. Parameter of the e-function and of the proposed WRC, shown inTable 5, were obtained by maximizing R^2.

	PPCT13	PPCT22	PPCT16
c_1		1.40E-04	
c_2		-42.2	
e_0	0.528	0.513	0.466
e_c		0.425	
related to	FWPT2	FWPT3	FWPT6
R^2		0.976	

(a)

$a_{FX\_e'}$	55.9
ε_a	-3.81
$n_{FX\_e'}$	4.90
ε_n	2.76
$m_{FX\_e'}$	0.303
ε_m	0.89
Cr	197
R^2	0.992

(b)

Table 5. Parameters (a) of the e-function (Equation 11) of PPCT13, PPCT22 and PPCT16 (Huang, et al., 1998) and (b) the proposed WRC model (Equation 21), all obtained using R^2 maximization

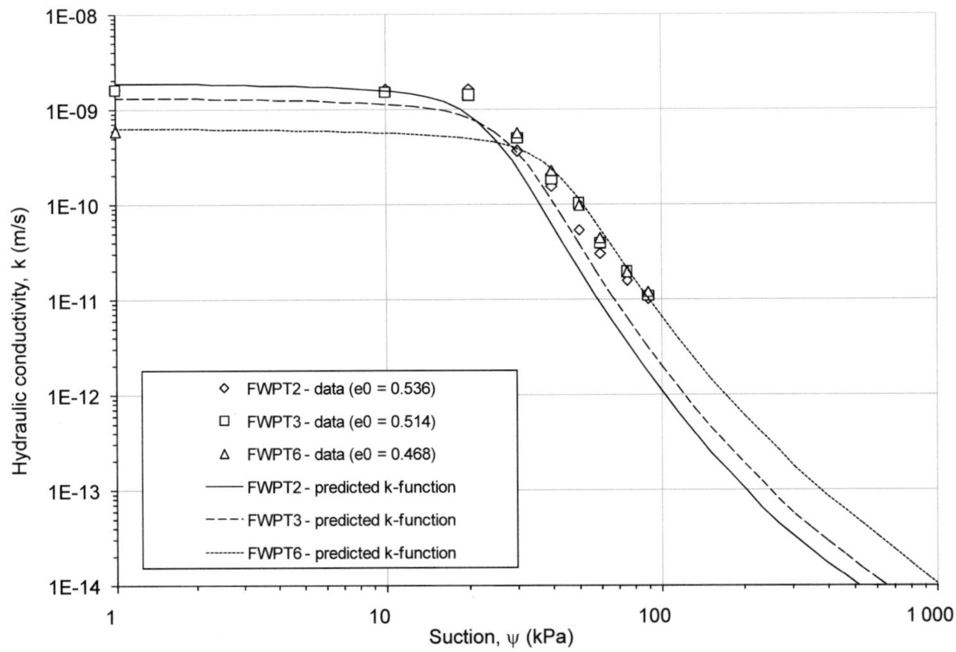

Fig. 16. Hydraulic conductivity function for Huang (1994)'s silty sand

The *k*-function of the silty sand was estimated using Equation 15 and Equation 16 based on the WRC determined using the proposed model. Parameters for equation Equation 15 are taken from Huang (1994). Since the derivative of the proposed model (θ' ; Equation 16) is rather complex to determine, the symbolic computation program *Maxima*[4] was used. The integrations were performed using quadratures in a *Scilab*[5] environment.

[4]http://maxima.sourceforge.net/

[5]Institut national de recherche en informatique et en automatique, France, Rocquencourt. http://www.scilab.org/

It can be observed that the *k*-functions are independent of each other, which is not the case of the WRC from which they derive. This is due to the intrinsic bias in Equation 16 associated with the supposition that the soil structure is incompressible (see section 2.3). However, the proposed WRC gives coherent *k*-function curves, the graphs being parallel on the log-log scale. The results in Fig. 16 show a rather good agreement between experimental data for FWPT6, but a much less good agreement for tests FWPT2 and FWPT3. Test FWPT6 underwent little shrinkage, which was not the case with FWPT2 and FWPT3. Equation 16 could be adapted for volume changes, using shrinkage factors, and a WRC model that considers volume changes, like the one proposed in this chapter. However, the development of a *k*-function model that considers shrinkage is beyond the scope of this report. The next section deals with application of the proposed model to the results of suction tests with DBP.

4.1.1.2 Application of the proposed water retention curve model to deinking by-products

4.1.1.2.1 The water retention curve of deinking by-products

Fig. 17 presents the WRC data from suction tests with DBP performed in order to obtain the WRC (tests MCT1 to MCT4). The degrees of saturation data presented in Fig. 17 were obtained considering volume changes in the data reduction process. The proposed WRC model was employed to fit the four sets of experimental data. As shown in Fig. 12, the trend of parameter a_{FX} against void ratio is closely related to whose of the air-entry value, both slopes being visually parallel. In Fig. 17, it can be observed that a good agreement results (R^2 =0.875). It is shown in Table 6 that no variation was needed for parameters n_{FX} , its slope being null.

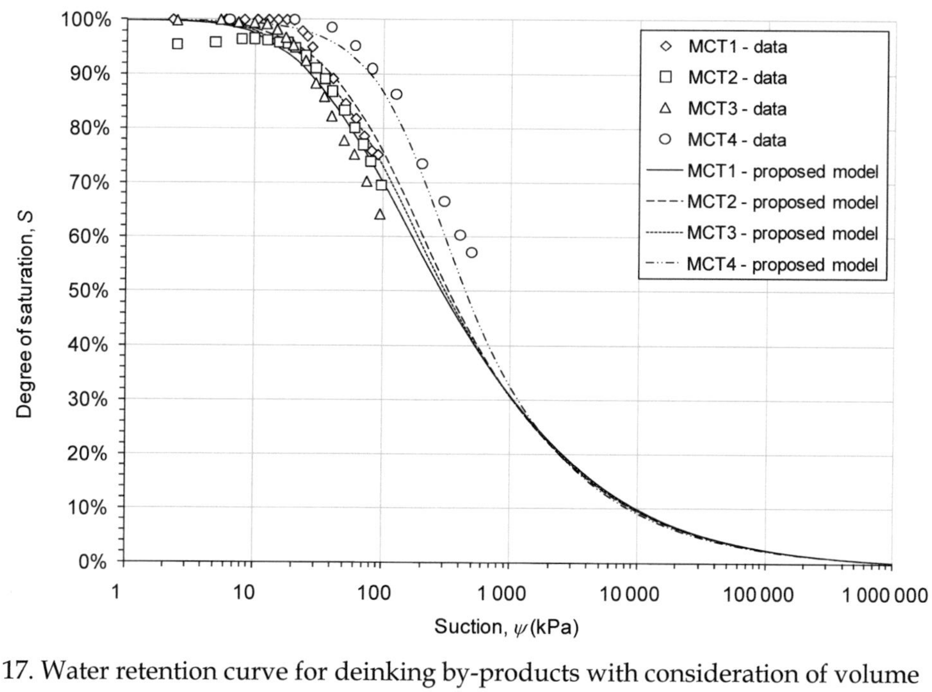

Fig. 17. Water retention curve for deinking by-products with consideration of volume change

a_{FX_ec}	1316
ε_a	-0.607
n_{FX_ec}	2.19
ε_n	0.00
m_{FX_ec}	1.32
ε_m	-0.36
Cr	1316
R^2	0.875

Table 6. Parameters of the fitted WRCs

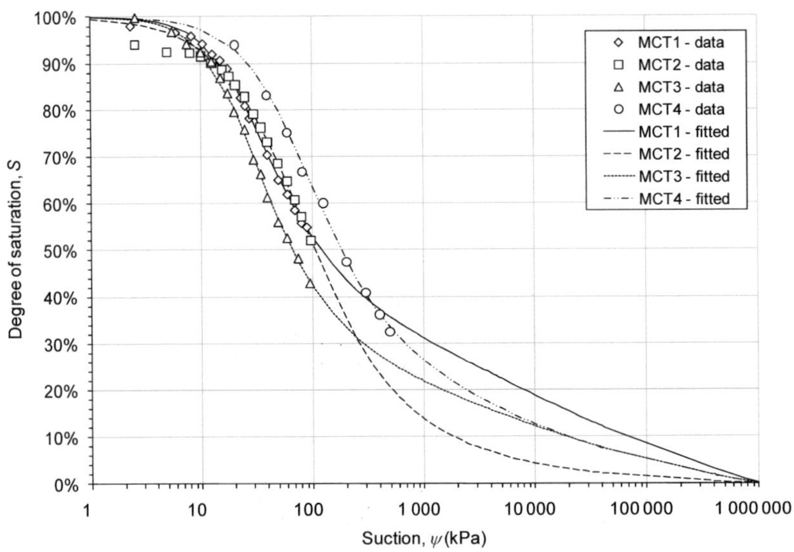

Fig. 18. Water retention curve for deinking by-products without consideration of volume change

	MCT1	MCT2	MCT3	MCT4	Average	Standard deviation
a_{FX}	19.9	60.3	19.0	51.9	37.8	21.4
n_{FX}	1.81	1.14	1.83	1.50	1.6	0.3
m_{FX}	0.570	1.615	0.742	0.857	0.9	0.5
Cr	3000	3000	3000	3000	3000.0	0.0
R^2	0.998	0.999	0.999	0.997		

Table 7. Parameters for the Fredlund & Xing (1994) water retention model for deinking by-products samples without consideration of volume change

If consideration is made that DBP do not undergo volume changes during application of suction, then the pore structure of the material would consequently remain unaltered. In this case, samples MCT1 to MCT4, which were consolidated to different initial void ratios, would behave as totally different materials. Fig. 18 presents suction test data for tests MCT1 to MCT4. The Fredlund & Xing (1994) model was used to fit experimental data, for which corrections due to volume changes in the data reduction process were not applied. Since volume is considered to be constant, the sets of data were treated independently, i.e. the relevant parameters were optimized in an independent manner. The values of the several parameters, their average values and their standard deviations are shown in Table 6.

The experimental results in Fig. 19 clearly show that volumetric water contents are significantly underestimated if volume changes are not considered, particularly at high suction levels. For example, test MTC1, at approximately $\psi = 20\ kPa$, consideration of volume changes lead to a degree of saturation 14% greater than the value obtained if volume changes were not considered. At $\psi = 90\ kPa$, the difference increases to 20%.

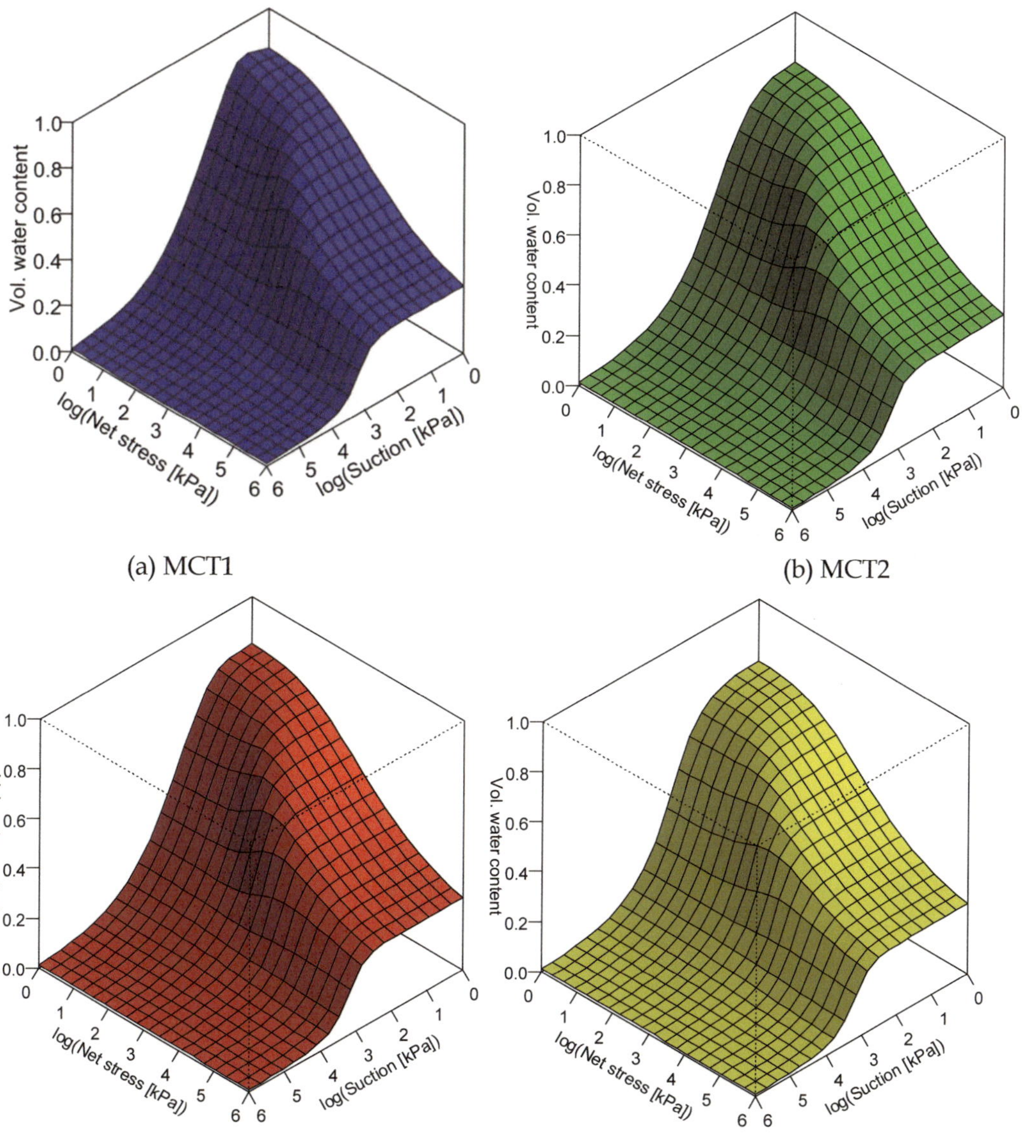

(a) MCT1 (b) MCT2

(c) MCT3 (d) MCT4

Fig. 19. Isometric representations of the water retention planes for tests MCT1 to 4

Since the void ratio is a function of both net stress and suction, and since the degree of saturation is a function of void ratio and suction, the degree of saturation can be plotted as a function of both net stress and suction in order to obtain water retention planes. Such three-dimensional representations may lead practitioners to better understandings of pore compression phenomena in unsaturated porous media.

4.1.1.2.2 Hydraulic conductivity functions for deinking by-products

Fig. 20 shows the hydraulic conductivity function (*k*-function) for tests MCT1 to MCT4. The curves were obtained based on their respective WRC that, in turn, were determined using the proposed model (Equation 21). Fig. 21 presents the *k*-functions for DBP based on the Fredlund & Xing (1994) WRC model (Equation 9 – no volume change), whose parameters are presented in Table 6. The value of k_{sat} was determined based on the initial void ratio (e_0) of each test (Table 2).

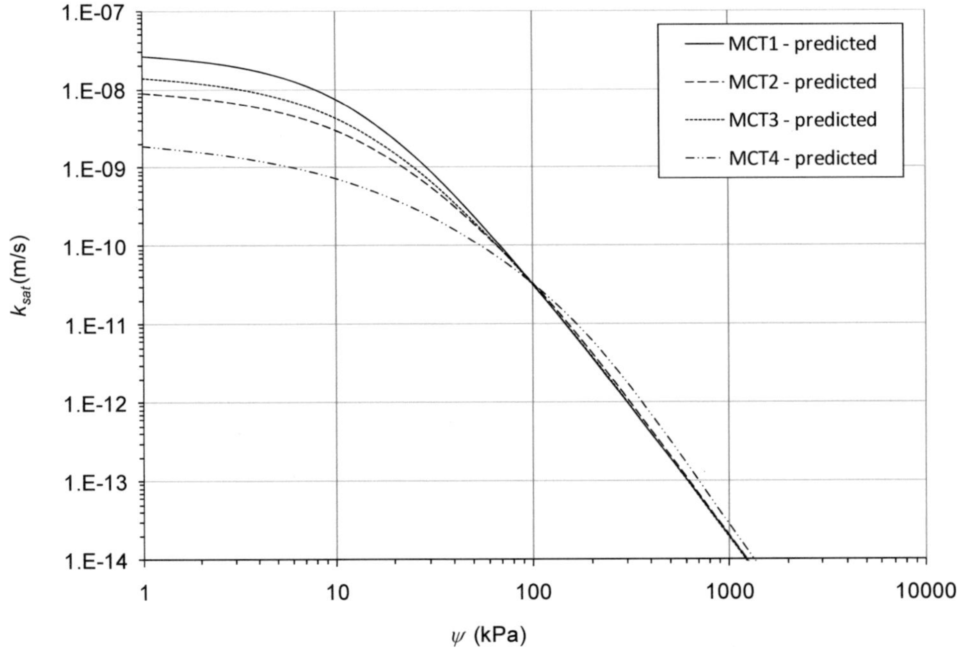

Fig. 20. Hydraulic conductivity functions for deinking by-products derived from models considering volume changes (proposed water retention curve model)

The different *k*-functions computed from the proposed model, plotted in Fig. 20, are closed to superimpose onto a single branch around a suction of 100 kPa, although the tendency of the void-ratio function (Fig. 15) and the degree of saturation (Fig. 18) shows a convergence around 1000 kPa. The bias in the computation of *k*-functions for HCMs (section 2.3) is still visible in Fig. 18, although barely apparent, perhaps because of the low standard deviation of porosity of DBP. This behavior is not observed in Fig. 21, where the four curves represent four independent samples Fig. 20 and Fig. 21 shows that the general trend of the *k*-function computed from the WRCs of respectively Fig. 18 and Fig. 19 is similar for both scenarios, although the *k*-function computed with the WRC that considers volume changes (Fig. 20) is more coherent with the theory (Fig. 5).

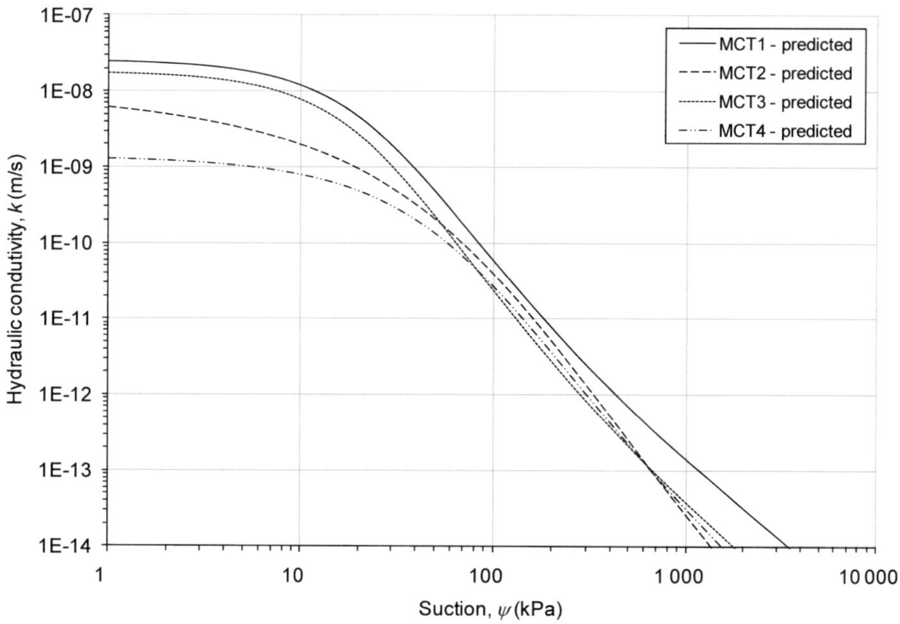

Fig. 21. Hydraulic conductivity functions for deinking by-products derived from models *not* considering volume changes (Fredlund & Xing (1994) water retention curve model)

4.1.1.3 Model to predict the hydraulic conductivity function of a highly compressible material based on tests with saturated samples

The k-function predictive model presented below has the advantage over Fredlund & Xing (1994)'s model (Equation 16) to assume compressible structure and to be based on tests performed with saturated samples.

4.1.1.3.1 The conceptual approach

Fig. 22 can be utilized to explain the concept behind the proposed k_{sat}-ψ_{aev} model. The k-function is described here by two curves. The first (denoted as curve 1) describes how the saturated hydraulic conductivity varies with suction for a sample of a highly compressible material consolidated to an initial void ratio e_{0i}. Using a void ratio function (e-function; proposed later), it is possible to predict the void ratio e_i for any suction value ψ and, using Equation 15, to predict the saturated hydraulic conductivity, k_{sati}, corresponding to e_i and ψ. Although the curve in Fig. 15 extends beyond the AEV, obtainment e vs. ψ curve beyond the AEV is not necessary for the purpose of determining curve 1 in Fig. 22. This is due to the fact that the value of e_c does not affect significantly the value of e_i for suctions lower than ψ_{aevi}, whether it is determined by a least square optimization technique or simply imposed based on a known residual value (as is the case in Fig.15). Using Equation 11 and the e-function, the AEV for test i, ψ_{aevi}, is then determined. The void ratio corresponding to ψ_{aevi} is e_{aevi}.

The second curve (denoted as curve 3 in Fig. 22) can also be estimated based on saturated hydraulic conductivity tests and the WRC. For the same test i as above, consolidated to an initial void ratio e_{0i}, when suction exceeds ψ_{aevi}, the void ratio decreases to a value e_{unsati} lower than e_{aevi}. Accordingly, the associated hydraulic conductivity decreases as expected to a value k_{unsati} lower than k_{sati}. This new hydraulic conductivity value, k_{unsati}, corresponds to the unsaturated hydraulic conductivity for sample *i* when its void ratio is e_{unsati}.

According to Equation 15, it is possible to obtain a saturated hydraulic conductivity corresponding to a void ratio e_{unsati}. In this case, another test *ii* (curve 2 in Fig. 22) must be performed using the same material, but the sample is consolidated to an initial void ratio $e_{0ii} < e_{0i}$. Suction is then applied to bring the sample to a void ratio e_{aevii} equal to e_{unsati} (Fig. 22).

Since, as shown schematically in Fig. 5, various WRCs superimpose into a single virgin desaturation branch Toll (1988) (this is equivalent to the consolidation behavior of clayey soils) and that the slope of the *k*-function can be calculated from the slope of the WRC, it may be inferred that *k*-functions also superimpose onto the same curve. Therefore, the curve obtained by joining the points (ψ_{aev};k_{sat}) for tests *i*, *ii*, etc., *i.e.* curve 3 and 4 in Fig. 22, would describe the *k*-function of the compressible material for suctions greater than the AEV.

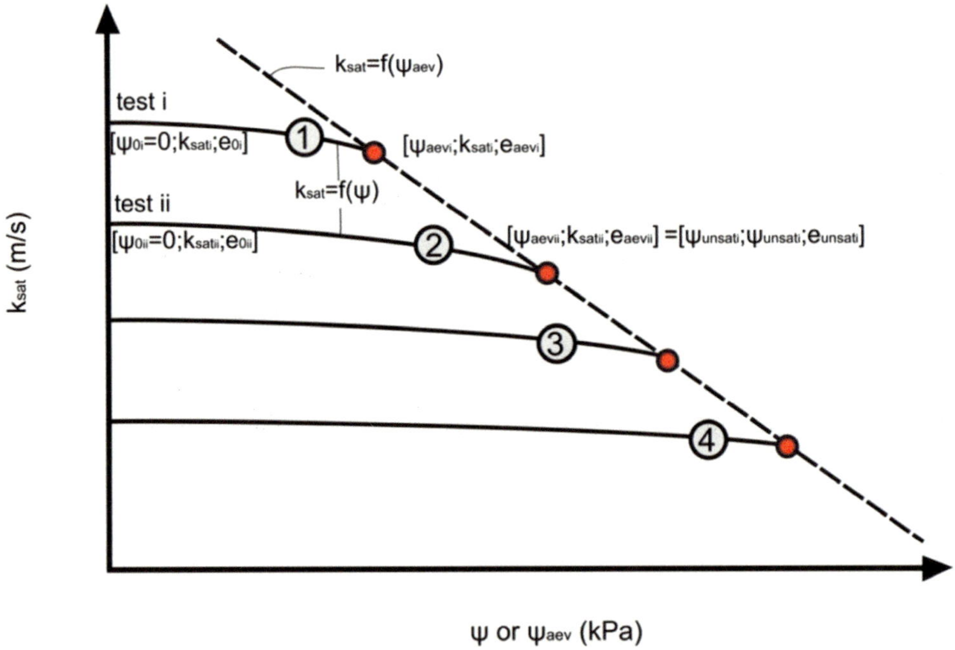

Fig. 22. Schematic representation of saturated hydraulic conductivity as a function of air-entry value

4.1.1.3.2 The proposed procedure

By substituting *e* in Equation 15 by e_i obtained from Equation 20, the following model for curve 1 (or curve 2) is obtained as follows:

$$k_{sat_i} = k_{sat_i} 10^{b(e_c+(e_{oi}-e_c)(1+c_1 e_{oi}\sigma')^{c_2}-e'_{aev})} \quad \text{for } \psi < \psi_{aev_i} \tag{22}$$

Curve 3 (Fig. 22) is obtained by replacing the parameters *e* and *e′* in Equation 15 by, respectively, e_{aev} and e'_{aev}, by isolating e_{aev} in Equation 11 and finally by substituting Equation 11 into Equation 15. Equation 23 is then used to establish a linear relationship on a *log*−*log* scale between k_{sat} and ψ_{aev}:

$$k_{sat} = k_{sat_{e'}} \left(\frac{\psi_{aev}}{\psi'_{aev}}\right)^{\frac{b}{\epsilon_a}} \quad \text{for } \psi_{aev} > \psi_{aev_i} \tag{23}$$

4.1.1.3.3 Validation and discussion

In order to validate the proposed procedure, Equation 22 and Equation 23 were applied to experimental data published by Huang et al. (1998). The results, presented in Fig. 23, are compared to unsaturated hydraulic conductivity measurements, also reported by Huang et al. (1998).

Hydraulic conductivity, suction and void ratio values obtained from flexible-wall permeability tests allowed for the determination of the parameters in Equation 22.

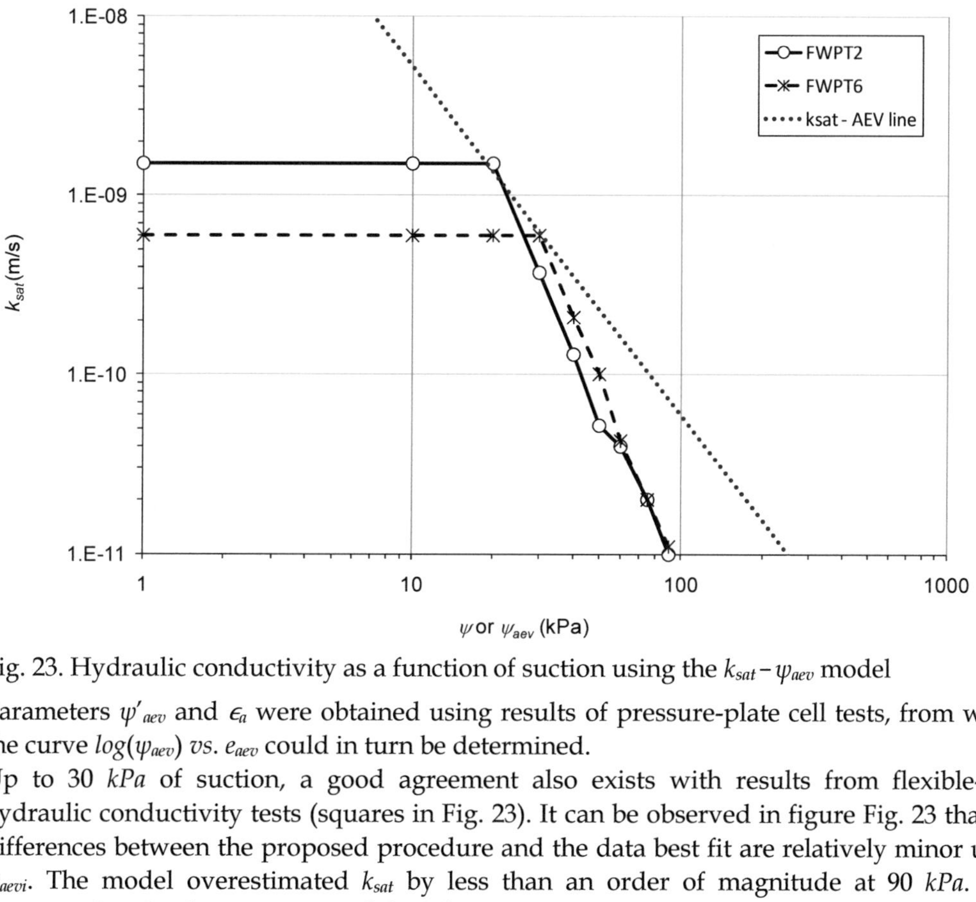

Fig. 23. Hydraulic conductivity as a function of suction using the $k_{sat} - \psi_{aev}$ model

Parameters ψ'_{aev} and ϵ_a were obtained using results of pressure-plate cell tests, from which the curve $log(\psi_{aev})$ *vs.* e_{aev} could in turn be determined.

Up to 30 *kPa* of suction, a good agreement also exists with results from flexible-wall hydraulic conductivity tests (squares in Fig. 23). It can be observed in figure Fig. 23 that the differences between the proposed procedure and the data best fit are relatively minor up to ψ_{aevi}. The model overestimated k_{sat} by less than an order of magnitude at 90 *kPa*. This suggests that the determination of the *k*-function can be made based on known relationships between k_{sat} and void ratio, and ψ_{aev} and void ratio with reasonable confidence.

4.1.1.3.4 Application of the k_{sat}-ψ_{aev} model to deinking by-products

Data from tests performed for DBP (Fig.8, Fig.15 and Fig. 13) were used to apply the k_{sat}-ψ_{aev} model. Results are shown is Fig. 24 for tests MCT1 and MCT4. For test MCT1, at 10 *kPa*, the hydraulic conductivity obtained using the proposed WRC and Fredlund & Xing (1994)'s model is twice as large as the one obtained using the k_{sat}-ψ_{aev} model. For test MCT4, hydraulic conductivities predicted using both models are similar for suction values lower than 400 *kPa*; for higher suction values, hydraulic conductivities obtained using the k_{sat}-ψ_{aev} model are higher than the ones obtained using the proposed WRC and Fredlund & Xing (1994)'s model, the *k*-function slope being steeper for the latter model than for the k_{sat}-ψ_{aev} model.

The discrepancies between the two predictions may be caused by the lack of data available. It can be observed in Fig. 20 and Fig. 23 that the k_{sat}-ψ_{aev} model overestimates the data and the prediction from the proposed WRC model.

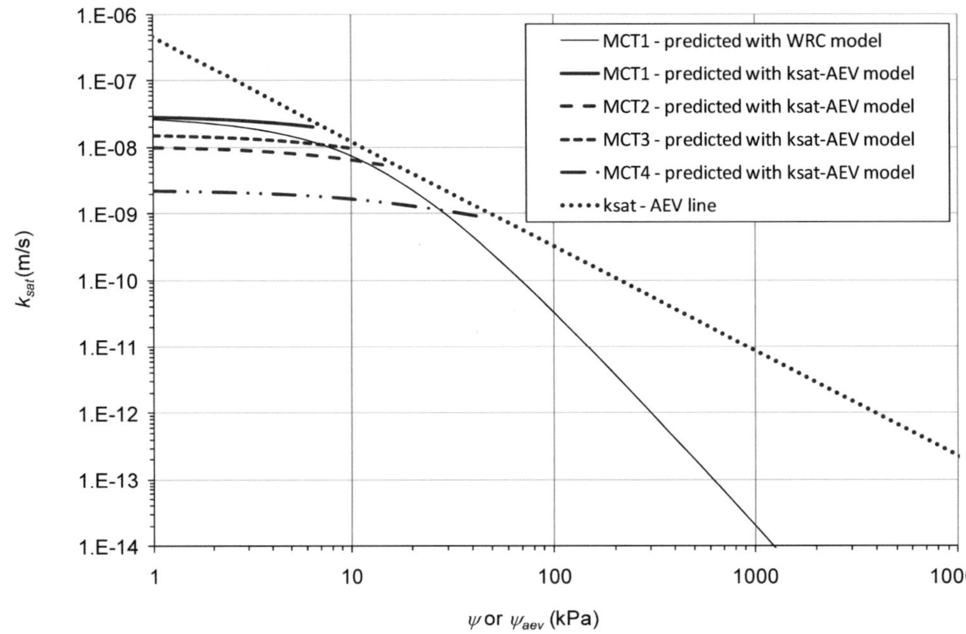

Fig. 24. Hydraulic conductivity functions for deinking by-products derived from the k_{sat}-ψ_{aev} model

5. Conclusion

The hydraulic properties of DBP were determined using the experimental procedure described by Cabral et al. (2004) and water retention curve (WRC) and hydraulic conductivity function (*k*-function) models considering suction-induced shrinkage developed by the authors Parent et al. (2006; 2004).

5.1 The compression energy concept

Although energy units are used in soil compaction studies, the concept of compression energy is not used often in geotechnics. A stress is a concentration of energy describing the action of compression exerted on the pore network. The energy paradigm links the WRC, the porosity function and Bishop (1959)'s parameter χ. It was demonstrated in this chapter that the capillary induced compression energy may be important even for suction values larger than the air-entry value (AEV). As a result, materials like DBP shrink over a large range of suction values beyond the AEV.

5.2 Model to determine the water retention curve of a highly compressible material

A model was proposed to describe the WRC of highly compressible materials (HCMs). The input parameters needed for the model were obtained directly from water retention tests. The experimental procedure used allowed to determine WRCs of materials undergoing significant volume changes during application of suction, i.e. HCM. Volume change in specimen was monitored during suction application, so that volumetric water contents can be continuously calculated.

The proposed WRC model was validated using published experimental data from tests performed with a compressible silty sand from Saskatchewan, Canada. Hydraulic conductivity functions (*k*-functions) based on the proposed WRC model fitted hydraulic conductivity values obtained from unsaturated permeability testing with this silty sand only for the data set that underwent no significant volume change, verifying the model bias of Fredlund et al. (1994)'s model (Equation 16) for HCM explained in section 2.3 As a result, there is a need for an accurate model able to predict the *k*-function of a HCM.

The proposed WRC model was applied to experimental data from representative tests on DBP. The proposed model fits experimental data with good accuracy (R^2=0.902). Volumetric water contents were significantly underestimated if volume change was eluded in the data reduction process.

Void ratio of DBP specimens tended to converge to the same value as suction increase. Consequently, their *k*-functions should also superimpose. Based on their respective WRC curve parameters, the *k*-functions for several tests were predicted using the Fredlund et al. (1994) model coupled with function that allowed variation in saturated hydraulic conductivity with void ratio. The *k*-functions obtained when the WRC model accounted for volume change converged to a single value at 10 000 *kPa*, even though the Huang et al. (1998) model was found to be inaccurate for HCM. On the other hand, if volume change was not accounted for, several independent *k*-functions were obtained.

We expect that the proposed WRC model could be applied to other compressible materials and that reliable *k*-functions could be derived using an appropriate *k*-function model. The appropriate parameters for the WRC must be obtained based on an experimental procedure such as the one presented in this paper. Further studies should also take into account the influence of hysteresis.

5.3 A model to predict the hydraulic conductivity function with saturated samples

A procedure to determine the *k*-function based on relationships between saturated hydraulic conductivity and void ratio, and between AEV and void ratio was developed and applied to DBP. A comparison between the *k*-function obtained by applying this procedure to experimental data reported in the literature (for a Saskatchewan silty sand) and actual unsaturated hydraulic conductivity data for the same silty sand shows a good agreement up to a suction value in the vicinity of 30 *kPa*. For higher suctions a reasonable agreement (less than one order of magnitude) is still obtained.

The use of the proposed procedure to determine the *k*-function requires suction and saturated hydraulic conductivity testing on samples consolidated to different initial void ratios.

However, these tests are more expeditious than direct determination of *k*-functions. Hence, the k_{sat}-ψ_{aev} procedure may be a valuable and cost-effective solution in many situations.

6. References

Abdolahzadeh, A., Vachon, B., & Cabral, A. (2008). Hydraulic barrier and its impact on the performance of cover with double capillary barrier effect. *61e Conférence géotechnique canadienne.* Edmonton.

Abdolahzadeh, A., Vachon, B., & Cabral, A. (2010). submitted. Assessment of the design of an experimental cover with capillary barrier effect using four years of field data. *Geotechnical and Geological Engineering Journal* .

Arya, L., & Paris, J. (1981). A physicoempirical model to predict the soi1 moisture characteristic from particle-size distribution and bulk density data. *Soil Science Society of America Journal , 45,* pp. 1023-1030.

Aubertin, M., Mbonimpa, M., Bussiere, B., & Chapuis, R. (2003). A model to predict the water retention curve from basic geotechnical properties. *Fourth International Conference on Acid Rock Drainage, 2,* pp. 731-746. Vancouver, Canada.

Audet, C., Lefebvre, G., Cabral, A., & Burnotte, F. (2002). State of development in the valorization of deinking by-products as an alternative to fine grained soils. *TAPPI.* Montréal, Canada.

Barbour, S. (1998). Nineteenth canadian geotechnical colloquium: The soil-water characteristic curve: a historical perspective. *Canadian Geotechnical Journal , 30* (5).

Baumgartl, T., & Kock, B. (2004). Modeling volume change and mechanical properties with hydraulic models. *Soil Science Society of America Journal , 68* (1), pp. 57-65.

Bédard, D. (2005). *Effet du fluage à long terme des sous-produits de désencrage dû à la perte de masse et son effet sur la compression et la conductivité hydraulique (Effect of creep of DBP due to mass loss and impact on the compressibility and hydraulic conductivity).* Master thesis, Université de Sherbrooke, Sherbrooke.

Bishop, A. (1959). The principle of effective stress. *Teknisk Ukeblad , 106,* pp. 859-863.

Bloemen, G. W. (1983). Calculation of the hydraulic conductivities and steady-state capillary rise in peat soils from bulk density and solid matter volume. *Zeitschrift für Pflanzenernährung und Bodenkunde , 146* (5), pp. 460-473.

Boivin, P., Garnierb, P., & Vauclin, M. (2006). Modeling the Soil Shrinkage and Water Retention Curves with the Same Equations. *Soil science society of America journal , 70,* pp. 1082-1093.

Brandyk, T., Szatylowicz, J., Oleszczuk, R., & Gnatowski, a. T. (2003). Water-related physical attributes of organic soils. In L. Parent, & P. Ilnìckì, *Organic Soils and Peat Material for Sustainable Agriculture* (pp. 33-66). CRC Press.

Brooks, R., & Corey, A. (1964). Hydraulic properties of porous media. *Hydrology paper, Colorado State University , 3.*

Burnotte, F., Lefebvre, G., Cabral, A., Audet, C., & Veilleux, A. (2000). Use of deinking residues for the final cover of a MSW landfill. *53rd Canadian Geotechnical Conference, 1,* pp. 585-591. Montréal, Canada.

Cabral, A., Burnotte, F., Lefebvre, G., Amyot, G., & Lacasse, G. (1999). Design construction and monitoring of a waste rock cover using pulp and paper residues.

Cabral, A., El Ghabi, B., Parent, S.-É., & Marineau, L. (2007). Design and performance of an experimental double capillary barrier cover placed in a municipal solid waste landfill. *Proc. 11th Intern. Waste Mgmt and Landfill Symp., CD-Rom.* Cagliari.

Cabral, A., Lefebvre, G., Burnotte, F., Panarotto, C., & Pastore, E. (1999). Use of pulp and paper residues as an alternative cover material to landfill and to acid generating tailings. *Seminario de Meio Ambiente em Industrias de Processo,* (pp. 56-70). Sao Paulo, Brazil.

Cabral, A., Planchet, L., Marinho, F. A., & Lefebvre, G. (2004). Determination of the soil water characteristic curve of highly compressible materials: Case study of pulp and paper by-product. *Geotechnical Testing Journal* .

Childs, E., & Collis-George, G. (1950). The permeability of porous materials. *Royal Society of London , Series A,* pp. 392-405.

DelAvanzi, E. (2004). *Unsaturated flow under increased gravitational field.* Boulder: University of Colorado at Boulder.

D'Onza, F., Gallipoli, D., Wheeler, S., Casini, F., Vaunat, J., Khalili, N., et al. (2010). In press. Benchmarking of constitutive models for unsaturated soils. *Géotechnique* .

Ettala, M. (1993). Quality of deinking sludge. *Journal of Environmental Science and Health , A28* (4), pp. 923-932.

Fredlund, D. (1967). *Comparison of soil suction and one-dimensional consolidation characteristics of a highly plastic clay.* National Research Council Canada, Division of Building Research.

Fredlund, D., & Morgenstern, N. (1976). Constitutive relations for volume change in unsaturated soils. *Canadian Geotechnical Journal 13(3)* , 261-276.

Fredlund, D., & Rahardjo, H. (1993). *Soil mechanics for unsaturated soil.* New York, U.S.A.: Wiley.

Fredlund, D., & Xing, A. (1994). Equations for the soil-water characteristic curve. *Canadian Geotechnical Journal , 31* (4), pp. 521-532.

Fredlund, D., Xing, A., & Huang, S. (1994). Predicting the permeability function for unsaturated soils using the soil-water characteristic curve. *Canadian Geotechnical Journal , 31* (4), pp. 533-546.

Fredlund, M., Wilson, G., & Fredlund, D. (2002). Use of the grain-size distribution for estimation of the soil-water characteristic curve. *Canadian Geotechnical Journal , 39* (5), pp. 1103-1117.

Gallipoli, D., Gens, A., Sharma, R., & Vaunat, J. (2003). An elasto-plastic model for unsaturated soil incorporating the effect of suction and degree of saturation on mechanical behavior. *Géotechnique , 53* (1), pp. 123-135.

Holtz, R., & Kovacs, W. (1981). *Introduction to Geotechnical Engineering.*

Huang, S. (1994). *Evaluation and laboratory mesurment of the coeficient of permeability in deformable unsaturated soils.* Saskatoon, Department of civil engineering, University of Saskatchewan, Saskatchewan, Canada.

Huang, S., Barbour, S., & Fredlund, D. G. (1998). Development and verification of a coefficient of permeability function for a deformable unsaturated soil. *Canadian Geotechnical Journal , 35* (3), pp. 411-425.

Jardine, R., Standing, J., & Kovacevic, N. (2004). Lessons learned from full-scale observations and practical application of advanced testing and modelling. *Deformation Characteristics of Geomaterials. 2,* p. 45. IS-Lyon: Editions Di Benedetto.

Kamon, M., Inazumi, S., Rajasekaran, G., & Katsumi, T. (2002). Evaluation of waste sludge compatibility for landfill cover application. *Soils and Foundations* (42), pp. 13-27.

Kawai, K., Karube, D., & Kato, S. (2000). The model of water retention curve considering effects of void ratio. In H. Rahardjo, D. Toll, & E. Leong (Ed.), *Unsaturated Soils for Asia,* (pp. 329-334).

Kennedy, G., & Price, J. (2005). A conceptual model of volume-change controls on the hydrology of cutover peats. *Journal of Hydrology*, *302*, pp. 13-27.

Khalili, N., & Khabbaz, M. (1998). A unique relationship for chi for the determination of shear strength of unsaturated soils. *Geotechnique*, *48* (5), pp. 681-688.

Khalili, N., Geiser, F., & Blight, G. (2004). Effective stress in unsaturated soils: Review with new evidence. *International journal of Geomechanics*, *4* (2), pp. 115-126.

Kraus, J., Benson, C., Maltby, C. V., & Wang, X. (1997). Laboratory and field hydraulic conductivity of three compacted paper mill sludges. *Journal of Geotechnical and Geoenvironmental*, *123* (7), pp. 654-662.

Lapierre, C., Leroueil, S., & Locat, J. (1990). Mercury intrusion and permeability of louiseville clay. *Canadian Geotechnical Journal*, *27*, pp. 761-773.

Latva-Somppi, J., Tran, H. N., Barham, D., & Douglas, M. A. (1994). Characterization of deinking sludge and its ashed residue. *Pulp and paper Canada*, 2 (82), pp. 382-385.

Leong, E. C., & Rahardjo, H. (1997). Review of soil-water characteristic curve equations. *Journal of Geotechnical and Geoenvironmental Engineering*, *123* (12), pp. 1106-1117.

Leong, E., & Rahardjo, H. (1997). Permeability functions for unsaturated soils. *Journal of Geotechnical and Geoenvironmental Engineering*, *123* (12), pp. 1118-1126.

McCartney, J., & Zornberg, J. (2005). The centrifuge permeameter for unsaturated soils. In A. Tarantino, E. Romero, & Y. Cui (Ed.), *Advanced Experimental Unsaturated Soi1 Mechanics - Proceedings of an International Symposium.* Trento, Italy.

Moo-Young, H., & Zimmie, T. (1996). Effects of freezing and thawing on the hydraulic conductivity of paper mill sludges used as landfill covers. *Canadian Geotechnical Journal*, *33*, pp. 783-792.

Mualem, Y. (1976). A new mode1 for predicting the hydraulic conductivity of unsaturated porous media. *Water Resources Research*, *12*, pp. 513-522.

Nemati, M., Caron, J., & Gallichand, J. (2000). Using paper de-inking sludge to maintain soil structural form: Field measurements. *Soil Science Society of America Journal*, *66* (2), pp. 367-373.

Nemati, M., Caron, J., Banton, O., & Tardif, P. (2002). Determining air entry value in peat substrates. *Soil Science Society of America Journal*, *64* (1), pp. 275-285.

Ng, C., & Pang, Y. (2000). Influence of stress state on soil-water characteristics and slope stability. *Journal of Geotechnical and Geoenvironmental Engineering*, *126* (2), pp. 157-166.

Nuth, M., & Laloui, L. (2008). Advances in modelling hysteretic water retention curve in deformable soils. *Computers and Geotechnics*, *35* (6), pp. 835-844.

Nuth, M., & Laloui, L. (2008). Effective Stress Concept in Unsaturated Soils :Clarification and Validation of an Unified Framework. *International Journal of Numerical and Analytical Methods in Geomechanics*, *32*, pp. 771-801.

Panarotto, C., Cabral,. A., Burnotte, F., Pastore, E., & Lefebvre, G. (1999). Utilisation des résidus de désencrage du papier comme recouvrement pour le contrôle du DMA: capacité de neutralisation de l'acidité résiduelle. *Congrès APGGQ*, (pp. 31-44). Rouyn-Noranda, Quebec, Canada.

Panarotto, C., Cabral, A. R., & Lefebvre, G. (2005). Environmental, geotechnical, and hydraulic behaviour of a cellulose-rich by-product used as alternative cover material. *Journal of environmental engineering and science*, *4*, pp. 123-138.

Paquet, J., & Caron, J. (1993). In situ determination of the water desorption characteristics of peat substrates. *Canadian Journal of Soil Science*, *73* (3), pp. 329-339.

Parent, S. (2006). *Hydraulic and geotechnical aspects of capillary barrier design using a highly compressible recylced material.* Sherbrooke: Ph.D. Thesis.

Parent, S., Cabral, A., Dell'Avanzi, E., & Zornberg, J. (2004). Determination of the Hydraulic Conductivity Function of a Highly Compressible Material Based on Tests with Saturated Samples. *Geotechnical Testing Journal , 27* (6), pp. 1-5.

Planchet, L. (2001). Utilisation des résidus de désencrage comme barrière capillaire et évapotranspirative (ÉT) pour les parcs à résidus miniers producteurs de DMA. Sherbrooke: Université de Sherbrooke.

Price, J., & Schlotzhauer, S. (1999). Importance of shrinkage and compression in determining water storage changes in peat : the case of a mined peatland. *Hydrological Processes , 13* (16), pp. 2591-2601.

Ratkowski, D. (1990). *Handbook of nonlinear regression model.* New York: M. Dekker.

Robart, G. (1998). *Étude de la perméabilité et de la compressibilité des résidus de désencrage (Studyof the permeability and compressibility of deinking residues).* Sherbrooke: Université de Sherbrooke.

Rode, P. C. (1990). Transient caluculation of moisture migration using a simplified description of hysteresis in the sorption isotherms. *Proceedings of the second symposium on building physics in the nordic countries.* Norwegian University of Science and Technology, Trondheim.

Roscoe, K., & Burland, J. (1968). *On the generalized stress-strain framework of "wet" clay* (Heymann and Leckie ed.). Cambridge: Cambridge University Press.

Salager, S., El Youssoufi, M., & Saix, C. (2010). Definition and experimental determination of a soil-water retention surface. *Canadian Geotechnical Journal , 47* (5), pp. 609-621.

Schlotzhauer, S., & Price, J. (1999). Soil water flow dynamics in a managed cutover peat field, Quebec: Field and laboratory investigations. *Water Resources Research , 35* (12), pp. 3675-3683.

Simms, P., & Yanful, E. (2002). Predicting soil-water characteristic curves of compacted plastic soils from measured pore-size distributions. *Geotechnique , 54* (4), pp. 269-278.

Smith, K., & Mullins, C. (2001). *Soil and environmental analysis : physical methods, 2nd edition.* M. Dekker.

Terzaghi, K. (1936). A Fundamental Fallacy in Earth Pressure Computations. *Journal of Boston Society of Civil Engineers , 23*, pp. 71-88.

Toll, D. (1995). A conceptual model for the drying and wetting of soils. In *Unsaturated Soils* (Vol. 2).

Toll, D. (1988). *The behaviour of unsaturated compacted naturally occurring gravel.* University of London: Ph.D. Thesis.

Tripathy, S., Rao, K., & Fredlund, D. (2002). Water content - void ratio swell-shrink paths of compacted expansive soils. *Canadian Geotechnical Journal , 39* (4), pp. 938-959.

van Genuchten, M. T. (1980). A closed-form equation for pre- dicting the hydraulic conductivity of unsaturated soils. *Soil Science Society of America Journal , 44*, pp. 892-898.

van Genuchten, M., Leij, F., & Yates, S. (1991). *The RETC Code for Quantifying the Hydraulic Functions of Unsaturated Soils, Version 1.0.* Riverside, California: EPA Report 600/2-91/065, U.S. Salinity Laboratory, USDA, ARS.

Vanapalli, S., Fredlund, D., & Pufahl, D. E. (1999). The influence of soil structure and stress history on the soil-water characteristics of a compacted till. *Geotechnique , 49* (2), pp. 143-159.

Vlyssides, A. G., & Economides, D. G. (1997). Characterization of wastes from a newspaper wash deinking process. *Frenesius Environ. Bull. , 6*, pp. 734-739.

Weiss, R., Alm, J., Laiho, R., & Laine, J. (1998). Modeling moisture retention in peat soils. *Soil Science Society of America Journal*, *62* (2), pp. 305-313.

Zhou, J., & Yu, J.-l. (2005). Influences affecting the soil-water characteristic curve. *Journal of Zhejiang University SCIENCE, 6A* (8), pp. 797-804.

Zhuang, J., Jin, Y., & Miyazaki, T. (2001). Estimating water retention characteristic from soil particle-size distribution using a non-similar media concept. *Soil Science, 166* (5), pp. 308-321.

Part 2

Empirical Approaches to Estimating Hydraulic Conductivity

4

Correlations between Hydraulic Conductivity and Selected Hydrogeological Properties of Rocks

Stanisław Żak
Wrocław University of Technology
Poland

1. Introduction

Solving problems related to the occurrence, accumulation, discharge and flow of groundwaters requires knowledge of many rock properties. Their determination is laborious and costly. Therefore, for a long time scientists have been looking for and formulating relationships between particular parameters. Literature contains plenty of relations representing mutual correlations between different hydrogeological properties of rocks. This pertains in particular to determination of hydraulic conductivity (Bear, 1972; Arya et al., 1999; Kasenow, 2002). Formulae obtained empirically are the commonest, but there are also those based on an adopted soil model. In this case the range of application is usually broader, and application limitations are directly connected with the application range of the model.

The discussion presented in this chapter concerns the relationship between hydraulic conductivity and selected hydrogeological properties of rocks, based on a rock model in the shape of a bundle of tortuous capillaries, known in literature (Carman, 1956). The properties which have the paramount importance for this model are specific surface area and porosity. As specific surface area is very often determined based on gradation analysis, these issues have received more detailed attention. A broader discussion of these issues is also connected with a clear formulation of all the assumption and simplifications comprised in the used formulae.

The best known relation based on a rock model in the form of a bundle of tortuous capillaries is a formula known as Kozeny-Carman-equation (Olsen, 1960; Liszkowska, 1996; Mauran et al. 2001; Chapuis & Aubertin, 2003; Carrier, 2003). However, based on this model, one may look for correlations with other hydrogeological parameters too (Petersen et al. 1996).

This chapter presents the results of such investigations in relation to most parameters used in hydrogeological calculations. In particular, this refers to mutual relations between hydraulic conductivity, specific surface area, effective grain diameter, effective capillary diameter, specific yield, specific retention, porosity and capillary rise height. One should emphasize that a satisfactory attempt to present such relationships could contribute to a significant reduction in the range of necessary analyses connected with soil identification.

An important element of the described research is verification of theoretically determined correlations between different parameters. Therefore, the results of experimental examinations will be presented and then compared with calculation results obtained from the derived relations.

2. Theoretical correlations between hydrogeological properties of rocks

Theoretical correlations between various properties of rocks always refer to a particular rock model. Therefore, it should be strictly defined. The model adopted in this work, presenting rock as a bundle of capillaries, is well-known and used to determine hydraulic conductivity. However, as its range of usefulness has been extended to include a possibility to define other rock properties, it will be discussed here more broadly.

2.1 Rock model

The discussion will cover a model presenting rock as undeformable material containing a bundle of tortuous capillaries with identical cross-sections. (Fig. 1).

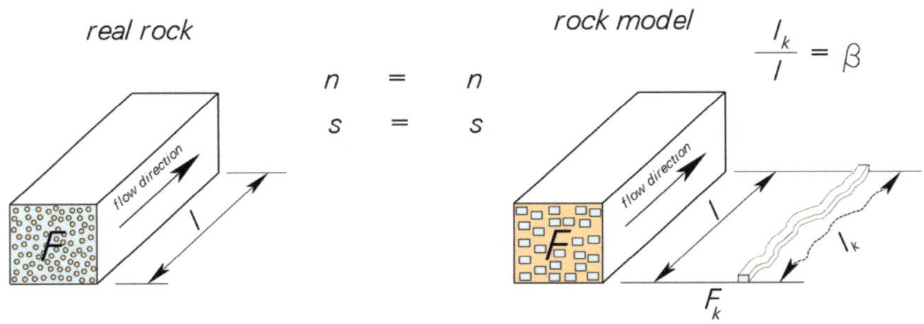

Fig. 1. Model of rock as a bundle of tortuous capillaries (s - specific surface area, n - porosity).

Capillaries are arranged in such a way that the line joining their beginnings and endings is parallel to a potential direction of a fluid flow in real soil. The horizontal cross-section of capillaries will be normally adopted as circle-shaped, although in some considerations concerning determination of hydraulic conductivity, any other shape can be adopted. Capillary tortuousness is characterised by the ratio of the length of a capillary along its axis to the length along a straight line between its beginning and ending. One can adopt different values of capillary tortuousness along different directions, thus allowing for anisotropic properties of soils.

The characteristic feature of the discussed rock model is the fact that its specific surface area s and porosity n are the same as specific surface area and porosity of real rock.

2.2 Specific surface area

Specific surface area is a very important parameter, on which the structure of the adopted rock model is based. For the needs of hydrogeology, it is very often determined based on soil gradation analysis, especially sieve analysis. Spherical grain shape is usually adopted then. If grain shape is more complex, specific surface area can be determined more precisely by considering three dimensions of the grain, i.e. the largest, the smallest and medium. These dimensions can be obtained by analysing grain shape in a small, randomly chosen sample or subjecting it to laser analysis using devices produced specially for this purpose. Hence, further considerations concerning determination of specific surface area based on soil gradation analysis will be covered more extensively, with the assumption that we have more detailed knowledge of the shape of rock-building grains.

Specific surface area s will refer to the ratio between the total area of grains and particles ΣF_z (the boundary surface of grain skeleton) contained in total rock volume V to this volume. For rock built of grains with identical sizes and shapes, specific surface area is

$$s = \frac{\sum F_z}{V} = \frac{N\,F_z}{V} = \frac{\frac{V_s}{V_z} F_z}{V} = \frac{\frac{V(1-n)}{V_z} F_z}{V} = \frac{(1-n)F_z}{V_z} = A\,\frac{(1-n)}{b} \qquad (1)$$

where: N - number of grains in total rock volume V, F_z - area of a an individual grain, V_s - volume of grain skeleton in volume V, V_z - volume of an individual grain, n - porosity, A - grain shape factor, b - largest grain dimension.

Shape factor can be determined from the boundary surface F_z of a grain, its volume V_z and the largest dimension b.

$$A = b\,\frac{F_z}{V_z} \qquad (2)$$

For example, if grains have the shape of a sphere with diameter d, then $b = d$ and $A = 6$. The value of grain shape factor is also 6 if a grain has the shape of a cube with side b or a cylinder with base diameter b and height b. However, if grains are elongated or oblate, value A clearly rises (Fig. 2).

sphere with diameter b = d	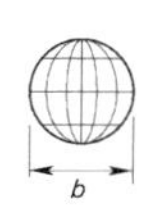	$A = 6$
cube with edge b	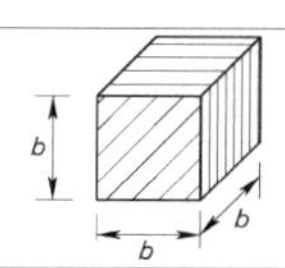	$A = 6$
cylinder with diameter b and height b	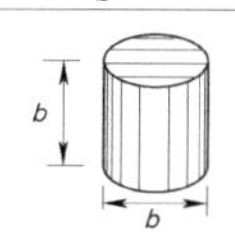	$A = 6$
cuboid with sides b, b, b/2	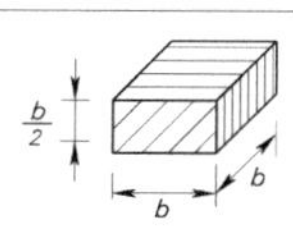	$A = 8$
cuboid with sides b, b/2, b/2	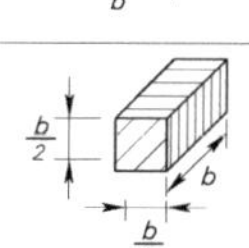	$A = 10$

Fig. 2. Shape factor A for different grain shapes.

Grains in sedimentary rocks are normally rounded. One can often assume that they have the shape of a sphere, a spheroid or another smooth solid. Fig. 3 illustrates shape factor relation for different ellipsoid dimensions a and c, where c – the smallest dimension and b – the largest dimension.

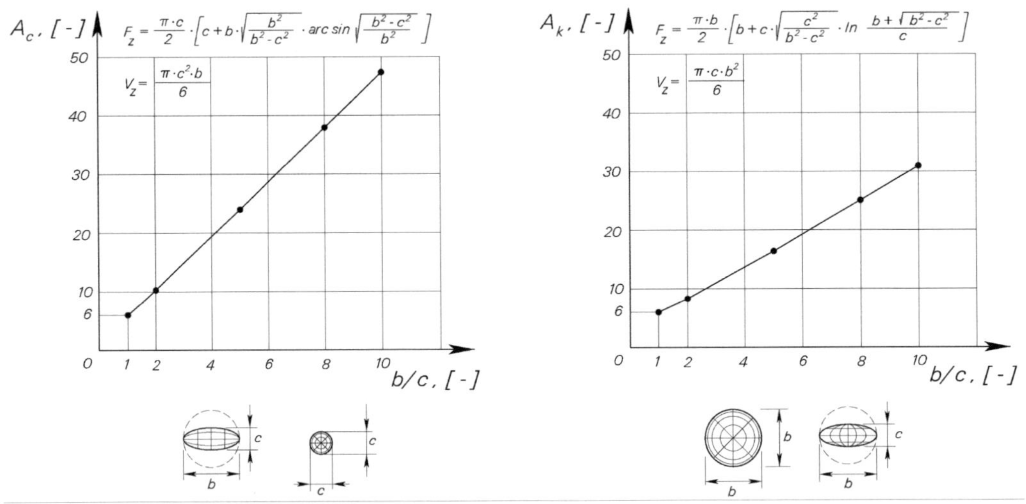

Fig. 3. Shape factor of a cigar-shaped (A_c) and disc-shaped (A_k) spheroid in relation to the ratio of its biggest to smallest dimension (b/c).

If a grain is rounded (its surface is smooth) and has three different dimensions a, b and c, then its volume V_z can be adopted as:

$$V_z = \frac{\pi\, a\, b\, c}{6} \tag{3}$$

The value of shape factor A for such shapes can be calculated with the assumption that A changes linearly from value A_k for a disc to value A_c for a cigar, with mean size a ranging from c to b. Then the shape factor is:

$$A = A_k + \frac{A_c - A_k}{\frac{b}{c} - 1}\left(\frac{b}{a} - 1\right) \tag{4}$$

Changes in shape factor are illustrated by Fig. 4.

For a soil type with varied grain sizes, determination of specific surface area s depends on the chosen method of gradation analysis. In the case of measuring the mass of different fractions building a rock sample, specific surface area can be calculated from the formula:

$$s = \frac{\sum F_z}{V} = \frac{\sum_i (N_i\, F_{zi})}{V} = \frac{\sum_i \left(\frac{m_{si}}{\rho_{si}\, V_{zi}}\, F_{zi}\right)}{\frac{m_s}{\rho_s\,(1-n)}} = \rho_s\,(1-n) \sum_i \left(\frac{g_i\, A_i}{\rho_{si}\, b_{ei}}\right) = \sum_i s_i \tag{5}$$

where: N_i – number of i-sized grains, F_{zi} – area of an average i-sized grain, ρ_{si} – density of i-fraction soil skeleton, m_{si} – mass of i-fraction soil, V_{zi} – volume of an average i-sized grain, m_s

– total mass of soil used for sieve analysis, ρ_s – density of the soil skeleton of all the sample, A_i – shape factor of a typical i-sized grain, b_{ei} – the largest dimension of effective grain size of i-fraction, $g_i = m_{si}/m_s$ - mass proportion of i-fraction, s_i – specific surface area component resulting from the proportion of i-fraction, $g_i = m_{si}/m_s$ – proportion of i-fraction.

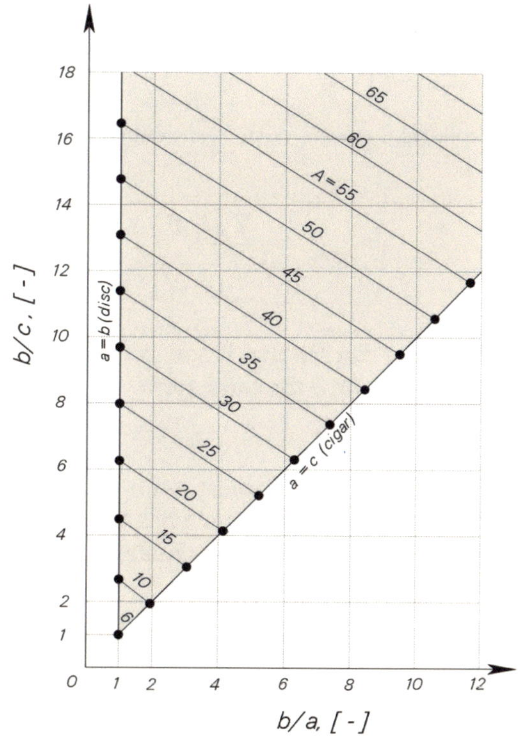

Fig. 4. Shape factor values for rounded grains, according to dimensions (a – medium, b – the largest and c – the smallest).

For a soil with spherical grains, whose soil skeleton density does not change with grain size, specific surface area, according to the above formula, will be

$$s = 6(1-n)\sum_i \frac{g_i}{d_{ei}} \tag{6}$$

where d_{ei} denotes the effective diameter of i- size.

Another method of determining specific surface area is connected with determining frequency f_i of grain occurrence in a soil sample with specific dimensions (the largest b_i, the smallest c_i and medium a_i). The occurrence frequency is normally expressed in %.

$$f_i = \frac{N_i}{\sum_i N_i} 100 = \frac{N_i}{N} 100 \tag{7}$$

where: N_i – number of grains with specific size i,
N – number of all grains in the tested sample.

Given grain dimensions and occurrence frequency, specific surface area can be determined from the formula:

$$s = \frac{\sum F_z}{V} = \frac{\sum_i (N_i F_{zi})}{V} = \frac{\sum_i \left(\frac{f_i N}{100} F_{zi}\right)}{\frac{\sum_i m_i}{\rho_s (1-n)}} = \frac{\rho_s (1-n) \frac{N}{100} \sum_i (f_i F_{zi})}{\sum_i (N_i V_{zi} \rho_{si})} = \frac{\rho_s (1-n) \sum_i \left(f_i \frac{V_{zi} A_i}{b_i}\right)}{\sum_i (f_i V_{zi} \rho_{si})} \quad (8)$$

In the case of a mixture of spherical grains with identical thickness and different diameters $d_{i,,}$ the specific surface area is:

$$s = 6\,(1-n) \frac{\sum_i (f_i d_i^2)}{\sum_i (f_i d_i^3)} \quad (9)$$

Analysing formulae (6) and (9), one can see that if soil is composed of spherical grains with identical diameters, both formulae assume the same form.

2.3 Effective grain size

Effective grain size b_e will be defined as the largest grain dimension in imaginary material built of grains with identical sizes and shapes, which has the same specific surface area and the same porosity as real soil. The grain shape in this material is the same as that of a typical grain in real soil.

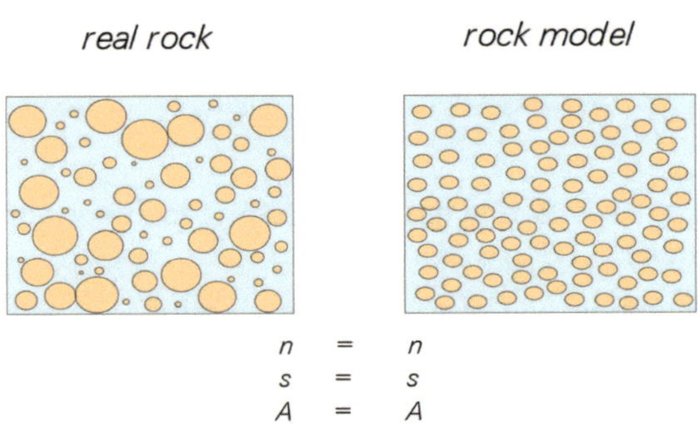

Fig. 5. Modelling of real rock as material composed of grains with identical shape and size (n - porosity , s - specific surface area, A - shape factor).

Comparing the specific surface area of well-graded real soil with that of its counterpart with uniform-size grains, one can define the effective grain size. When determining the mass of various sizes building a rock sample, we obtain:

$$\rho_s (1-n) \sum_i \left(\frac{g_i A_i}{\rho_{si} b_{ei}}\right) = \frac{A_e (1-n)}{b_e} \quad (10)$$

Assuming that

$$A_e = \sum_i g_i A_i \quad (11)$$

one can define effective grain size b_e.

$$b_e = \frac{\sum_i g_i A_i}{\rho_s \sum_i \left(\frac{g_i A_i}{\rho_{si} b_{ei}} \right)} \tag{12}$$

If all grains in a rock sample, regardless of their size, have similar shapes i.e. $A_1 = A_2 = \ldots = A_e$ and the density of soil skeleton does not change either ($\rho_{s1} = \rho_{s2} = \ldots = \rho_s$), then the effective grain size is

$$b_e = \frac{1}{\sum_i \frac{g_i}{b_{ei}}} \tag{13}$$

Certainly, for a soil whose grains are spherical, b_{ei} values are equal to the diameters of particular grains sizes d_{ei}, and b_e is equal to effective grain diameter d_e. Hence we get the commonly used relation

$$d_e = \frac{1}{\sum_i \frac{g_i}{d_{ei}}} \tag{14}$$

When it comes to determining the frequency f of grain occurrence in a soil sample with defined dimensions, what we get when comparing the specific surface area of well-graded rock with that of rock with uniform grade size is:

$$\frac{(1-n)\sum_i \left(f_i \frac{V_{zi} A_i}{b_i} \right)}{\sum_i (f_i V_{zi})} = \frac{A_e (1-n)}{b_e} \tag{15}$$

Assuming in this case that

$$A_e = \sum_i \left(\frac{f_i}{100} A_i \right) \tag{16}$$

the effective grain size will be

$$b_e = \frac{\sum_i \left(\frac{f_i}{100} A_i \right) \sum_i (f_i V_{zi})}{\sum_i \left(f_i \frac{V_{zi} A_i}{b_i} \right)} \tag{17}$$

If the shape of a grain does not depend on its size, i.e. $A_1 = A_2 = A_3 = \ldots = A_N = A_e$, then

$$b_e = \frac{\sum_i (f_i V_{zi})}{\sum_i \left(f_i \frac{V_{zi}}{b_i} \right)} \tag{18}$$

For rounded grains, whose grain volume could be adopted in accordance with formula (3), we get:

$$b_e = \frac{\sum_i f_i\, a_i\, b_i\, c_i}{\sum_i f_i\, a_i\, c_i} \tag{19}$$

For rock whose grains are spherical, the values of b_i are equal to the diameters of particular grain sizes d_i and b_e is equal to effective grain diameter d_e. Hence we get:

$$d_e = \frac{\sum_i \left(f_i\, d_i^3\right)}{\sum_i \left(f_i\, d_i^2\right)} \tag{20}$$

One can see then that effective grain size will be defined according to different formulae, depending on the chosen method of rock gradation analysis.

2.3.1 Effective grain size in a particular fraction

The term rock fraction refers to the isolated part of mineral skeleton of rock whose grains are contained within a specified size range. For instance, in the case of sieve analysis, a fraction will be rock composed of grains that will pass through the upper sieve with mesh openings D_{i-1}, and will be retained on a sieve with openings D_i.
If grain size distribution curve (Fig. 6) is known, then the value adopted as effective fraction value will be the one enabling correct determination of its specific surface area.
Therefore, if b_{ei} denotes the effective size of fraction i, then specific surface area component resulting from the proportion of this fraction, according to formula (5), is

$$s_i = g_i \frac{A_i\,(1-n)}{b_{ei}} \tag{21}$$

The specific surface area component calculated in this way should be equal to the specific surface area of this fraction treated as well-graded material, which it actually is.
In order to calculate this component of specific surface area, one must adopt specific grain size distribution. To achieve this aim, it was accepted that gradation graph between equivalent grain diameters D_{i-1} and D_i has the shape of a straight line segment (Fig. 6). One should emphasize here that with sufficiently dense breakdown into fractions, the gradation curve can be represented as a curve composed of straight line segments. Moreover, it was assumed that soil mass in the range between D_{i-1} and D_i depends on the mean dimension of the grain. Then, dividing any line segment i into N equal sections (Fig. 7) and assuming that density, grain shape and the ratio of the largest to the mean grain dimension $\frac{b_{i,j}}{a_{i,j}} = \lambda_i$ within each fraction do not change, one can determine the specific surface area component, in accordance with formula, (3) in the form

$$s_i = A_i\,(1-n)\sum_{j=1}^{N}\frac{g_{i,j}}{b_{i,j}} = \frac{A_i}{\lambda_i}(1-n)\sum_{j=1}^{N}\frac{g_{i,j}}{a_{i,j}} \tag{22}$$

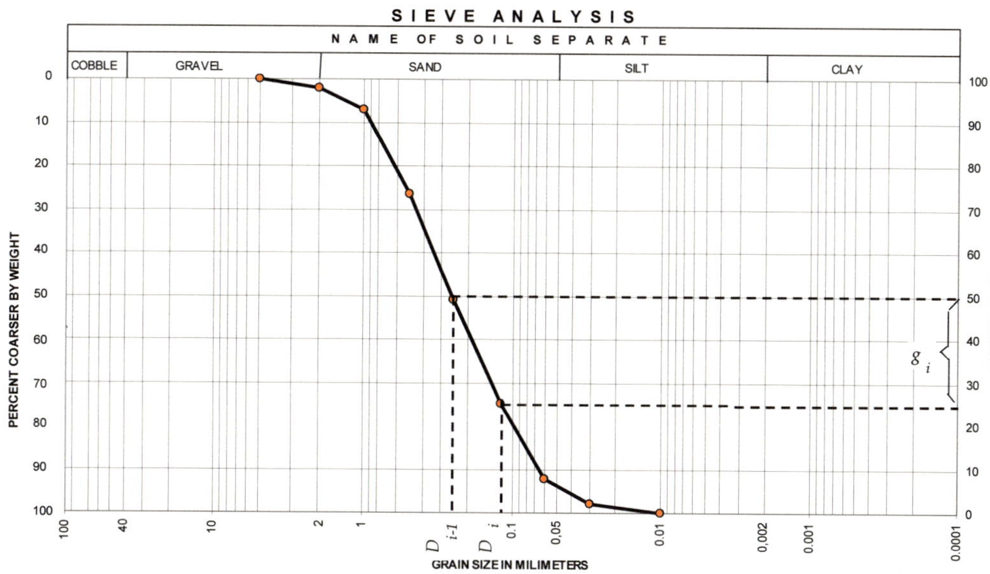

Fig. 6. Typical grain size distribution graph.

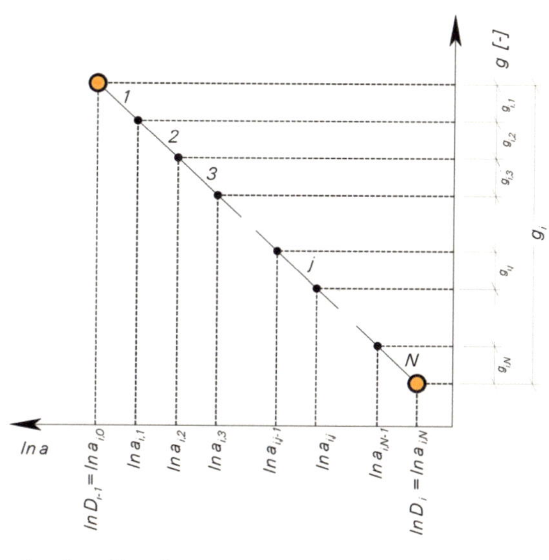

Fig. 7. A fragment of grain size distribution curve.

where $b_{i,j}$ – effective size of part j of the fraction, and

$$g_{i,j} = \frac{g_i}{N} \tag{23}$$

Assuming that division figure N of line segment i approaches infinity, the sum in equation (22) can be replaced by an integral and then

$$s_i = \frac{A_i}{\lambda_i}(1-n)\int_{D_i}^{D_{i-1}} \frac{dg}{a} \tag{24}$$

Since we have assumed that g is a linear function (Fig. 7),

$$g = \alpha\, x + \beta \tag{25}$$

And

$$dg = \frac{dg}{dx}\, dx = \alpha\, dx \tag{26}$$

where α is the slope of a line

$$\alpha = \frac{g_i}{\ln(D_{i-1}) - \ln(D_i)} \tag{27}$$

The x-axis on the gradation curve is in logarithmic coordinates, so $x = \ln a$ and

$$dx = \frac{1}{a}\, da\,.$$

Hence, after substituting appropriate data in equation (24), we obtain

$$s_i = \frac{A_i}{\lambda_i}(1-n)\int_{D_i}^{D_{i-1}} \frac{dg}{a} = \frac{A_i}{\lambda_i}(1-n)\int_{D_i}^{D_{i-1}} \frac{\alpha}{a^2}\, da = \frac{A_i}{\lambda_i}(1-n)\, g_i \frac{\frac{1}{D_i} - \frac{1}{D_{i-1}}}{\ln \frac{D_{i-1}}{D_i}} \tag{28}$$

Comparing formulae (21) and (28), one can determine the effective size of fraction b_{ei}

$$b_{ei} = \lambda_i \frac{\ln \frac{D_{i-1}}{D_i}}{\frac{1}{D_i} - \frac{1}{D_{i-1}}} \tag{29}$$

It should be emphasized that effective grain size b_{ei} calculated from formula (29) refers to the situation when the mass of grains contained between D_{i-1} and D_i depends on the mean grain dimension. Such a case should occur if the gradation curve is constructed based on aerometric analysis. If, on the other hand, sieve analysis is performed, then it is mostly the largest grain dimension that determines what is left on a particular sieve. Consequently, for results obtained from sieve analysis, one should adopt $\lambda_i = 1$.

With the assumption that grains are spherical, b_{ei} is equal to the effective diameter of i-fraction, d_{ei}.

2.4 Effective capillary diameter

Effective capillary diameter will be defined as the size of capillaries in material with the same specific surface area and the same porosity as real soil, but with all capillaries having the same constant cross-section. In order to define the size of these capillaries, let us cut out, theoretically, a soil cuboid with volume $V = F \cdot l$ (Fig. 1). Let us assume that instead of soil we have the same volume of material containing sinuous capillaries with any, but identical, cross-section F_k. Let us designate the cross-section circumference of an individual capillary as U_k. The boundary surface of capillaries inside volume V of such material is

$$s\ V = U_k\ \Sigma l_k, \tag{30}$$

where Σl_k denotes the total length of all capillaries, and l_k – the length of an individual capillary. On the other hand, the volume of capillaries inside volume V is

$$n\ \ V = F_k\ \Sigma l_k. \tag{31}$$

Dividing equations (31) and (30) by sides, we get

$$\frac{n}{s} = \frac{F_k}{U_k} \tag{32}$$

The ratio F_k/U_k is defined as hydraulic radius R_h. If we accept that capillaries have circular cross-section, then the cross-section diameter will be defined as effective Φ_e and, according to the above equation, it will be

$$\Phi_e = \frac{4\,n}{s} \tag{33}$$

Taking specific surface area into account, we obtain a correlation between effective grain size b_e, and effective capillary diameter Φ_e

$$\Phi_e = \frac{4\,n}{A_e(1-n)}\,b_e \tag{34}$$

If we assume that grains are spherical, then b_e will be equal to sphere diameter d_e and $A_e = 6$, and

$$\Phi_e = \frac{2\,n}{3(1-n)}\,d_e \tag{35}$$

In this way, known correlations between effective capillary diameter, effective grain diameter, specific surface area and porosity have been derived.

2.5 Capillary rise height

If we know the effective diameter of capillaries in rock, we can determine capillary rise height h_k.

$$h_k = \frac{4\,\sigma\cos\theta}{\Phi_e\,\rho\,g} \cong \frac{4\,\sigma}{\Phi_e\,\rho\,g} \tag{36}$$

where: σ – surface tension, ρ - density, g – gravitational acceleration, θ – contact angle.
Taking into account the above relationships between the effective capillary diameter and the specific surface area of a grain, we get

$$h_k = \frac{4\,\sigma}{\Phi_e\,\rho\,g} = \frac{\sigma\,s}{n\,\rho\,g} = \frac{6(1-n)\,\sigma}{n\,\rho\,g\,d_e} \tag{37}$$

After substitution of appropriate data for rock composed of quartz sand, the rise height h_k of water at the temperature of 10°C becomes

$$h_k = 0{,}462\,\frac{1-n}{n}\,\frac{1}{d_e} \tag{38}$$

where the values of h_k and d_e are expressed in cm.
An analogical relationship was presented by Polubarinova-Kochina (1962)

$$h_k = 0{,}45\,\frac{1-n}{n}\,\frac{1}{d_{10}} \tag{39}$$

In this case, d_{10} stands for grain diameter below which grains make 10% of soil mass, and is also expressed in cm.
Apparently, the numerical coefficients in formulae (38) and (39) are very comparable. A bigger difference in capillary rise values could be related to the method of determining d_e and d_{10}.

2.6 Hydraulic conductivity

The hydraulic conductivity of soil will be defined with the assumption that a flow in soil is equivalent to a flow through a bundle of capillaries with a constant cross-section. (Fig. 8).
If the length of filtration path is l, then the length of an individual capillary, owing to its tortuousness, is higher and amounts to $l_k = l \cdot \beta$, where coefficient β expresses capillary tortuousness. Volumetric flow rate Q_k of an individual capillary, with the assumption of a flow of incompressible fluid, can be expressed by the formula Hagen-Poisseuille

$$Q_k = F_k\,v_{śr} = F_k\,\frac{R_h^2}{\alpha\,\eta}\,\frac{\Delta p}{l_k} = F_k\,\frac{\rho\,g\left(\frac{F_k}{U_k}\right)^2}{\alpha\,\eta\,\beta}\,\frac{\Delta H}{l} \tag{40}$$

where: $v_{śr}$ – mean capillary flow velocity, R_h – hydraulic radius of a capillary, α – coefficient related to capillary shape, η- dynamic viscosity, Δp – pressure difference at flow path length l, ΔH – hydraulic head difference at flow path length l.
By defining the number of capillaries N contained in a material

$$N = \frac{\sum l_k}{l_k} = \frac{s\,V}{U_k\,l\,\beta} = \frac{s\,F}{U_k\,\beta} \tag{41}$$

one can define the total volumetric flow rate Q through a bundle of capillaries

$$Q = Q_k\,N = \frac{F_k\,s\,F\,\rho\,g\left(\frac{F_k}{U_k}\right)^2}{U_k\,\alpha\,\eta\,\beta^2}\,\frac{\Delta H}{l} = F\,\frac{\rho\,g\,n^3}{\alpha\,\beta^2\,\eta\,s^2}\,\frac{\Delta H}{l} \tag{42}$$

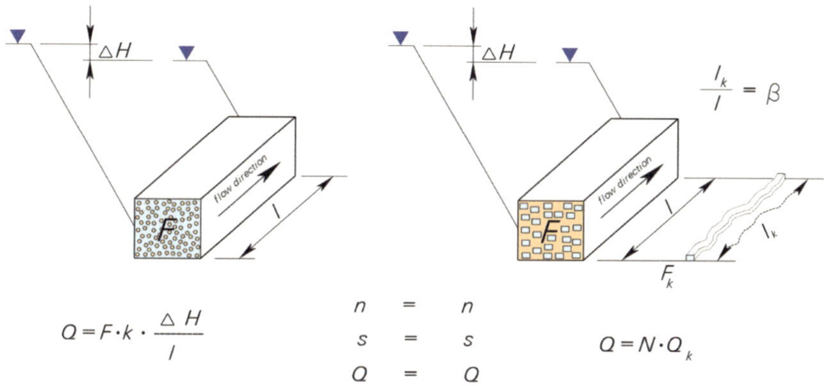

Fig. 8. Assumptions for a flow through real soil and a soil model (n – porosity , s – specific surface area, Q - volumetric flow rate).

Comparing the above relation to volumetric flow rate defined according to Darcy's law, we get a formula for hydraulic conductivity k

$$k = \frac{\rho\, g\, n^3}{\alpha\, \beta^2\, \eta\, s^2} \tag{43}$$

The performed tests have revealed that the value of product $\alpha' = \alpha\ \beta^2$ is equal to c. 5 (Carman, 1956) and the extreme variation interval of this product ranges from 3,5 to 5,8 (Richardson et al., 1991).
Consequently, if the shape of soil grains is close to a sphere, then after substituting value 5 for α and expressing specific surface area s by effective soil diameter, we obtain

$$k = \frac{\rho\, g\, n^3}{5\, \eta\, s^2} = \frac{\rho\, g\, n}{80\, \eta}\, \Phi_e^2 = \frac{1}{180}\, \frac{n^3}{(1-n)^2}\, \frac{\rho\, g}{\eta}\, d_e^2 \tag{44}$$

The above formula is known in literature as Carman - Kozeny formula (Olsen, 1960; Liszkowska, 1996; Mauran et al. 2001; Chapuis & Aubertin, 2003; Carrier, 2003).

2.7 Specific yield

As a result of drainage process, soil retains bound waters, which are held by forces stronger than gravitation. Bound waters can form a film around soil grains and particles. As capillaries have different diameters, additional amounts of water can be held at grain contact. The idea of determining specific yield consists of adopting the mean thickness of bound water layer present on the surface of capillaries formed by pores in soil. Assuming that capillaries have circular cross-section with effective diameter Φ_e, and the mean thickness of bound waters on the boundary surface of soil equals δ, one can define specific yield from the formula

$$\mu = \frac{V_w}{V} = \frac{\pi\, (\Phi_e - 2\,\delta)^2 \sum l_k}{4\, V} = \frac{n\, (\Phi_e - 2\,\delta)^2}{\Phi_e^2} \tag{45}$$

where V_w stands for the value of water drained by gravity off the total volume V of rock fully saturated with water, and Σl_k - the total length of capillaries in volume V.

Equation (45) contains the notion of bound water thickness related to a capillary with circular cross-section. In order to define it, the author used the results of thorough and reliable research conducted by other scholars, including the results of relationships between hydraulic conductivity and specific yield (Drainage Manual, 1984). Based on them, the relationship between effective capillary diameter Φ_e [m], and bound water thickness δ [m] was defined. It has the following form:

$$\delta = 7.54 \cdot 10^{-7} + \frac{8.67 \cdot 10^{-6} \cdot \Phi_e}{0.000032 + \Phi_e} \quad (46)$$

The author suggests employing this equation for natural soils containing dust and clay particles on grain surface.
For sorted washed soils, the author used the research results produced by C.F. Tolman, defining the relationship between grain diameter and specific yield (Fig. 9).

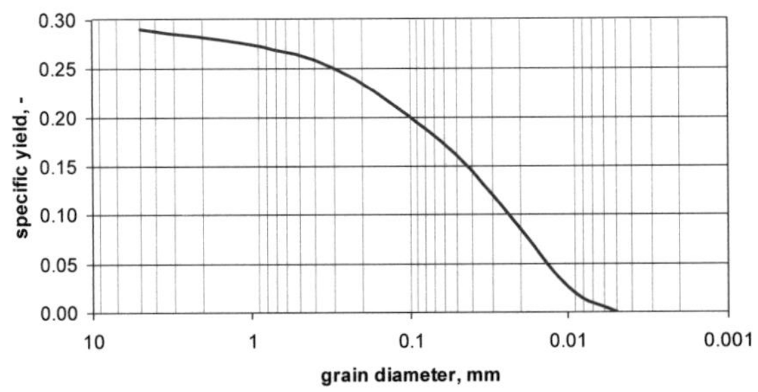

Fig. 9. Correlation between specific yield and grain diameter in non-cohesive soils with porosity n=0.3 (Tolman 1937).

For these soils, the relationship between effective capillary diameter Φ_e [m] and thickness of bound water film δ [m] was obtained in the form:

$$\delta = 8.88 \cdot 10^{-7} + \frac{7.15 \cdot 10^{-6} \cdot \Phi_e}{0.0001 + \Phi_e} \quad (47)$$

Certainly, both in formula (46) and (47), condition $2\delta \geq \Phi_e$ must be met, so the highest film thickness may not be higher than half the diameter of a capillary. Assuming that $\delta = 0.5\,\Phi_e$, one can calculate the effective capillary diameter below which soil pores are completely filled with bound water, and water drainage from the soil will not take place. For homogeneous sorted soils, this is 2.99 μm, and for others - 2.07 μm. Consequently, film thickness in a capillary with circular cross-section would be 1.5 μm and 1.04 μm respectively. Comparing this thickness to film thickness on a flat surface, we can observe that it will be exactly twice as thinner and will come to 0.75 μm or 0.52 μm. This results from the balance of mass contained in pores, as equation $\pi\,\Phi_e\delta = \frac{\pi}{4}\Phi_e^2$ should be fulfilled.

In compliance with relations (46) and (47), the mean thickness of bound water film δ remaining in capillaries with circular cross-section may reach the maximum value of c. 8 - 9.5 micrometres. The pattern of these changes is shown in Fig. 10.

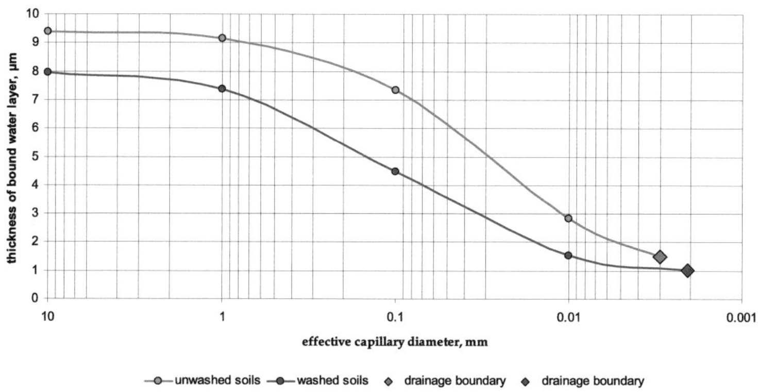

Fig. 10. Bound water layer thickness depending on effective capillary diameter.

It should be emphasized here that both curves are well-matched to measurement points. The point is that specific yields calculated by means of formula (45), with the assumed mean thickness of water layer, according to formulae (46) and (47), are very consistent with the results described by Tolman and Dranaige Manual.

One should add here that adopting the described rock model (a bundle of capillaries with identical diameters) for describing drainage process is a result of broader investigations undertaken by the author. Many variants have been considered, in which a layer of bound waters remains on the surface of grains and soil particles after drainage process. It has been assumed that its thickness on a flat surface is the same as that on the surface of spherical and ellipsoid-shaped grains. It has also been assumed to form a film on the surface of variously-shaped capillaries. In particular, it has been assumed that capillaries have the shape of adjoining balls or the shape of a void formed when spherical soil grains are arranged in such a way that lines joining the centres of these balls form cubes and rhombi. For all these cases, the relation between the thickness of this layer and specific yield was determined theoretically in a way analogical to formula (45). Then, as before, specific yields calculated based on these formulas were compared with the results of the above tests. Based on this comparison, the mean thickness of bound waters was determined. In all the discussed cases, it turned out that film thickness δ was lower for small grains and higher for large ones. Therefore, the simplest geometry of the capillaries was adopted for further investigations, i.e. that with constant circular-shaped cross-section. Adopting such capillary geometry enables employing a uniform rock model.

2.8 Specific retention

Specific retention s_r is defined as the ratio of volume V_z of bound water i.e. water retained in rock by a force higher than gravity, to total rock volume V.

$$s_r = \frac{V_z}{V} \tag{48}$$

Volume V_z is the volume of water which remains in rock after the drainage process. When it comes to drainage of rock fully saturated with water, the following relation between specific retention s_r, specific yield μ and porosity n takes place:

$$s_r = n - \mu \tag{49}$$

Noticeably, specific retention, as well as other parameters discussed before, can be determined from two other known rock parameters.

3. Description of laboratory tests

The aim of the tests was to verify correlations between rock properties that had been obtained based on a rock model in the shape of a bundle of capillaries. Some of them, as already mentioned, had been investigated before, had been accepted and are now often used. This is true about correlations between hydraulic conductivity, porosity, specific surface area and effective grain diameter known as Carman- Kozeny formulas. Therefore in his research, the author focused mainly on verifying mutual correlations between hydraulic conductivity, specific yield and other rock parameters.

The tests were conducted on washed quartz sand with well-rounded grains. Based on microscope observation of a studied soil sample, the mean values of the ratios of the highest to the mean grain dimension and of the largest to the smallest dimension were determined. These values come to 1.1 and 1.2 respectively. The calculated shape factor for such grains, according to equation (4) is 6.3. The sands used for investigations were characterised by grain sizes in the ranges 1.5-1.02 mm, 1.02-0.5 mm, 0.5-0.25 mm, 0.35-0.06 mm and 2.0-0.06 mm. The first three sand fractions were specially selected so that all the grains composing them did not vary a lot in diameter. This was to ensure good homogeneousness of the drained soil. Sands with grain sizes 0.35-0.06 mm and 2.0-0.06 mm corresponded to well-graded soil. The gradation curve was determined for these sands.

Each fraction was subjected to preliminary research, which covered grain skeleton density ρ_s and capillarity (capillary rise height) h_k. Capillarity was defined as the height of a water column allowing the air to break through a soil sample placed in a glass funnel.

The analyses of hydraulic conductivity and specific yield were conducted by placing soil in high columns. The soil was first washed, then dried at the temperature of 105-110°C, and cooled in a dessicator. After cooling it down to the temperature of c. 21°C, the soil was placed and thickened in batches in a thermally insulated column with the height of 1.75 m and diameter of 23.23 cm^2. The amount of soil placed in the column was weighed in order to determine its porosity. The porosity was calculated from the formula

$$n = \frac{V - \frac{m_s}{\rho_s}}{V} \tag{50}$$

where: V- column volume, m_s – soil mass in the column and ρ_s - soil skeleton density.

Then, very slowly, de-aerated water with the temperature of 21±1°C was fed from the bottom. After saturation with water, the flow of de-aerated water through the column was realised with the assigned hydraulic head drop of c. 0.2. The flow was carried out until gas bubbles were dissolved in pores and constant volumetric flow rate was achieved. Based on the volumetric flow rate of the flowing water, hydraulic conductivity was determined. After defining the hydraulic conductivity, the determination of specific yield was started. At the initial stage of the investigation, constant volumetric flow rate was forced at the column base until the air broke through the soil sample.

Then, gravitational outflow of water was enabled. During the drainage process, the mass of drained-off water was measured. The research lasted from 2 months to three years. Stable temperature was maintained for all this time. The devices mounted at the base of the column enabled the analysis and determination of suction height $\Psi_{s,}$. The results of laboratory tests are shown in Tab. 1.

Rock No	Gradation range	n	h_k	Ψ_s	k	μ
	mm	-	cm	cm	cm/s	-
1	1.5 - 1.02	0.352	4.5	5.2 do 7.6	0.765	0.337
2	1.02 - 0.5	0.317	9.0	9.8 do 14.7	0.180	0.285
		0.329			0.206	0.298
3	0.5 - 0.25	0.315	23.0	23.9 do 34.0	0.0340	0.278
		0.334			0.0502	0.298
4	0.35 - 0.06	0.297	34.0	34.7 do 53.6	0.0160	0.259
		0.308			0.0183	0.271
5	2.0 - 0.06	0.289	25.4	25.0 do 35.3	0.0405	0.245

Table 1. Results of laboratory tests of hydrogeological properties of rocks.

As suction height Ψ_s changed during the fall of water table, Table 1 presents its change range.

4. Comparison of calculation results and laboratory tests

All the formulas presented in the article, concerning determination of various hydrogeological properties of rocks, are interrelated. The analysis of these formulas demonstrates that if any two properties are known, then any remaining one can be determined.

In the first stage of the comparative analysis it was assumed that laboratory results of sieve analysis and porosity n are known. Based on them, seven different rock parameters, namely effective grain diameter, specific surface area, effective capillary diameter, capillary rise, hydraulic conductivity, specific yield and specific retention were calculated. For rocks No. 1, 2 and 3, it was assumed that the gradation curve (Fig. 5) is composed of one straight line segment. Consequently, the effective diameter was calculated for them directly from formula (29). For rock designated as 5, the effective grain diameter was determined from formula (14). As the gradation curves for all the studied soils were obtained as a result of sieving, in formula (29) λ=1 was adopted. The specific surface area was calculated from formula (5), adopting grain shape factor equal to 6.3. The effective capillary diameter, capillary rise height and specific yield were calculated from formulas (33), (36) and (45) respectively. The thickness of bound water layer found in formula (45) was adopted as for washed soils, i.e. from formula (47). The specific retention was determined from formula (49), and the results of all calculations were compiled in Table 2.

When comparing the corresponding results in Tables 1 and 2, one can observe good accordance of specific yield and hydraulic conductivity. The conformity of suction height with capillary rise height can be also regarded as high. Slightly bigger differences occur in relation to passive capillarity. However, one should remember that passive capillarity has

been determined for small rock samples placed in a funnel, i.e. for poorly condensed samples. Moreover, determining passive capillarity is influenced by increased rock porosity next to the funnel walls.
Tab. 3., in turn, presents properties of sandy soil calculated based on the knowledge of hydraulic conductivity and porosity.

Rock No	Data	Calculation results							
	Gradation range	n	s	b_e	Φ_e	h_k	k	μ	s_r
	mm	-	mm^{-1}	mm	mm	cm	cm/s	-	-
1	1.5 - 1.02	0.352	3.32	1.23	0.424	7.0	0.781	0.333	0.019
2	1.02 - 0.5	0.317	6.15	0.699	0.206	14.4	0.166	0.288	0.029
		0.329	6.05	0.699	0.218	13.6	0.193	0.300	0.029
3	0.5 - 0.25	0.315	12.45	0.347	0.101	29.3	0.0398	0.271	0.044
		0.334	12.11	0.347	0.110	26.9	0.0502	0.289	0.045
5	2.0 - 0.06	0.289	10.64	0.421	0.109	27.3	0.0421	0.250	0.039

Table 2. Rock properties determined based on sieve analysis and porosity.

Rock No	Gradation range	Data		Calculation results					
		k	n	s	b_e	Φ_e	h_k	μ	s_r
	mm	cm/s	-	mm^{-1}	mm	mm	cm	-	-
4	0.35 - 0.06	0.0160	0.297	17.99	0.246	0.0660	44.9	0.247	0.0504
		0.0183	0.308	17.76	0.245	0.0694	42.8	0.257	0.0513

Table 3. Rock properties calculated based on hydraulic conductivity and porosity.

In his case, the specific surface area was calculated from formula (44), the effective capillary diameter - from formula (33), and the effective grain parameter - from formula (34). The capillary rise height, specific yield, and specific retention were calculated analogically by means of the same relations as those shown in Table 2. One can observe that also in this case, i.e. for rock No. 4 with non-uniform gradation, high consistence of the calculated parameters with parameters obtained from direct laboratory tests was obtained.

5. Conclusions

In this chapter, the author has presented mutual relationships between hydrogeological properties of rocks. They are related to hydraulic conductivity, specific surface area, porosity, effective grain diameter, effective capillary diameter, capillary rise, specific yield and specific retention. All the relations were derived based on a rock model in the form of a bundle of tortuous capillaries with constant circular-shaped cross-section. The analysis of thus obtained relations demonstrates that if any two rock properties are known, one can define all the remaining ones. Hence, knowing two properties, one can determine six or seven other properties of rock. Some of the formulas included in the chapter have already been known, verified and widely used for practical problems for a long time. This pertains in particular to Carman-Kozeny formulas. Except for formerly known formulas, a new one, related to determining specific yield, has been proposed. It is based on adopting the mean thickness of bound water in capillaries with circular cross-section. The presented formula for bound water thickness in homogeneous sorted soils is based on research results from Tolman (Tolman, 1937), and in other soils - on results published in Drainage Manual (Kasenow, 2002). Its form also enables determining boundary effective capillary diameter, below which only bound waters occur and water is no longer drained off rock. Introducing a formula for specific yield also has enabled linking rock properties with specific retention.

Determining different rock parameters requires different amounts of work. One of the simplest tests, almost always performed when recognising soils, is gradation analysis. Therefore a lot of attention has been paid to using its results for determining other hydrogeological properties of rocks. Complete derivation of formulas enabling determination, based on gradation analysis, of specific surface area and effective grain size, considering complex size of soil-building grains and non-uniform density of soil skeleton has been presented.

A very important element of this research is verification of presented relations between different rock properties. To achieve this aim, long-term laboratory tests were launched. The key issue was the investigation of specific yield and hydraulic conductivity. They were determined on samples placed in high columns. Such a method produces good results but determining specific yield is very time-consuming. As a result of many years of investigations, few results were obtained. This is why investigations connected with verification of presented relations should be continued.

Notwithstanding, it should be emphasised that the presented results have demonstrated satisfactory accuracy of the described relations. In particular, positive results connected with using new formulas for determining specific yield and its relations with hydraulic conductivity have been obtained.

6. References

Arya, L.M.; Leij F.J.; Shouse, P.J. & Genuchten, M. Th. (1999). Relationship between the hydraulic conductivity function and the particle-size distribution. *Soil Sci.Soc. Am. J.* Vol. 63, (September-October 1999), pp. (1063-1070), ISSN 0361-5995

Bear, J. (1972). *Dynamics of fluids in porous media*, Dover Publications, ISBN 0-486-65675-6, New York USA

Carman, P.C. (1956). *Flow of Gases through Porous Media*. Butterworths Scientific Publications, London.

Carrier, W.D. (2003). Goodbye, Hazen; Hello, Kozeny-Carman. *Journal of Geotechnical and Geoenviromental Engineering*, Vol. 129, No 11 (November 2003), pp1054-1056), ISSN 1090-0241

Chapuis, R.P. & Aubertin, M. (2003). On the use of the Kozeny-Carman equation to predict the hydraulic conductivity of soils. *Can, J. Geotech./Rev. Can. Geotech.* Vol. 40 No 3 pp. 616-628, ISSN 1208-6010

Drainage Manual (1984). US Department of the Interior. U.S. Government Printing Office, Denver, Colorado

Richardson, J. F.; Harker, J. H. & Backhurst J. R. (1991). *Coulson and Richardson's Chemical Engineering. Volume 2. Particle Technology and Separation Processes*, Butterworth Heineman, ISBN 0-7506-4445-1 Oxford

Kasenow, M. (2002). *Determination of hydraulic conductivity from grain size analysis*, Water Resources Publications, ISBN 1-887201-31-9, Colorado USA

Liszkowska, E. (1996). Wzór Carmana-Kozeny uniwersalnym wzorem na obliczanie współczynnika filtracji, *Geologos*, No 1 (1996) (193-202)

Mauran, S.; Rigaud, L. & Coudevylle O. (2001). Application of the Carman-Kozeny correlation to high-porosity and anisotropic consolidated medium: The compressed expanded natural graphite, *Transport in Porous Media,* Vol. 43, pp. 355-376, ISSN 0169-3913

Olsen, H.W. (1960). Hydraulic flow through saturated clays, *Clays and Clay Minerals,* Vol. 9, No 1, pp. 131-161, ISSN 00009-8604

Petersen L.W.; Moldrup P.; Jacobsen O.H. & Rolston D.E. (1996). Relations between specific surface area and soil physical and chemical properties, *Soil Science* Vol. 161, No 1, (January 1996) pp. 9-21, ISSN 0038-075X

Polubarinova-Kochina P. (1962). *Theory of Ground Water Movement*, Princeton Univ. Pr. ISBN 0691080488, New Jersey USA

Tolman C.F. (1937). *Ground Water*, Mc Graw-Hill, New York.

5

Rock Mass Hydraulic Conductivity Estimated by Two Empirical Models

Shih-Meng Hsu[1], Hung-Chieh Lo[1], Shue-Yeong Chi[1] and Cheng-Yu Ku[2]
[1]*Sinotech Engineering Consultants, Inc.*
[2]*National Taiwan Ocean University,*
Taiwan

1. Introduction

Undertaking engineering tasks such as tunnel construction, dam construction, mine development, the abstraction of petroleum, and slope stabilization require the estimation of hydraulic conductivity for fractured rock mass. The understanding of hydraulic properties of fractured rock mass, which involves the fluid flow behaviour in fractured consolidated media, is a critical step in support of these tasks. To obtain hydraulic properties of fractured rock mass, double packer systems can be adopted (NRC 1996). They can be used to determine the hydraulic conductivity in a portion of borehole using two inflatable packers. Although this type of test can directly measure the hydraulic parameter, costs of the testing are fairly high. Several studies (Snow, 1970; Louis, 1974; Carlsson & Olsson, 1977; Burgess, 1977; Black, 1987; Wei et al., 1995;) have proposed the estimation of rock mass hydraulic conductivity using different empirical equations, which were based on the concept that rock mass permeability decreases with depth, as shown in Table 1. These empirical equations provide a great feature for characterizing rock mass hydraulic properties quickly and easily. However, the applicability of these equations is very limited because depth is not the only significant variable on the prediction of rock mass permeability. Hydraulic properties of rock mass may vary with geostatic stress, lithology and fracture properties, including fracture aperture and frequency, fracture length, fracture orientation and angle, fracture interconnectivity, filling materials, and fracture plane features (Lee & Farmer, 1993; Sahimi, 1995; Foyo et al., 2005; Hamm et al., 2007). Thus, a more applicable empirical equation for estimating hydraulic conductivity of rock mass possibly must include the aforementioned factors.

This chapter proposes two empirical models to estimate hydraulic conductivity of fractured rock mass. The first empirical model was based on the rock mass classification concept. The study developed a new rock mass classification scheme for estimating hydraulic conductivity of fractured rocks. The new rock mass classification system called as "HC-system" based on the following four parameters: rock quality designation (RQD), depth index (DI), gouge content designation (GCD), and lithology permeability index (LPI). HC-values can be calculated from borehole image data and rock core data. The second empirical model was simply based on results of borehole televiewer logging, flowmeter logging and packer hydraulic tests. Three borehole prospecting techniques for fractured rock mass hydrogeologic investigation were performed to explore various hydrogeologic characteristics, such as fracture width, fracture angle, flow velocity and hydraulic

Equation	Reference
$k = az^{-b}$	Black (1987) a and b are constants. z is the vertical depth below the groundwater surface.
$\log K = -8.9 - 1.671 \log Z$	Snow (1970) K (ft^2) is the permeability. z (ft) is the depth.
$K = 10^{-(1.6 \log z + 4)}$	Carlson and Olsson (1977) K (m/s) is the hydraulic conductivity. z (m) is the depth.
$K = K_s e^{(-Ah)}$	Louis (1974) K (m/s) is the hydraulic conductivity. K_s is the hydraulic conductivity near ground surface. H (m) is the depth. A is the hydraulic gradient.
$\log K = 5.57 + 0.352 \log Z$ $-0.978(\log Z)^2 + 0.167(\log Z)^3$	Burgess (1977) K (m/s) is the hydraulic conductivity. Z (m) is the depth.
$K = K_i[1 - Z / (58.0 + 1.02Z)]^3$	Wei et al. (1995) Z is the depth. K is the hydraulic conductivity. K_i (m/s) is the hydraulic conductivity near ground surface.

Table 1. Diverse approximations for estimating rock mass hydraulic conductivity

conductivity. By adopting a correlation analysis, the dependence between hydraulic conductivity and other prospecting data can be identified. The consequence results can be used to determine rock mass hydraulic conductivity.

2. The first empirical model

The classical rock mass classification systems have gained wide attention and are frequently used as powerful design aids in rock engineering. A great feature of the existing systems is that the characterization of rock mass properties for specific engineering purposes can be quickly obtained at a relatively low cost. There are six common systems used for engineering purposes, including Rock load (Terzaghi, 1946), Stand-up time (Lauffer, 1958), RQD (Deere et al., 1967), RSR (Wickham et al., 1972), RMR (Bieniawski, 1973), and Q-system (Barton et al., 1974). The above systems or other available systems were designed on the geomechanical assessment of rock mass. However, there is very limited study on rock mass permeability assessments (Bieniawski, 1989). Because permeability of rock mass is related to groundwater seepage into excavations for tunnels, mines, and other construction sites, Gates (1997) proposed the hydro-potential (HP) value as a new rock mass classification, semi-quantitative technique employed to evaluate the potential for developing groundwater in bedrock. The HP-value technique is a modification of the engineering rock mass quality designation (Q) originally developed for evaluation of rock competency in tunnel design and seismic rock fall susceptibility.

To reduce the cost of estimating the hydraulic parameter, the objective of the study presented herein is to propose a new application of the rock mass classification concept on the estimation of hydraulic conductivity of fractured rocks. The new rock mass classification system will be verified by in situ hydraulic test data from two hydrogeological investigation programs in three boreholes to demonstrate its rationality in predication of hydraulic conductivity of fractured rocks. Besides, the model verification using another borehole data with four additional in-situ hydraulic tests from similar geologic rocks was also conducted to further verify the feasibility of the proposed empirical HC model.

2.1 Components of new rock mass classification system

Prior to describing the new rock mass classification system, potential factors, including rock quality designation (RQD), depth index (DI), gouge content designation (GCD), and lithology permeability index (LPI), that may affect the degree of permeability should be considered. In addition, the rating approach for each factor that represents the magnitude of permeability is also described as follows.

2.1.1 Rock Quality Designation (RQD)

In rock engineering, from the mechanical point of view, the degree of fracturing stands for rock quality. This provides a simple index to judge the engineering quality of the rock. From the hydrogeological point of view, fractures reflect the ability of a formation to transmit water through fractures themselves. Thus, the degree of fracturing may be regarded as a factor in evaluating rock mass permeability.

To assess the influence of the fracture characteristic on permeability, the rock quality designation (RQD) index, which was developed by Deere et al. (1967), can be adopted. The RQD index was introduced over 40 years ago as an indicator of rock mass conditions. The RQD value is defined as the cumulative length of core pieces longer than 100 mm in a run (R_S) divided by the total length of the core run (R_T) and can be obtained from the following equation.

$$RQD = \frac{\sum Length\ of\ Intact\ and\ Sound\ Core\ Pieces\ >\ 100\ mm}{Total\ Length\ of\ Core\ Run,\ mm} \times 100\% = \frac{R_s}{R_T} \times 100\% \quad (1)$$

In this study, a core run for calculating a RQD value is herein defined as a selected zone of a hydraulic test. Equation (1) may be utilized to identify rock mass permeability.

2.1.2 Depth Index (DI)

Many researchers (for example Lee & Farmer, 1993; Singhal & Gupta, 1999) pointed out that rock mass permeability may decrease systematically with depth. The decrease in permeability with depth in fractured rocks is usually attributed to reduction in fracture aperture and fracture spacing. The reduction is due to the effect of geostatic stresses, and thereby the permeability of fractured rocks will reduce. The depth may be considered as a factor in evaluating rock mass permeability.

To assess the influence of the depth on permeability, a depth index called as DI was defined as the following equation.

$$DI = 1 - \frac{L_c}{L_T} \quad (2)$$

in which L_T is the total length of a borehole; L_c is a depth which is located at the middle of a double packer test interval in the borehole. The value of DI is always greater than zero and less than one. The greater the DI value, the higher the permeability.

2.1.3 Gouge Content Designation (GCD)

In general, the permeability of clay-rich gouges has extremely low values (Singhal & Gupta, 1999). The RQD value may decrease by an increase of fractures in a core run. If the fractures contain infillings such as gouges, permeability of the fractures will reduce. To assess the influence of the gouge materials on permeability, a gouge content designation index called as GCD was defined as the following equation.

$$GCD = \frac{R_G}{R_T - R_S} \tag{3}$$

in which R_G is the total length of gouge content. The value of GCD is always greater than zero and less than one. The greater GCD value stands for the more gouge content in a core run, and thereby it will reduce the permeability of the core run.

2.1.4 Lithology Permeability Index (LPI)

Lithology is the individual character of a rock in terms of mineral composition, grain size, texture, color, and so forth. For an intact rock, the magnitude of permeability depends largely on the individual character of the rock. It may be affected by the average size of the pores, which in turn is related to the distribution of particle sizes and particle shape. In sedimentary formations grain-size characteristics are most important because coarse-grained and well-sorted material will have high permeability as compared with fine-grained sediments like silt and clay. Thus, the lithology may be regarded as a factor in evaluating rock mass permeability.

To assess the influence of lithology on permeability, a lithology permeability index called as LPI was defined as Table 2.

2.2 Rock mass permeability system

As stated in Section 2.1, the rock mass permeability may be dependent on the following four parameters: rock quality designation (RQD), depth index (DI), gouge content designation (GCD), and lithology permeability index (LPI). However, the permeability is not simply affected by only one factor. It is possibly affected by any two factors, three factors or even all four factors. Thus, a rock mass classification scheme was applied to establish the rock mass permeability system. The new rock mass classification scheme is the product of the four parameters. It can account for the synthetic effect from the four parameters on permeability. The new rock mass classification system called as "HC-system" can be given by the following equation:

$$HC = \left(1\text{-}\frac{RQD}{100}\right)(DI)(1\text{-}GCD)(LPI) \tag{4}$$

The value of each parenthesis at the right hand side of Equation (4) is always greater than zero and less than one depending on the values assigned to the four parameters. The greater the value of each parenthesis, the higher the permeability. Thus, the system performs a

Lithology	Hydraulic conductivity (m/s)				Range of rating	Suggested Rating
	Reference[1]	Reference[2]	Reference[3]	$K_{average}$		
Sandstone	10^{-6}~10^{-9}	10^{-7}~10^{-9}	10^{-7}~10^{-9}	$10^{-7.5}$	0.8-1.0	1.00
Silty Sandstone	—	—	—	—	0.9-1.0	0.95
Argillaceous Sandstone	—	—	—	—	0.8-0.9	0.85
S.S. interbedded with some Sh.	—	—	—	—	0.7-0.8	0.75
Alternations of S.S & Sh.	—	—	—	—	0.6-0.7	0.65
Sh. interbedded with some S.S.	—	—	—	—	0.5-0.7	0.60
Alternations of S.S & Mudstone	—	—	—	—	0.5-0.6	0.55
Dolomite	10^{-6}~$10^{-10.5}$	10^{-7}~$10^{-10.5}$	10^{-9}~10^{-10}	10^{-8}	0.6-0.8	0.70
Limestone	10^{-6}~$10^{-10.5}$	10^{-7}~10^{-9}	10^{-9}~10^{-10}	10^{-8}	0.6-0.8	0.70
Shale	10^{-10}~10^{-12}	10^{-10}~10^{-13}	—	$10^{-10.5}$	0.4-0.6	0.50
Sandy Shale	—	—	—	—	0.5-0.6	0.60
Siltstone	10^{-10}~10^{-12}	—	—	10^{-11}	0.2-0.4	0.30
Sandy Siltstone	—	—	—	—	0.3-0.4	0.40
Argillaceous Siltstone	—	—	—	—	0.2-0.3	0.20
Claystone	—	10^{-9}~10^{-13}	—	10^{-11}	0.2-0.4	0.30
Mudstone	—	—	—	—	0.2-0.4	0.20
Sandy Mudstone	—	—	—	—	0.3-0.4	0.40
Silty Mudstone	—	—	—	—	0.2-0.3	0.30
Granite	—	—	10^{-11}~10^{-12}	$10^{-11.5}$	0.1-0.2	0.15
Basalt	10^{-6}~$10^{-10.5}$	10^{-10}~10^{-13}	—	$10^{-11.5}$	0.1-0.2	0.15

[1]B.B.S. Singhal & R.P. Gupta (1999)

[2]Karlheinz Spitz & Joanna Moreno (1996)

[3]Bear(1972)

Table 2. Description and ratings for lithology permeability index numerical assessment of the rock mass permeability using the four parameters. However, it should be noted that if (1−RQD) is zero, the value of 0.01 in the term of (1−RQD) is suggested to avoid the HC-value to be zero. Currently, the study took the same weight for each factor in Equation (4). While collecting more observed data, a further study can be considered to assign a different weight for each factor to giver better correlation between the hydraulic conductivity and HC-value. The rationality of Equation (4) must be verified by observed data through in situ hydraulic tests.

2.3 Study on correlation of hydraulic conductivity and HC-system

To verify rationality of Equation (4), the study collected in situ hydraulic test data to determine a relationship between hydraulic conductivity and HC-values. The needed data include rock core logs, BHTV (Borehole Acoustic Televiewer) image logs, locations of hydraulic tests, and hydraulic conductivity results. The hydraulic test results of boreholes were used to perform the dependence of HC on hydraulic conductivity.

2.3.1 Description of study area, boreholes, BHTV investigation and hydraulic test data

The study area is located in Taiwan. Taiwan's strata are distributed in long and narrow strips, almost parallel to the island's axis. Metamorphic rock lies under the Central and Snow Mountain Ranges. Sedimentary rock forms part of the island-wide piedmonts and coastal plains as well as the Coastal Mountain Range. The island of Taiwan has three geological zones divided by longitudinal faults: the Central Range, Western Piedmont and Eastern Coastal Mountain Range zones (Figure 1).

About 26 hydraulic conductivity measurements and borehole image scanning using BHTV were conducted in four boreholes in Western Piedmont, primarily at three sites: Da-Keng, Shang-Ming, and Caoling (Figure 1) in which borehole HB-94-01 is in the Da-Keng site, boreholes HB-95-01 and HB-95-02 are in the Shang-Ming site, and borehole CH-04 is in the Caoling site. Besides, the Da-Keng and Caoling sites are in central Taiwan and the Shang-Ming site is in southern Taiwan. The dominant rock strata of the Shang-Ming site include Miocene sedimentary rock with layers of sandstone or shale or their alternation. The major structures consist of a series of parallel easterly inclined thrust faults and folds, which often form local fractured zones, including geological structures such as the Pingshi fault, the Biauhu fault, and the Chin-Shan fault. In addition, borehole HB-94-01 in the Da-Keng site and borehole CH-04 in the Caoling site also have similar rock strata but without geological structures. Based on the loggings and geological analysis, HB-95-01 and HB-95-02 are strongly influenced by the faults; nevertheless, HB-94-01 and CH-04 are not. Rock core photos (Figure 2(a)) indicated soft and cohesive gouges are extensive in borehole HB-95-02; while the borehole CH-04 is not influenced by the faults (Figure 2(b)).

The depth of the boreholeHB-94-01 is 110 m. The principal lithologic units for the borehole are sandstone and siltstone. The interval of 36 m to 44 m is a fractured zone compared to other depths in the borehole. A total of 8 hydraulic tests using a double packer system were carried out to determine hydraulic conductivity (Sinotech Engineering Consultants, LTD., 2005). The strategy of the test design from the drilling work was to determine hydraulic properties from different geological structures such as no fracture, a single fracture, or multiple fractures at different depths. The drilling depths of HB-95-01 and HB-95-02 are 250 m and 350 m, respectively. The principal lithologic units for HB-95-01 are sandstone, argillaceous sandstone, and sandy mudstone. The principal lithologic units for HB-95-02 are

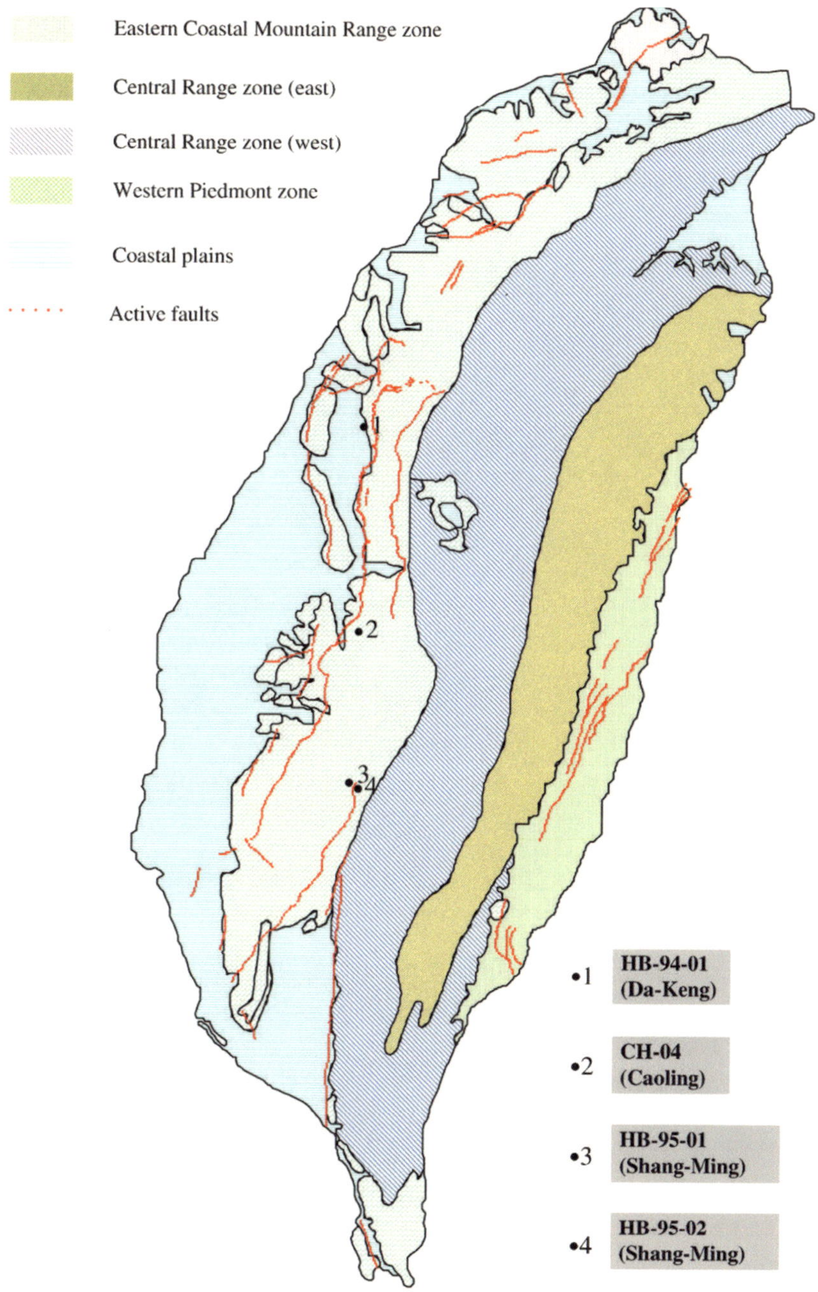

Fig. 1. Location of major faults and four boreholes for this study in Taiwan

sandstone, argillaceous sandstone, and sandstone mixed with some mudstone. HB-95-01 and HB-95-02 are close to the Biauhu fault and Pingshi fault, respectively. Rock core data indicated soft and cohesive gouges are extensive in both boreholes (Figure 2) in which the hydraulic properties of fault-related rocks can be studied. The study completed 3 and 14 hydraulic tests in HB-95-01 and HB-95-02, respectively (Sinotech Engineering Consultants,

LTD., 2007). The strategy of the test design was to determine hydraulic conductivity in more permeable zones and clay-rich gouge zones. Besides, the drilling length of borehole CH-04 is 120 m. The principal lithologic units of the borehole CH-04 are mainly sandstone, shale, and sandstone with some thin shale. Four different intervals were sealed by double packers for conducting the hydraulic tests. Those hydraulic test data were used for the model verification and it is described in Section 2.3.3.

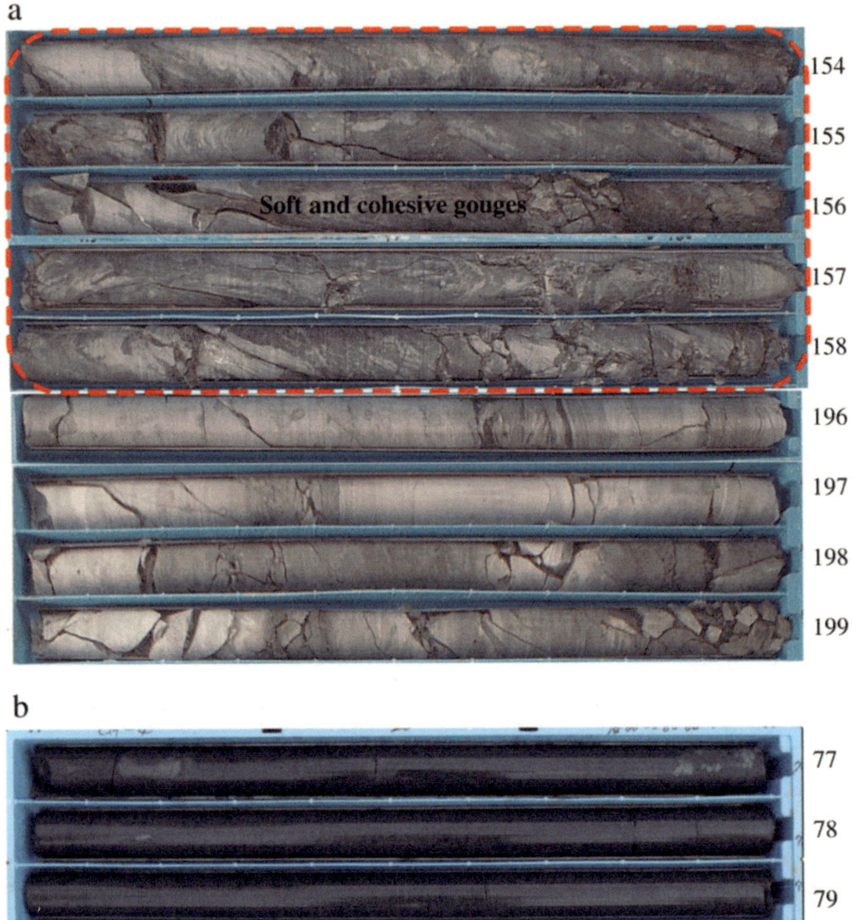

Fig. 2. (a). Rock core photos of borehole HB-95-02 with fault influence; (b). Rock core photos of borehole CH-04 without fault influence

BHTV image logs were gathered to characterize lithology, gouge and fractures for each borehole and essential to the proper design of hydraulic testing. Figure 3 shows two hydraulic test zones and corresponding BHTV images from different boreholes. It is obvious that fractures or shear band can be identified clearly using the high-resolution BHTV. The BHTV image in Figure 3(a) shows multiple fractures. This information with a double packer system (Figure 4) can be used to investigate hydraulic property of a water-bearing zone of subsurface. The BHTV image in Figure 3(b) shows shear band. This information can be utilized to investigate hydraulic property of fault-related rocks.

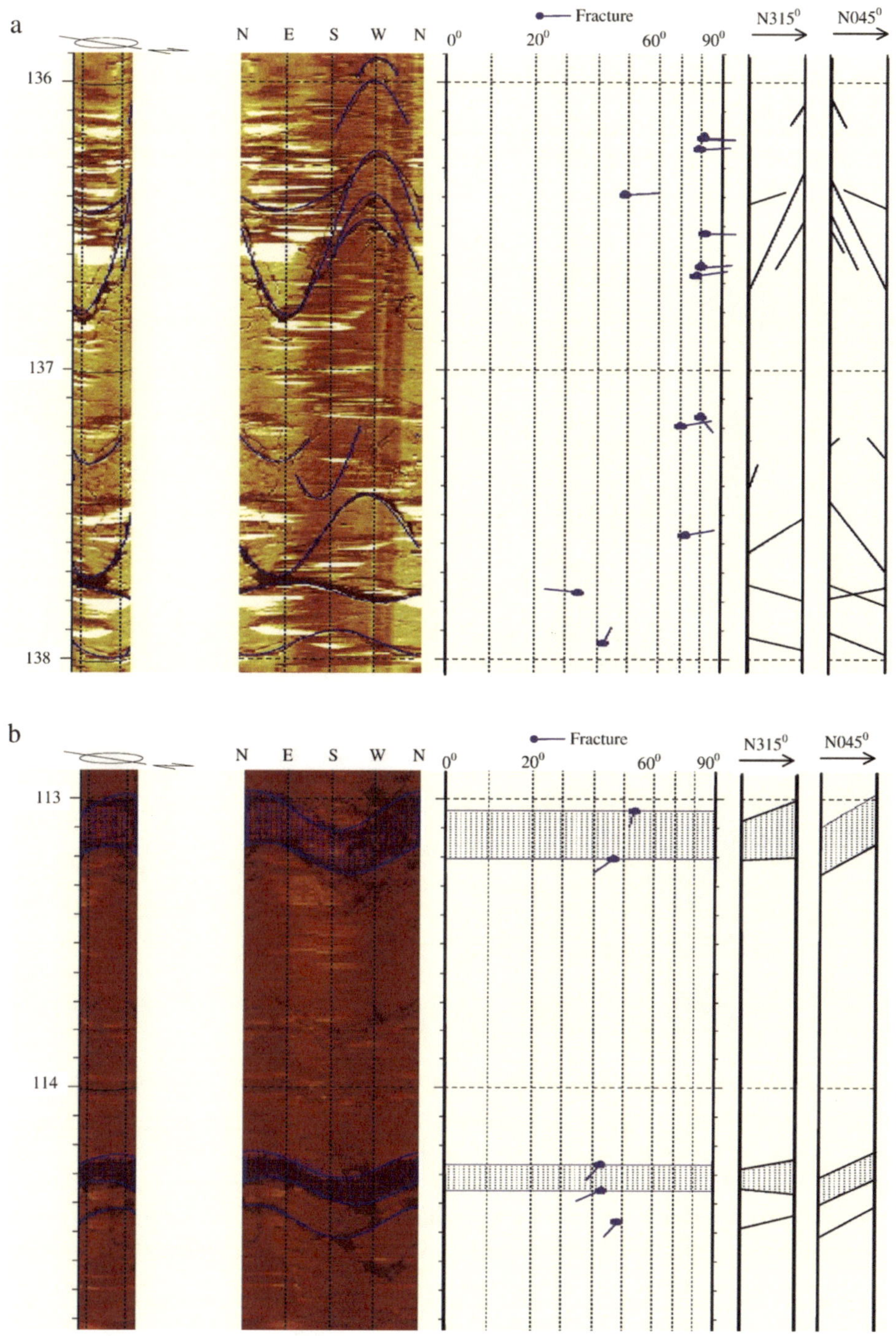

Fig. 3. (a). hydraulic test interval and corresponding BHTV image (depth 136 m-138 m, HB-95-01); (b). hydraulic test interval and corresponding BHTV image (depth 113 m-115 m, HB-95-02)

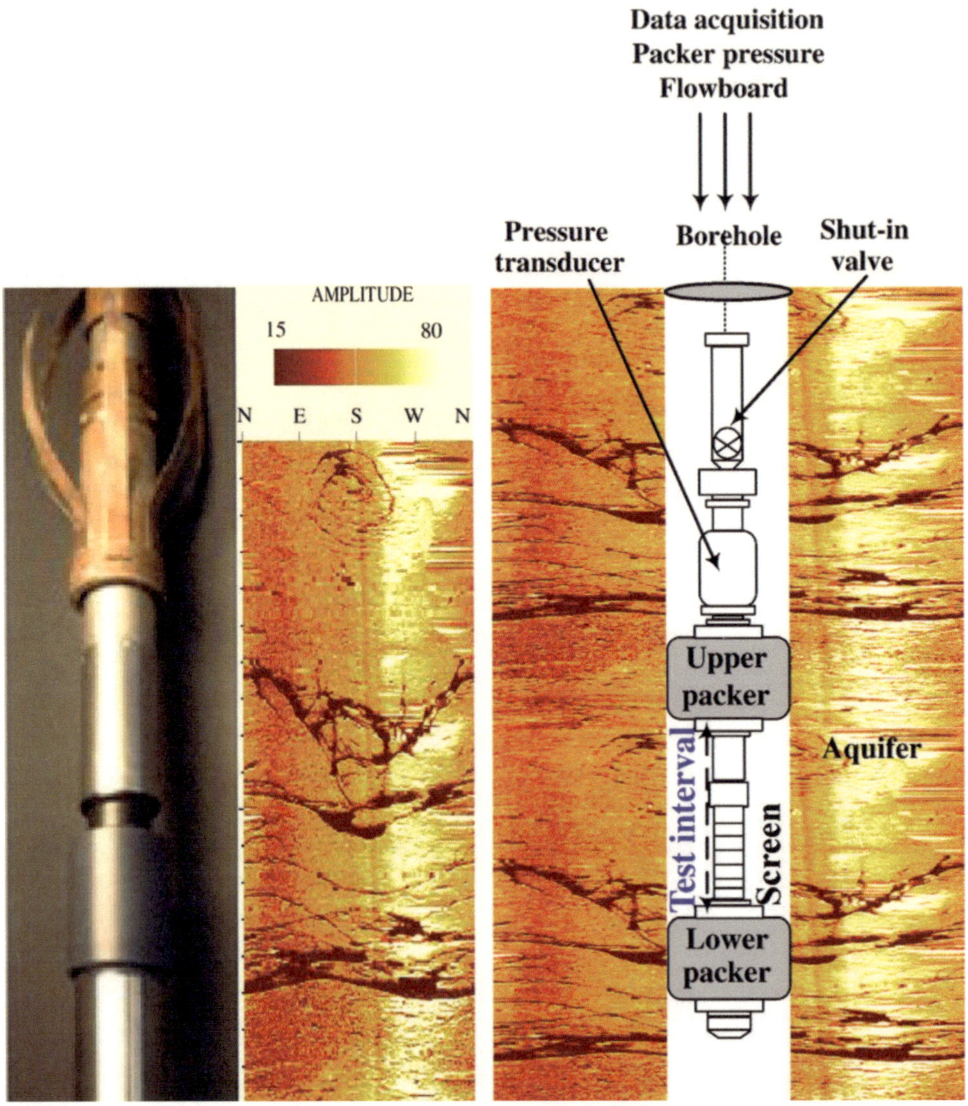

Fig. 4. Schematic drawing of BHTV, acoustic image of borehole, and double packer system

In this study, 26 hydraulic test data collected during hydraulic tests can be analyzed by analytical methods. Water pressure and discharge rate measurements with time for each hydraulic test were collected. The data analysis was performed using a professional version of the AQTESOLVE test analysis software, which enables both virtual and automatic type curve matching (Duffield, 2004). The quantitative evaluation of hydraulic parameters was carried out as an iterative process of the best-fit theoretical response curves based on the measured data of the hydraulic test. Figure 5 shows the evaluation of hydraulic parameters using AQTESOLVE. For the test interval of 118.5 m to 121.7 m, although three fractures and a fracture zone of approximately 7.25 cm thickness were seen on the borehole image, lack of interconnectivity of fractures and soft and cohesive gouges existing at the fractures may reduce the permeability of rock masses.

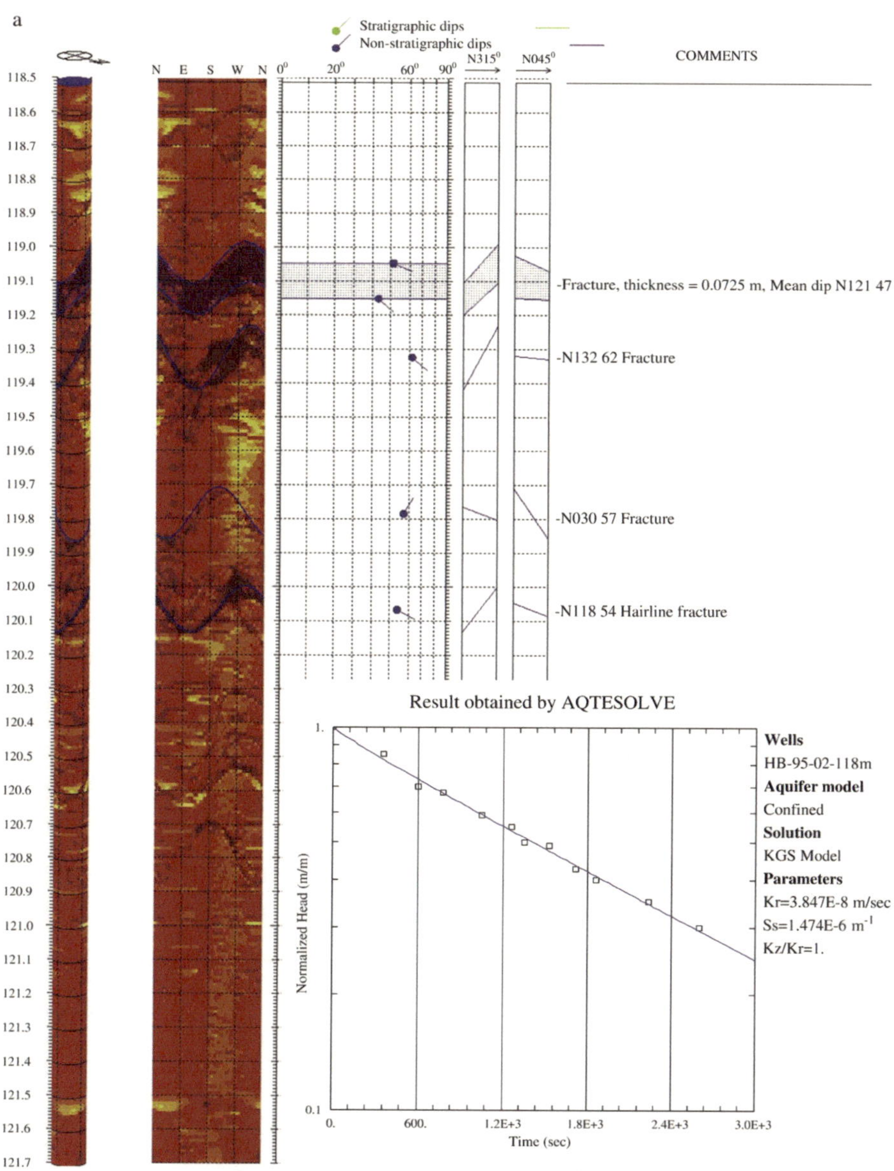

Fig. 5. Evaluation of hydraulic parameters using AQTESOLVE and BHTV images at pack-off zones of 118.5 m to 121.7 m in borehole HB-95-02

2.3.2 Relationship between hydraulic conductivity and HC-system

Regression analysis was performed to estimate the dependence of HC on hydraulic conductivity. A total of 22 hydraulic test data (borehole CH-04 data not included) were applied to the study. HC-values for the hydraulic tests can be computed from borehole image data and rock core data, in which the values of RQD and GCD at each test interval can be calculated from borehole image data and rock core data with Equations (1) and (3), respectively. The value of DI can be calculated using Equation (2). The value of LPI for each test zone can be obtained from rock core data and Table 2. Table 3 shows the calculated results for the HC-system based on the verified data. The regression results indicated that a power law relationship exists between the hydraulic conductivity and HC values with a coefficient of determination of 0.866 as shown in Figure 6. The empirical HC model is obtained as shown in Equation 5.

$$K = 2.93 \times 10^{-6} \times (HC)^{1.380}, R^2 = 0.866 \quad (5)$$

If only HB-94-01 testing data were adopted, a better correlation with the coefficient of determination of 0.905 can be obtained as shown in Equation (6).

$$K = 2.31 \times 10^{-6} \times (HC)^{1.342}, R^2 = 0.905 \quad (6)$$

It should be noted that the values of (1−GCD) inHB-94-01 borehole are all equal to 1. The results of Equation (6) demonstrate that the empirical HC model may also be more accurate for the estimation of the rock mass hydraulic conductivity if the fractures do not contain infillings. There are a few limitations that need to be noted for the use of Equation (5). The data used to develop the equation are limited in number and in the lithologies represented. From the definition of DI, DI cannot be determined for inclined boreholes because the data collected were from vertical boreholes.

Boreholes	Test intervals (m)	1-RQD/100	DI	1-GCD	LPI	HC	K (m/s)
HB-94-01	34.7-36.3	0.094	0.677	1.000	1.000	0.0635	7.06E-08
	36.4-38.0	0.438	0.662	1.000	1.000	0.2895	1.64E-06
	56.7-58.3	0.063	0.477	1.000	0.950	0.0283	1.53E-08
	74.6-76.2	0.500	0.315	1.000	0.400	0.0629	5.30E-08
	77.2-78.8	0.010	0.291	1.000	0.400	0.0012	4.22E-10
	82.6-84.2	0.125	0.242	1.000	0.400	0.0121	2.31E-09
	90.2-91.8	0.010	0.173	1.000	0.400	0.0007	2.86E-10
	94.2-95.8	0.500	0.136	1.000	0.400	0.0273	4.53E-09
HB-95-01	99.0-101.9	0.345	0.598	0.200	0.400	0.0165	9.80E-09
	117.2-120.1	0.690	0.526	1.000	0.850	0.3081	9.76E-07
	133.2-136.1	0.724	0.461	0.286	1.000	0.0954	4.68E-08
HB-95-02	88.6-91.4	0.071	0.743	1.000	0.600	0.0318	1.56E-07
	96.0-99.2	0.031	0.721	1.000	0.600	0.0135	2.42E-08
	118.5-121.7	0.219	0.657	0.071	0.700	0.0072	1.36E-09
	134.8-138.0	0.344	0.610	0.727	0.700	0.1068	1.17E-07
	154.8-158.0	0.938	0.553	0.103	0.700	0.0376	1.99E-08
	173.0-176.2	0.938	0.501	0.103	0.700	0.0340	9.08E-09
	189.8-193.0	0.594	0.453	1.000	0.700	0.1883	1.01E-06
	196.6-199.8	0.563	0.434	0.500	1.000	0.1220	6.00E-08
	213.2-216.0	0.679	0.387	1.000	1.000	0.2625	4.54E-07
	249.0-251.8	0.393	0.285	0.091	0.700	0.0071	4.03E-09
	272.0-274.8	0.214	0.219	1.000	0.700	0.0328	3.36E-08

Table 3. The calculated results for HC-system based on 22 hydraulic test data

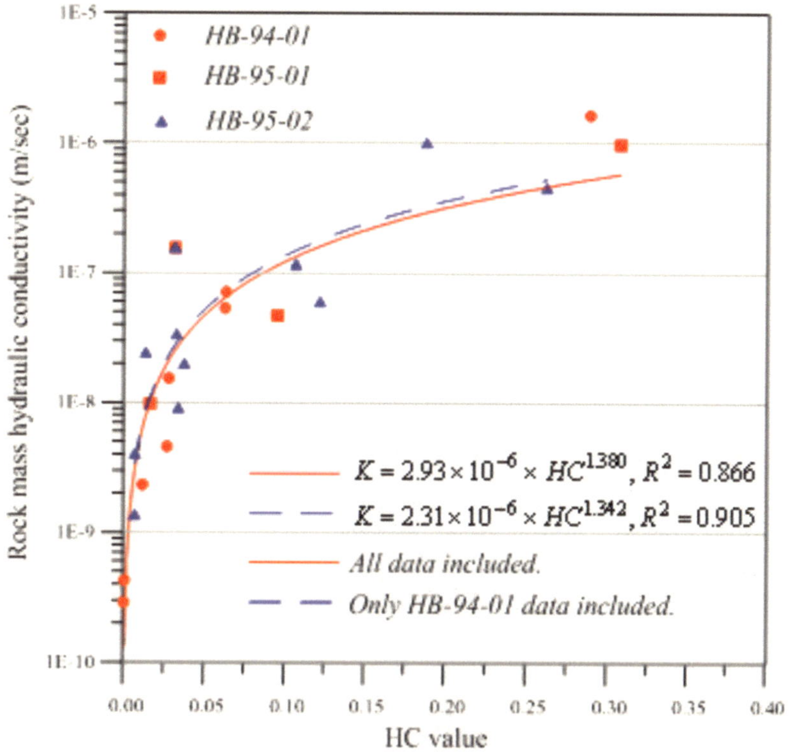

Fig. 6. Relationship between hydraulic conductivity and HC values

2.3.3 The empirical HC model verification

In order to further verify the feasibility of the proposed empirical HC model, the model verification is conducted. Borehole CH-04 data was adopted to verify the empirical HC model. The depth from 24.5 m to 26.6 m, 32.5 m to 34.1 m, 65.7 m to 67.8 m, and 77.8 m to 79.9 m were sealed by double packers for conducting the hydraulic tests. The quantitative evaluation of hydraulic parameters was then performed using AQTESOLVE. Table 4 shows four hydraulic test data for the model verification, in which $K_{HC\text{-}model}$ and $K_{in\text{-}situ}$ represent K obtained by Equation (5) and the in situ hydraulic test, respectively. Figure 7 shows that the comparison of the rock mass hydraulic conductivity obtained by in situ test and that from the estimation of the empirical HC model. Very good correlation can be found (Figure 7). This verification example demonstrates that the empirical HC model is able to determine the rock mass hydraulic conductivity for different sites in which the lithologic conditions are similar.

Test intervals (m)	RQD (%)	DI	1-GCD	LPI	HC	$K_{HC\text{-}model}$	$K_{in\text{-}situ}$
24.5-26.6	81.0	0.787	0.952	0.55	0.0785	9.06E-08	7.14E-08
32.5-34.1	43.8	0.723	0.975	0.55	0.2179	3.69E-07	1.11E-06
65.7-67.8	47.6	0.444	0.976	0.55	0.1248	1.71E-07	9.95E-08
77.8-79.9	95.2	0.343	1.000	0.55	0.0090	4.59E-09	9.09E-09

Table 3. Four hydraulic test data for the model verification

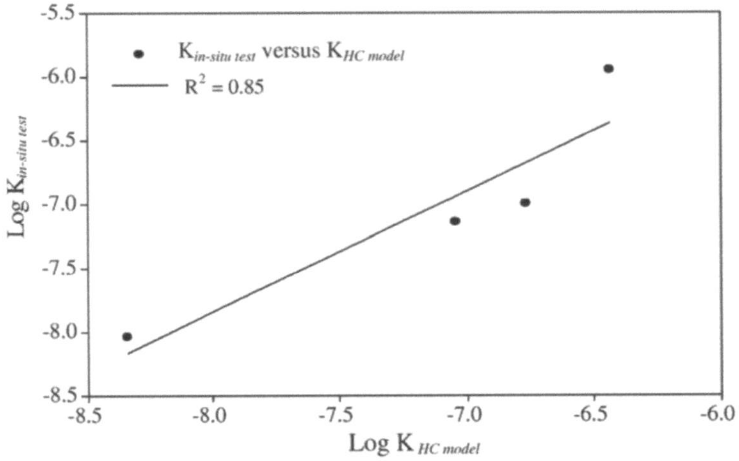

Fig. 7. Correlation between $K_{HC\text{-}model}$ and $K_{in\text{-}situ}$

2.3.4 Application of HC model

According to the above study results, the high correlation between hydraulic conductivity and HC implies that the new rock mass classification in the presented study is reasonable. It may provide two important applications in hydrogeology. The first application is that the regression equation (Equation (5) or Equation (6)) is capable of providing a useful tool to predict hydraulic conductivity of fractured rocks based on measured HC-values. By using this approach, hydraulic conductivity data in a given site can be directly acquired, which removes the cost on hydraulic testing. Secondly, for in-situ aquifer tests the HC-system is a valuable new rock mass classification system for preliminary assessment of the degree of permeability in a pack-off interval of a borehole. It is beneficial to the hydraulic test design.

3. The second empirical model

In recent years, many borehole hydrogeological investigations have been conducted to understand the relationship between hydraulic conductivity and fracture properties. For example, Tanaka and Miyakawa (1992) reported that some hydraulically conductive regions within a borehole existed in high-density fracture zone. Their results were supported by data acquired through packer tests and borehole televiewer logging. In addition, Gustafson et al., (1991) described that 44 to 61% of fractures in granite are non-conductive based on the field hydrological tests. By performing similar approaches, Hamm et al., (2007) demonstrated that the fracture aperture has stronger relationship to hydraulic conductivity than fracture frequency. They also proposed that the cubic fracture aperture model has close relationship to transmissivity with the highest correlation coefficient of 0.88.

In this study, a hydrogeologic investigation employing a series of subsurface exploration technologies was conducted at three active landslide sites in southern Taiwan. Each site was initially investigated with borehole televiewer logging to identify potentially significant fracture features and fracturing degree at depth and its hydrogeologic implications. Flowmeter logging was then performed to measure high permeability zones and fracture hydraulic connectivity along the borehole. Based on the prospecting results, test sections of hydraulic tests can be arranged. The hydraulic packer tests were carried out to further characterize the hydrogeologic system of the site and quantitatively determine the hydraulic properties of major hydrogeologic units or different geological structures. Finally, the borehole data were used in correlation and regression analyses to define the dependence and establish models between the fracture properties and hydraulic conductivity. The rationality of the regression results was carefully assessed in predicting fractured rock hydraulic conductivity.

3.1 Investigation technologies

A comprehensive hydrogeologic investigation on slopeland may include surficial geology investigations, borehole drilling, testing of soil and rock properties from rock cores, landslide mapping with light detecting and ranging (LIDAR), resistivity image profiling (RIP), double-ring infiltration tests, borehole televiewer logging, electrical logging, flowmeter logging, and packer tests. This study focuses on describing borehole televiewer logging, flowmeter logging, and packer tests with the purpose of obtaining relationships between fracture properties and their corresponding hydraulic conductivity. The techniques are described as follows.

3.1.1 Borehole televiewer logging

Using borehole televiewers to characterize fractured-rock properties has been adopted for many years. The tool acquires continuous, 360-degree images of the borehole wall while the probe moves along the length of the borehole. The results provide relevant geological and structural information needed to hydraulically analyze the subsurface, such as the location, orientation and angle of the fractures, fracture width, infilling material of fractures, and structural planar features. In addition, the borehole image is capable of clarifying the uncertainties of the traditional rock core drilling technique, including those derived from human errors for misplacing rock core samples from its original place, or interpretation for missing intervals (Williams & Johnson, 2004; Hsu et al., 2007). Generally, borehole televiewers are of two types, including: the acoustic televiewer (ACTV) and the optical televiewer (OPTV) (Figure 8). The ACTV uses a fixed acoustic transducer and rotating acoustic mirror to scan the borehole wall with a focused ultrasound beam. The amplitude and travel time of the reflected acoustic signal are recorded simultaneously as separate image logs. The OPTV system consists a ring of lights, a hyperboloidal mirror, and video camera housing in the transparent window. The OPTV is capable of providing real-time borehole images (Williams & Johnson, 2004). Each probe has its own suitable prospecting environments that depend on specific groundwater conditions. Therefore, both acoustic and optical televiewers were used in this study.

3.1.2 Heat-pulse flowmeter logging

The heat-pulse flowmeter (HPFM) system consists of a wire-grid heating element and two sensitive thermistors (heat sensors) located above and below the wire-grid. The wire-grid generates a sheet of heat in the water and the heat migrates towards one of the thermistors, depending on direction and rate of groundwater flow (Sloto & Grazul, 1989). The direction and rate of flow can be computed once the peak temperature has been detected by the thermistor (Figure 9). When the stationary measurements are conducted at several depths along a borehole, the distribution of groundwater velocity can be obtained. This not only provides useful information to characterize the aquifer permeability, but also brings good indications to identify the location of the flow path, water-producing/receiving zones, and fracture connectivity (Paillet, 1986; Miyakawa et al., 2000; Williams & Paillet, 2002). In addition, Miyakawa et al. (2000) pointed out that the HPFM is usually carried out by pumping water into or extracting water from the borehole, because it is difficult to detect hydraulic pathways in a natural state due to low groundwater velocities.

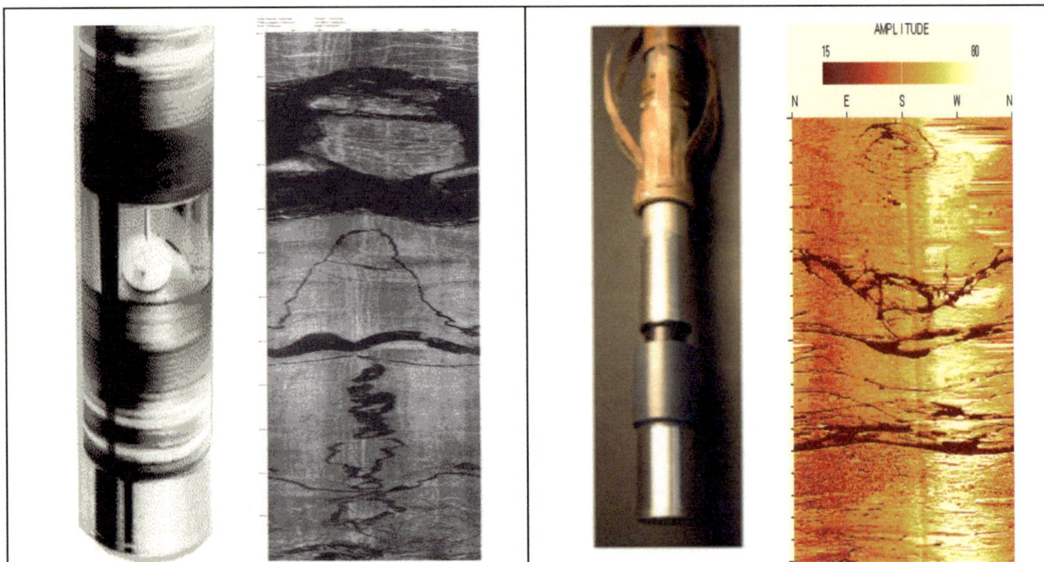

(a) Optical televiewer | (b) Acoustic televiewer

Fig. 8. The televiewer and scanned borhole images

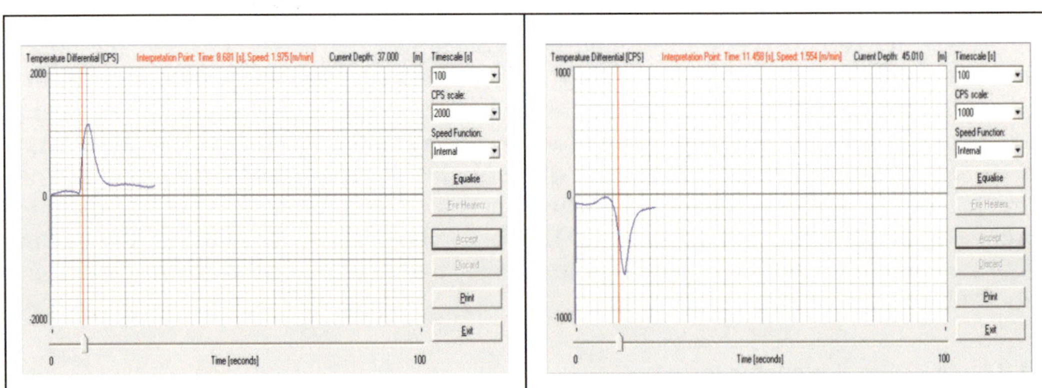

(a) The upward flow heat-pulse curve | (b) The downward flow heat-pulse curve

Fig. 9. The heat-pulse curve (the elapsed time in X-axis and the temperature difference measured by the thermistor in Y-axis)

3.1.3 Double packer hydraulic test

The double packer hydraulic test is one of the most common approached applied to determining the hydraulic conductivity and storage coefficient along discrete sections of a borehole. It is now recognized that this approach is capable of investigating the variability of a borehole as it intersects various hydrogeological units.

The double packer hydraulic test was conducted by isolating a section of borehole with a set of packers and measuring the rate of flow and/or pressure over a period of time. The system adopted in this study contained two inflatable rubber packers, a shut-in valve, flow meters, a submersible pump, and three transducers for measuring the piezometric pressure in the isolated interval and the areas above and below the packers. The rubber packers were inflated using nitrogen delivered through a polyethylene air line. The shut-in valve was used to open and close the hydraulic connection between the pipe string and the test section. The pumping or injection rate was measured at the land surface with a flow meter.

Four types of hydraulic tests can be applied to the double packer system including the pump test, injection test, slug test, and pressure pulse test. A pump test involves pumping at a constant or variable rate and measuring changes in water levels during pumping. In an injection test, fluid is injected into the test interval at a constant head. In the slug test, a known amount of water is delivered to the test interval and the changes in pressure are monitored as equilibrium conditions return. In a pressure head test, an increment of pressure is applied to the test interval and the pressure decay is monitored over time. Typically, the selection for type of test is based on the expected permeability of strata, the volume of rock to be sampled, and the availability of time and equipment. The data collected during the hydraulic test can be analyzed by analytical methods using professional software AQTESOLVE as stated in Section 2.3.1.

3.2 Case study and prospecting results

Borehole logging was conducted at three landslide sites in the south central portion of Taiwan (Figure 10). The purpose of the investigation was to determine the hydraulic properties from various geological structures, such as the degree of fracturing and hydraulic conductivity, to test hypotheses related to the causes of the landslides. The logging was conducted in six boreholes that ranged in depth from 70 to 80 m. Two boreholes from each of the following three sites were investigated: Tung-Chi (borehole FH-13 and FH-15), Bao-Long (FH-03 and FH-05), and Gi-Lu (borehole FH-21 and FH-23). The Tung-Chi and Bao-Long sites are located in Kaohsiung County, southern Taiwan. The principle lithologic units of the two sites are weathering slate with clay-rich gouges and fresh shale with thin layered sandstone, respectively. The Gi-Lu site is located in Pingtung County, southern Taiwan, where the dominated geologic unit composed of fresh slate with a minor amount of quartz and metamorphic sandstone.

The investigation first identified the position and degree of the fracturing using borehole televiewer logging. Second, the angle and width of fracture or fracture zone were calculated by adopting the post processing software. Lastly, flowmeter logging was used to determine the strata permeability and fracture connectivity. Based on the results of these preliminary logging data to determine favourable hydraulic conditions, test sections were selected for the double packer hydraulic test, and the results were used to determine the hydraulic conductivity for different geological structures (Figure 11). A total of 18 double packer hydraulic test sections were obtained in this study. Table 4 summarizes the results of the average fracture angle, fracture width, average flow velocity, hydraulic conductivity, and product of fracture width and flow velocity for each test section. These calculated data were used in correlation and regression analyses to define the dependence and establish models between the fracture properties and hydraulic conductivity.

Fig. 10. Location of three active landslide sites

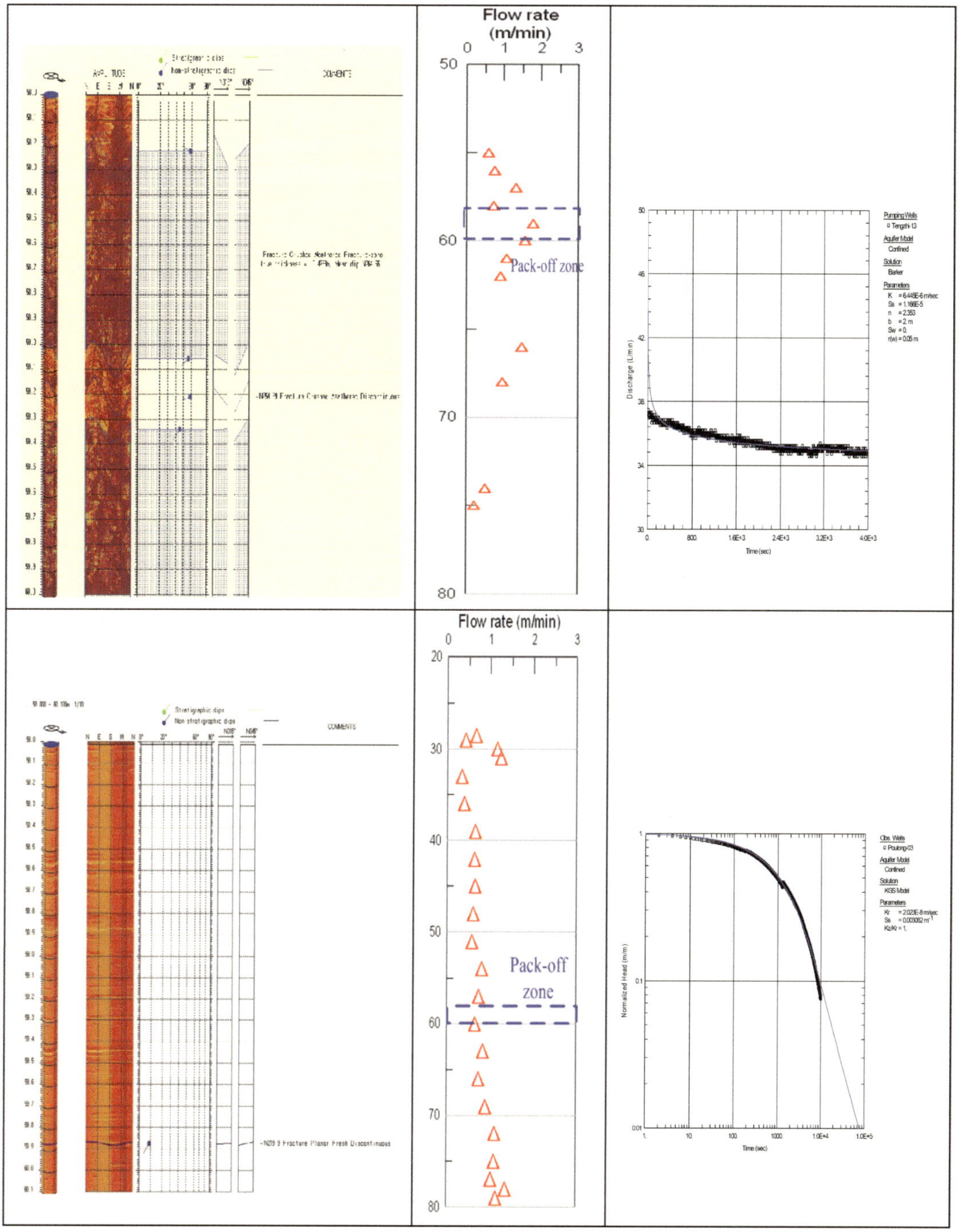

Fig. 11. The computed angle and width for single fracture or fracture zone, the measured groundwater velocity and hydraulic conductivity obtained from AQTESOLVE (from left to right) at 58.0-60.0m pack-off zones in Tung-Chi landslide site (above); and at 58.0-60.1m pack-off zones in Bao-Long landslide site (below)

Borehole	Test Intervals (m)	Average Fracture Angle, FA (deg)	Fracture (zone) Width, FW (m)	Average Flow Velocity, V (m/min)	Hydraulic Conductivity, K (m/sec)	Product of Fracture Width and Flow Velocity (FW×V)
FH-13	46.5-48.5	42.0	2.000	NA	3.59E-06	NA
	58.0-60.0	57.0	1.485	1.25	6.45E-06	1.856
	61.0-63.0	35.0	0.845	1.01	9.57E-06	0.845
	68.0-70.0	32.6	0.015	0.96	3.25E-07	0.014
FH-15	45.5-47.5	29.5	1.430	2.40	1.59E-05	3.432
	56.6-58.6	29.0	2.000	2.47	2.91E-05	4.940
	65.8-67.8	32.3	0.080	2.45	1.67E-06	0.196
FH-03	29.5-31.6	19.0	0.005	1.20	1.13E-08	0.006
	39.0-41.1	11.0	0.005	0.97	1.98E-10	0.005
	58.0-60.1	09.0	0.005	0.70	3.20E-09	0.004
FH-05	34.0-36.1	38.0	0.055	0.52	1.68E-07	0.029
	50.0-52.1	31.2	0.035	1.15	4.10E-07	0.040
FH-21	31.9-33.4	34.0	0.445	NA	8.40E-06	NA
	35.3-37.0	41.7	0.220	NA	2.54E-06	NA
	51.8-53.3	38.0	0.430	0.10	3.19E-07	0.043
	63.8-65.3	40.0	0.525	0.49	4.88E-07	0.257
FH-23	50.5-52.0	54.3	0.130	0.10	2.67E-08	0.013
	57.0-58.5	38.2	0.105	0.33	1.12E-07	0.035

NA: Non-detectable when test section above groundwater table

Table 4. The prospecting results

3.3 Data analysis

3.3.1 Correlation analysis

To define the dependence between the rock mass hydraulic conductivity and fracture angle, fracture width, and the flow velocity, a univariate correlation analysis was performed. The correlation coefficient between the different values can be computed by following equation:

$$r_{xy} = \frac{\sum_{i=1}^{n}(x_i - \overline{x})(y_i - \overline{y})}{S_x S_y} \tag{7}$$

where $\overline{x}$ and $\overline{y}$ are the sample means of x and y, S_x and S_y are the sample deviations of x and y, which are defined as:

$$S_x = \sqrt{\frac{\sum_{i=1}^{n}(x_i - \overline{x})}{n}} \tag{8}$$

$$S_y = \sqrt{\frac{\sum_{i=1}^{n}(y_i - \overline{y})}{n}} \tag{9}$$

The results of the correlation analysis are shown in Table 5. The table shows that the fracture width and flow velocity have a good correlation with hydraulic conductivity, with correlation coefficient of 0.75 and 0.69, respectively. In addition, the analysis indicates that the fracture angle and hydraulic conductivity were independent. Most noticeably, by multiplying fracture width and flow velocity, the products of two values shown strong positive correlation to hydraulic conductivity with correlation coefficient of 0.93. Since the borehole flow velocity is usually related to fracture connectivity, a strong correlation to hydraulic conductivity can be found when considering for both fractures width and connectivity.

	Correlation Coefficient			
	Average Fracture Angle, FA (deg)	Fracture (zone) Width, FW (m)	Average Flow Velocity, V (m/min)	Product of Fracture Width and Flow Velocity (FW×V)
Hydraulic Conductivity, K (m/sec)	-0.01	0.75	0.69	0.93

Table 5. The results of correlation analysis

3.3.2 Regression analysis

A regression analysis was used to establish the relationship between hydraulic conductivity and fracture angle, fracture width, flow velocity and the product of fracture width and flow velocity. Although the fracture angle and flow velocity were difficult to establish good relationship with hydraulic conductivity as shown in Table 6, a power low relationship exists between the hydraulic conductivity and fracture width in the semi-log scale with a coefficient of determination of 0.73. Furthermore, a better regression result, which the coefficient of determination of 0.83 was established between hydraulic conductivity and the product of fracture width and flow velocity in log-log scale (Figure 12) Therefore, it can be concluded that the rock mass hydraulic conductivity not only related to the fracture width, but also possesses strong relationship when their corresponding connectivity took into account. This regression results is desired to predict hydraulic conductivity of the study area based on borehole televiewer and flowmeter logging results, which removes cost of packer test for additional test intervals or boreholes.

Factors Related to Hydraulic Conductivity	Best Regression Equation	Coefficient of Determination (R^2)
Average Fracture Angle, FA (deg)	$K = 1.79E\text{-}13(FA)^{4.27}$	0.42
Fracture (zone) Width, FW (m)	$K = 5.42E\text{-}6(FW)^{1.30}$	0.73
Average Flow Velocity, V (m/min)	$K = 2.65E\text{-}8\exp(2.18V)$	0.28
Product of Fracture Width and Flow Velocity (FW×V)	$K = 6.29E\text{-}6(FW \times V)^{1.25}$	0.83

Table 6. The results of regression analysis

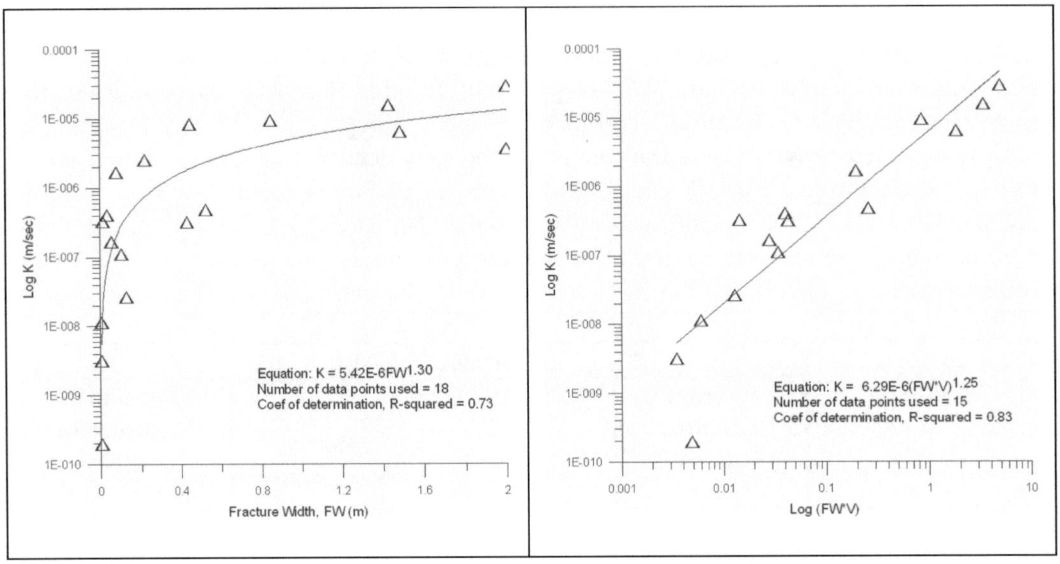

Fig. 12. Regression results of Fracture width (FW) with hydraulic conductivity (left); and the product of fracture width and flow velocity (FW×V) with hydraulic conductivity (right).

3.3.3 Model verification

In order to further verify the applicability of the regression result, the verification was conducted using another two in-situ borehole prospecting data obtained in a nuclear power plant. The geologic unit of the site is mainly composed of shale or shale with thin layered sandstone. Both televiewer and flowmeter loggings were performed, and five intervals were selected for packer test. The quantitative evaluation of hydraulic conductivity was then applied using AQTESOLV based on the data of packer test. Table 7 shows five test data for the model verification, in which K_{FW} and $K_{FW\times V}$ represent K obtained by corresponding equations in Table 6; $K_{in\text{-}situ}$ represents K obtained by the in situ hydraulic test. Figure 13 shows the comparison of rock mass hydraulic conductivity obtained from in-situ test data and from regression results. The coefficient of determination between in-situ hydraulic

Borehole	Test Intervals (m)	Fracture (zone) Width, FW (m)	Average Flow Velocity, V (m/min)	Product of Fracture Width and Flow Velocity (FW×V)	Predicted Hydraulic Conductivity, K_{FW}(m/sec)	Predicted Hydraulic Conductivity, $K_{FW\times V}$(m/sec)	Measured Hydraulic conductivity, $K_{in\text{-}situ}$ (m/sec)
BH-37	65.5-67.0	0.08	2.99	0.24	2.03E-07	1.05E-06	1.17E-06
	73.5-75.0	0.07	2.22	0.14	1.55E-07	5.59E-07	2.31E-07
BH-43	28.0-29.5	0.17	3.22	0.55	5.41E-07	2.96E-06	1.16E-05
	43.0-44.5	0.08	2.24	0.18	2.03E-07	7.33E-07	2.10E-07
	54.0-55.5	2.00	2.60	5.20	1.33E-05	4.94E-05	1.81E-05

Table 7. Five test data for model verification

conductivity and that predicted from the product of fracture width and velocity is 0.74, which is higher than the hydraulic conductivity predicted from fracture width only. Although the prediction still possesses some deviations when compared to real data (a predicted value is an order of magnitude lower than the measured value), the verification result demonstrated that the regression equation is capable of estimating the fractured rock hydraulic conductivity without packer testing, especially for the site with similar lithologic environments.

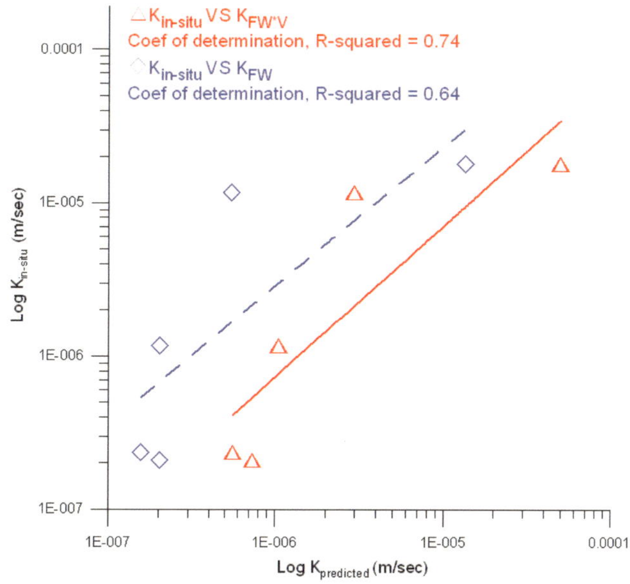

Fig. 13. Verification of regression results: hydraulic conductivity predicted from the product of fracture width and flow velocity (solid line); hydraulic conductivity predicted from fracture width only (dash line)

4. Conclusions

The estimation of rock mass hydraulic conductivity using feasible empirical equations possesses great advantages. This chapter proposes two empirical models to estimate hydraulic conductivity of fractured rock mass with the features of high efficiency and low cost.

The first empirical model was based on the rock mass classification concept. The study develops a new rock mass classification scheme for estimating hydraulic conductivity of fractured rocks. The new rock mass classification system called as "HC-system" based on the following four parameters: rock quality designation (RQD), depth index (DI), gouge content designation (GCD), and lithology permeability index (LPI). HC-values can be calculated from borehole image data and rock core data. To verify rationality of the defined HC-system, the study collected data from the results of two hydrogeological investigation programs in three boreholes with 22 in-situ hydraulic tests to determine a relationship between hydraulic conductivity and HC. Regression analysis was performed to estimate the dependence of HC on hydraulic conductivity. The field results indicated that the rock mass in the study area has the conductivity between the order 10^{-10} and 10^{-6} m/s at the depth between 34 m and 275 m below ground surface. The regression results demonstrated that a power law relationship exists between the two variables with a coefficient of determination of 0.866. Besides, the model verification using another borehole data with four additional in-situ hydraulic tests was also conducted to further verify the feasibility of the proposed empirical HC model. The regression equation provides a useful tool to predict hydraulic conductivity of fractured rocks based on measured HC-values. By using this regression equation, hydraulic conductivity data in a given site can be directly acquired, which removes the cost on hydraulic tests. For in-situ aquifer tests, the HC-system is a valuable new rock mass classification system for preliminary assessment of the degree of permeability in a packed-off interval of a borehole.

The second empirical model was simply based on the results of borehole televiewer logging, flowmeter logging and packer hydraulic tests at three active landslide sites in southern Taiwan. Three borehole prospecting techniques for hydrogeologic investigation of fractured rock mass were performed to explore various hydrogeologic characteristics, such as fracture width, fracture angle, flow velocity and hydraulic conductivity. By adopting a correlation analysis, the dependence between hydraulic conductivity and other prospecting data was identified. While the analysis revealed that the fracture width and flow velocity showed good correlation with hydraulic conductivity, the fracture angle and hydraulic conductivity were uncorrelated. In addition, by multiplying fracture width and flow velocity, the product of two values strongly correlated to hydraulic conductivity with the correlation coefficient of 0.93. The regression analysis also indicated that a power law relationship with a coefficient of determination of 0.83 existed between the hydraulic conductivity and the product of fracture width and flow velocity. Furthermore, the regression equation was verified using other borehole prospecting data. The results demonstrated that the regression equation is capable of predicting the hydraulic conductivity of fractured rock based on borehole televiewer and flowmeter logging results at the site with similar lithologic conditions. The study also demonstrated that such an approach is very constructive for a subsurface hydrogeologic assessment, particularly in the absence of packer test data due to a limited budget.

5. References

Barton, N.; Lien, R. & Lunde, J. (1974). Engineering classification of rock masses for the design of tunnel support, *Rock Mech.*, Vol. 6, 183-236.

Bear, J. (1972). *Dynamics of Fluids in Porous Media*, American Elsevier Publication Co., New York.

Bieniawski, Z. T. (1973). Engineering classification of jointed rock masses, *Trans. S. Afr. Inst. Civ. Eng.*, Vol. 15, 335-344.

Bieniawski, Z. T. (1989). *Engineering Rock Mass Classifications-A Complete Mannual for Engineers and Geologists in Minging, Civil, and Petroleum Engineering*, John Wiley & Sons, New York.

Black, J. H. (1987). Flow and flow mechanisms in crystalline rock, in Fluid Flow in Sedimentary Basins and Aquifers, *Geol. Soc. Special Publication No. 34*, pp. 186-200.

Burgess, A. (1977). *Groundwater Movements Around a Repository—Regional Groundwater Analysis*, Kaernbraenslesaekerhet, Stockholm, Sweden.

Carlsson, A. & Olsson, T. (1977). *Hydraulic properties of Swedish crystalline rocks-hydraulic conductivity and its relation to depth*, Bulletin of the Geological Institute, University of Uppsala, 71–84.

Deere, D. U.; Hendron, A. J.; Patton, F. D. & Cording, E. J. (1967). Design of surface and near surface construction in rock, *Proceedings of 8th U.S. Symposium. Rock Mechanics. AIME*, pp. 237-302, New York.

Duffield, G. M. (2004). *AQTESOLVE Version 4 User's Guide*, Developer of AQTESOLV HydroDOLVE, Inc., Reston, VA, USA.

Foyo, A.; Sa'nchez, M. A.; & Tomillo, C. (2005). A proposal for secondary permeability index obtained from water pressure test in dam foundations, *J. Geo. Eng.*, Vol. 77, 69-82.

Gates, W .C. B. (1997). The hydro-potential(HP) value: a rock classification technique for evaluation of the ground-water potential in fractured bedrock, *Environmental & Engineering Geoscience*, Vol. 3, No. 2, 251-267.

Gustafson, G.; Rhen, I. & Stanfor, R. (1991). *Evaluation and conceptual modeling based on the pre-investigations 1986-1990*, Aspo Hard Rock Laboratory Technical Report 91-22, Swedish Nuclear and Waste Management.

Hamm, S.; Kim, M.; Cheong, J.; Kim, J.; Son, M. & Kim, T. (2007). Relationship between hydraulic conductivity and fracture properties estimated from packer tests and borehole data in a fractured granite, *Engineering Geology*, 92, 73-87.

Hsu, S.; Chung, M.; Ku, C.; Tan, C. & Weng, W. (2007). An application of acoustic televiewer and double packer system to the study of the hydraulic properties of fractured rocks, 60th Canadian Geotechnical Conference & 8th joint CGS/IAH-CNC Groundwater Conference, pp. 415-422, V1, Ottawa, Canada.

Lauffer, H. (1958). Gebirgsklassifizierung für den stollenbau, *Geo. Bauwesen*, Vol. 74, 46-51.

Lee, C. H. & Farmer, I. (1993). *Fluid flow in discontinuous rocks*, Chapman&Hall, London, UK.

Louis, C., (1974). *Rock Hydraulics in Rock Mechanics (ed. L. Muller)*, Springer Verlag, Vienna.

Miyakawa, K.; Tanaka, K.; Hirata, Y. & Kanauchi, M. (2000). Detecting of hydraulic pathways in fractured rock masses and estimation of conductivity by a newly developed TV equipped flowmeter, *Engineering Geology*, Vol. 56, 19-27.

National Research Council. (1996). *Rock fractures and fluid flow: contemporary understanding and applications*, National Academy Press, Washington DC, USA.

Paillet, F. L. & Hess, A. E. (1986). *Geophysical well-log analysis of fractured crystalline rocks at east Bull Lake*, Ontario, Canada: U.S. Geological Survey Water Resources Investigations Report 86-4052.

Sahimi, M. (1995). Flow and transport in porous media and fractured rock, Wiley-VCH.

Singhal, B. B. S. & Gupta, R.P. (1999). *Applied hydrogeology of fractured rocks*, Kluwer Academic Publishers, The Netherlands, 400 p.

Sinotech Engineering Consultants, LTD. (2005). "Tseng-Wen transbasin diversion tunnel project-supplemental geology investigation," Southern Water Resources Office, Water Resources Agency, Ministry of Economic Affairs, Taiwan.

Sloto, R. A. & Grazul, K. E. (1989). *Results of borehole geophysical logging and hydraulic tests conducted in area D supply wells, former U.S. Naval Air Warfare Center; Warminster, Pennsylvania*, Water Resource Investigation Report 01-4263, U.S. Geological Survey.

Snow, D. T. (1969). Anisotropic permeability of fractured media, *Water Resources Research*, Vol. 5, No. 6, 1273-1289.

Spitz, K., and Moreno, J. (1996). "A practical guide to groundwater and solute transport modeling," John Wiley, New York, 480 p.

Tanaka, K. & Miyakawa, K. (1992). Application of borehole television system to deep underground survey, *Jpn. Soc. Eng. Geol.,* Vol. 32, 289-303.

Terzaghi, K. (1946). *Rock defects and loads on tunnel support,* Rock Tunneling with steel supports, ed. R. V. Proctor and T. White, Commercial Shearing Co., Youngstown, OH, 15-99.

Wickham, G.E., Tiedemann, H.R., and Skinner, E.H. (1972). "Support determineation based on geologic predictions," Proc. Rapid Excav. Tunnelinf Conference, AIME, New Yprk: 43-64.

Hess, A.E. (1986). "Identifying hydraulically conductive fractures with a slow velocity borehole flowmeter." Canadian Geotechnical Journal 23: 69-78.

Sloto, R.A. and Grazul, K.E. (1989). "Results of borehole geophysical logging and hydraulic tests conducted in area D supply wells, former U.S. Naval Air Warfare Center; Warminster, Pennsylvania" Water Resource Investigation Report 01-4263, U.S. Geological Survey.

Tanaka, K. and Miyakawa, K. (1992). "Application of borehole television system to deep underground survey." Jpn. Soc. Eng. Geol. 32: 289-303.

Wei, Z. Q.; Egger, P. & Descoeudres, F. (1995). Permeability predictions for jointed rock masses, *International Journal of Rock Mechanics, Mineral Science and Geomechanics,* Vol. 32, 251-26l.

Wickham, G. E.; Tiedemann, H. R. & Skinner, E. H. (1972). Support determination based on geologic predictions, Proc. *Rapid Excavation Tunnel,* AIME, pp. 43-64, New York.

Williams, J. H. & Paillet, F. L. (2002). Using flowmeter pulse to define hydraulic connections in the subsurface: a fractured shale example, *Journal of Hydrology,* Vol. 265, 100-117.

Williams, J. H. & Johnson, C. D. (2004). Acoustic and optical borehole-wall imaging for fractured-rock aquifer studies, *Journal of Applied Geophysics,* Vol. 55, 151-159.

6

Hydraulic Conductivity of Layered Anisotropic Media

Stanisław Żak
Wrocław University of Technology
Poland

1. Introduction

Calculating a flow in a multi-layered medium (material) requires entering geometric boundaries of all layers, boundary conditions in these areas and filtration parameters (Bear, 1972). In the case of numerous layers, entering these data is very laborious, and sometimes even impossible. What we do then is try to model the flow by adopting one medium with appropriate anisotropic filtration properties. In such a case, one should decide whether such an assumption is possible from a theoretical point of view, and what parameters of such a medium should be adopted. A full estimation of filtration properties of anisotropic media requires defining the values of hydraulic conductivity in the direction of the principal axis of permeability tensor (Bear, 1972; Batu, 1998; Rogoż, 2007). For two-dimensional orthotropic media, including media composed of parallel homogeneous isotropic layers, these will be two values - one in parallel direction ($k_{\|}$), and the other - perpendicular ($k_{\perp}$) to layer boundary (Snow, 1969; Cheng, 2000). These values can be determined by performing permeameter measurements, with a forced flow in the above directions. If this is not possible, measurements are performed by realising flow in the direction diagonal to layering, and then calculating $k_{\|}$ and $k_{\perp}$. However, this leads to errors in determining these coefficients. The aim of this work is to discuss the causes of these errors and estimate their value.

2. Theoretical basis

The possibility to replace a system of layers with different permeability parameters with one medium with specified parameters should be justified theoretically. Based on the flow model adopted for calculations, one should clearly state the conditions under which a system of layers can be treated as one medium, what parameters should be adopted for such a medium and how to determine them correctly. Hence, theoretical issues have received a lot of attention so that the problems of flow through layered media can be presented in a possibly full scope, which is essential for explaining the theses of the paper.

2.1 Flow through a single layer

Let us consider a simple case of an incompressible fluid flow through a homogeneous and anisotropic aquifer with constant thickness M (Fig. 1).

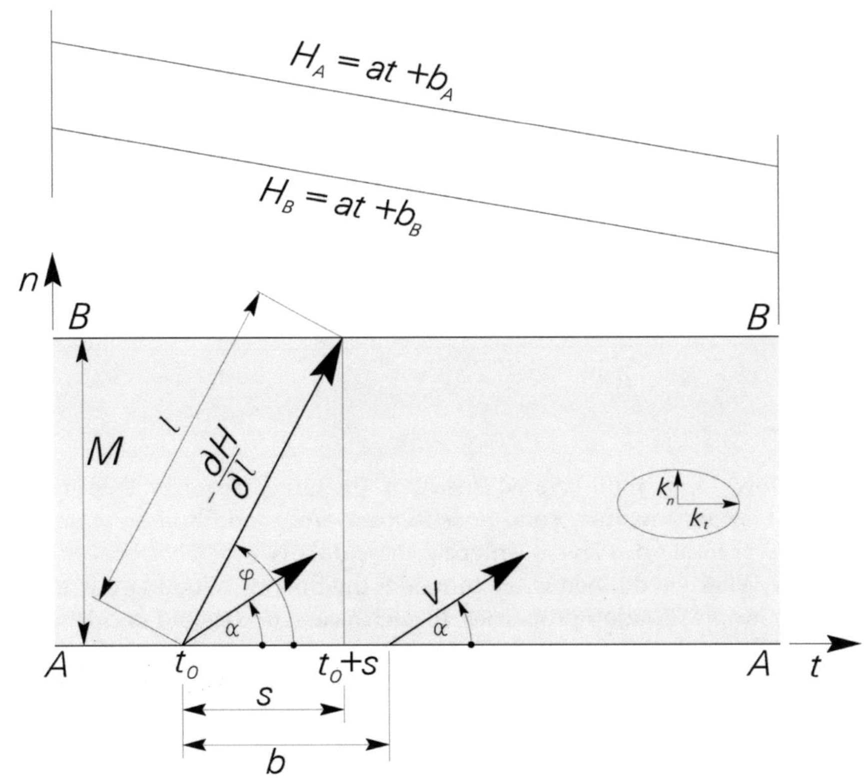

Fig. 1. Diagram of a flow through a single anisotropic layer.

Let us accept that the flow occurs according to Darcy's law. Let us designate the hydraulic conductivity in a plane tangential to the aquifer floor and roof as k_t, and in the direction normal to that plane as k_n. Let the hydraulic head change linearly along line A-A (axis t) according to the equation

$$H_A = at + b_A \, , \tag{1}$$

and along line B-B according to

$$H_B = at + b_B \, , \tag{2}$$

where a, b_A and b_B are constants, and $b_A > b_B$. Substituting $l = \frac{M}{\sin\varphi}$ and $s = \frac{M}{\tan\varphi}$, one can express hydraulic head drop along l in the form:

$$I = \frac{H_B(t_0 + s) - H_A(t_0)}{l} = -\frac{b_A - b_B}{M}\sin\varphi + a\cos\varphi \tag{3}$$

Filtration direction through an aquifer depends on the direction of hydraulic gradient. It can be determined by calculating the derivative $\frac{dI}{d\varphi}$ and comparing it to zero. Then

$$\varphi = \arctan\left(-\frac{b_A - b_B}{aM}\right) \tag{4}$$

The hydraulic head drop for the above angle is:

$$\begin{aligned} I_t &= a \\ I_n &= -\frac{(b_A - b_B)}{M} \\ I &= -\sqrt{\frac{(b_A - b_B)^2}{M^2} + a^2} \end{aligned} \tag{5}$$

while filtration velocity components are

$$\begin{aligned} v_t &= -k_t a \\ v_n &= k_n \frac{b_A - b_B}{M} \end{aligned} \tag{6}$$

The direction of filtration velocity can be determined from the formula

$$\tan\alpha = \frac{v_n}{v_t} = \frac{k_n \dfrac{\partial H}{\partial n}}{k_t \dfrac{\partial H}{\partial t}} = \frac{k_n \dfrac{\partial H}{\partial l}\sin\varphi}{k_t \dfrac{\partial H}{\partial l}\cos\varphi} = \frac{k_n}{k_t}\tan\varphi \tag{7}$$

Hence, for the discussed example

$$\alpha = \arctan\left(-\frac{k_n}{k_t}\frac{b_A - b_B}{aM}\right) \tag{8}$$

As we can see from relation (7), the more different the values of hydraulic conductivity k_n and k_t, the bigger the difference between hydraulic gradient direction and filtration velocity vector. If the layer is isotropic, angle φ will be equal to angle α.

2.2 Flow through layer boundary

If a flow occurs through the boundary between layers with different hydraulic conductivity values, streamline refraction occurs. Let us consider a flow through the boundary between two layers of soil media orthotropic in two dimensions (Fig. 2).

Two boundary conditions must be met in this area. The hydraulic head H_1 in layer 1 should be the same as H_2 in layer 2 and the normal velocity component to layer boundary v_{1n} in layer 1 and v_{2n} in layer 2 should be also the same.

$$\begin{aligned} H_1 &= H_2 \\ v_{1n} &= v_{2n} \end{aligned} \tag{9}$$

From the latter condition we obtain:

$$k_{1n}\frac{\partial H_1}{\partial n} = k_{2n}\frac{\partial H_2}{\partial n} \tag{10}$$

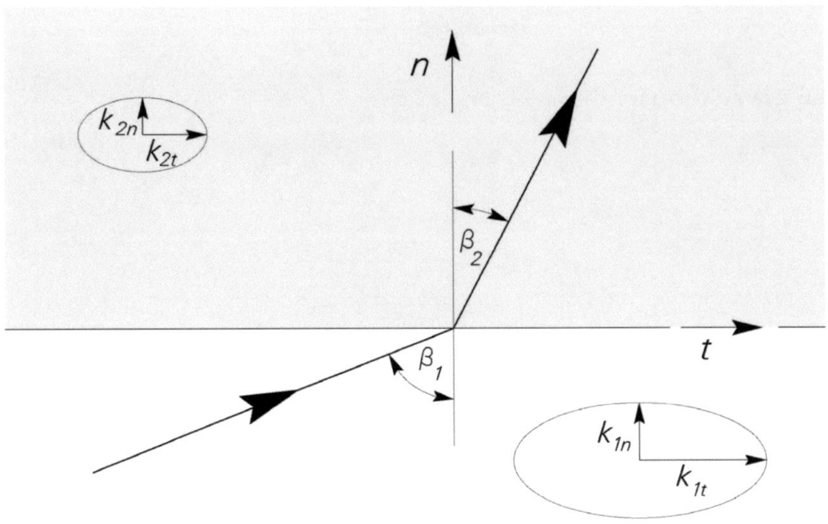

Fig. 2. Streamline refraction at the boundary of anisotropic layers.

But

$$\tan\beta_1 = \frac{k_{1t}\dfrac{\partial H_1}{\partial t}}{k_{1n}\dfrac{\partial H_1}{\partial n}} \quad \text{i} \quad \tan\beta_2 = \frac{k_{2t}\dfrac{\partial H_2}{\partial t}}{k_{2n}\dfrac{\partial H_2}{\partial n}} \tag{11}$$

After substituting relations (11), equation (10) assumes the form

$$\frac{k_{1t}\dfrac{\partial H_1}{\partial t}}{\tan\beta_1} = \frac{k_{2t}\dfrac{\partial H_2}{\partial t}}{\tan\beta_2} \tag{12}$$

Since H_1=H_2, we obtain

$$\frac{k_{1t}}{k_{2t}} = \frac{\tan\beta_1}{\tan\beta_2} \tag{13}$$

What emerges is that streamline refraction at the boundary of anisotropic media depends solely on the value of hydraulic conductivity in the direction tangential to layer boundary and does not depend on the values of hydraulic conductivity in the direction normal to layer boundary.

2.3 Flow through a multi-layered medium

Let us consider an incompressible fluid flow through an undeformable soil medium composed of N homogeneous and 2-D orthotropic parallel layers. Let us designate the hydraulic conductivity and thickness of the ith layer as k_{it} , k_{in} and M_i respectively. Let us assume that a steady-state flow diagonal to layering takes place through such a system. (Fig. 3).

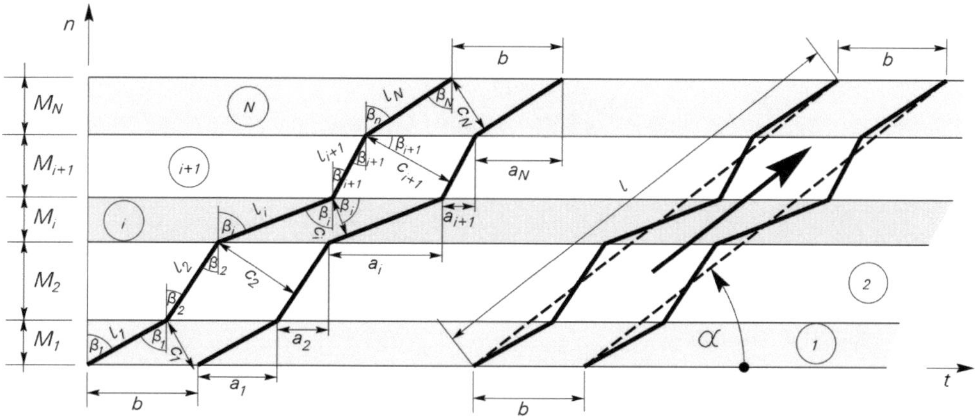

Fig. 3. Streamlines and filtration direction in a layered medium.

Let us also assume that the gradient of hydraulic head along lines A-A and B-B changes according to equations (1) and (2). In such a case, streamlines will take the shape of a polygonal chain composed of straight line segments. The differences between hydraulic heads along each streamline during the flow through particular layers will have the constant value ΔH_i. At layer boundaries, according to relation (13), streamline refraction will take place, which, after adopting the symbols from Fig. 3, will be expressed as

$$\frac{k_{it}}{k_{(i+1)t}} = \frac{\tan(\beta_i)}{\tan(\beta_{i+1})} \tag{14}$$

Water discharge between two streamlines through each layer must be the same. For a flow through the ith layer it is:

$$q = \sqrt{k_{it}^2\left(\frac{\partial H_i}{\partial t}\right)^2 + k_{in}^2\left(\frac{\partial H_i}{\partial n}\right)^2}\int_0^b \cos(\beta_i)\,dt = b\cos(\beta_i)\sqrt{k_{it}^2\left(\frac{\partial H_i}{\partial t}\right)^2 + k_{in}^2\left(\frac{\partial H_i}{\partial n}\right)^2} \tag{15}$$

Taking into account the relations

$$\frac{\partial H_i}{\partial t} = \frac{\Delta H_i}{l_i}\cos(\varphi_i) \tag{16}$$

$$\frac{\partial H_i}{\partial n} = \frac{\Delta H_i}{l_i}\sin(\varphi_i) \tag{17}$$

$$l_i = \frac{M_i}{\cos(\beta_i)} \tag{18}$$

We obtain the formula

$$q = b\cos(\beta_i)\frac{\Delta H_i}{l_i}\sqrt{k_{it}^2\cos^2(\varphi_i) + k_{in}^2\sin^2(\varphi_i)} \tag{19}$$

For the symbols used in Fig. 3, the relation corresponding to equation (13) between β_i and φ_i assumes the form

$$\tan(\varphi_i) = \frac{k_{it}}{k_{in}} \frac{1}{\tan(\beta_i)} \tag{20}$$

On can also prove that

$$\tan(\beta_i) = \frac{k_{it}}{k_{1t}} \tan(\beta_1) \tag{21}$$

Also, considering the relations

$$1 + \tan^2(\beta_i) = \frac{1}{\cos^2(\beta_i)} \tag{22}$$

And

$$\tan(\alpha) = \frac{\sum_{i=1}^{n} M_i}{\sum_{i=1}^{n} a_i} = \frac{\sum_{i=1}^{n} M_i}{\sum_{i=1}^{n} M_i \tan(\beta_i)} = \frac{k_{1t} \sum_{i=1}^{n} M_i}{\tan(\beta_1) \sum_{i=1}^{n} M_i\, k_{it}} \tag{23}$$

One can calculate the sum of hydraulic losses along all the flow path through N layers. It comes to

$$\Delta H = \sum_{i=1}^{N} \Delta H_i = \frac{q}{b \sin^2(\alpha)} \sum_{i=1}^{N} \frac{M_i}{k_{in}} \sqrt{\sin^2(\alpha) + \frac{k_{it}^2 \left(\sum_{i=1}^{N} M_i \right)^2}{\left(\sum_{i=1}^{N} M_i k_{it} \right)^2} \cos^2(\alpha)} \sqrt{\sin^2(\alpha) + \frac{k_{in}^2 \left(\sum_{i=1}^{N} M_i \right)^2}{\left(\sum_{i=1}^{N} M_i k_{it} \right)^2} \cos^2(\alpha)} \tag{24}$$

In order to determine the hydraulic conductivity of all the layered structure, let us replace the system of layers between the streamlines spaced at b intervals with one homogeneous layer with the same width b and length l, in accordance with Fig. 3. If we assume that the hydraulic conductivity of this layer in the flow direction determined by angle α is k_e, then the head loss will be

$$\Delta H = \frac{q \sum_{i=1}^{n} M_i}{k_e\, b \sin^2(\alpha)} \tag{25}$$

Comparing the size of hydraulic losses in both cases, one obtains the sought formula for the value of equivalent hydraulic conductivity k_e for a flow through a layered medium at any angle α.

$$\frac{1}{k_e}=\frac{\sum_{i=1}^{N}\frac{M_i}{k_{in}}\sqrt{\sin^2(\alpha)+\frac{k_{it}^2\left(\sum_{i=1}^{N}M_i\right)^2}{\left(\sum_{i=1}^{N}M_ik_{it}\right)^2}\cos^2(\alpha)}\sqrt{\sin^2(\alpha)+\frac{k_{in}^2\left(\sum_{i=1}^{N}M_i\right)^2}{\left(\sum_{i=1}^{N}M_ik_{it}\right)^2}\cos^2(\alpha)}}{\sum_{i=1}^{N}M_i} \tag{26}$$

For a flow parallel to layering (α = 0° or α = 180°), hydraulic conductivity k_e will be designated as $k_{\parallel}$ and, according to the above formula it is:

$$k_{\parallel}=\frac{\sum_{i=1}^{n}k_{it}M_i}{\sum_{i=1}^{n}M_i}, \tag{27}$$

while in perpendicular direction (α = 90° or α = 270°), hydraulic conductivity k_e will be designated as $k_{\perp}$ and in this case, according to formula (26), it is

$$k_{\perp}=\frac{\sum_{i=1}^{n}M_i}{\sum_{i=1}^{n}\frac{M_i}{k_{in}}} \tag{28}$$

It is worth emphasizing that equations (27) and (28) can be easily obtained by directly analysing a flow in two directions: parallel and perpendicular to layering.
After substituting relation (27) in equation (26), we obtain

$$\frac{1}{k_e}=\frac{\sum_{i=1}^{N}\frac{M_i}{k_{in}}\sqrt{\sin^2(\alpha)+\frac{k_{it}^2}{k_{\parallel}^2}\cos^2(\alpha)}\sqrt{\sin^2(\alpha)+\frac{k_{in}^2}{k_{\parallel}^2}\cos^2(\alpha)}}{\sum_{i=1}^{N}M_i} \tag{29}$$

It follows from equation (29) that in order to determine the equivalent hydraulic conductivity of a medium composed of anisotropic layers for diagonal direction α, it is essential to know all the values of hydraulic conductivity k_{it} and k_{in} of particular layers and their thickness M_i. Therefore, from the theoretical point of view, it is not possible to model a flow through such a structure like that through a single layer.
On the other hand, in the case when particular layers in this structure have isotropic properties, i.e. one can adopt $k_{it}=k_{in}=k_i$ for all the layers, relation (29) can be expressed in the following form (Bear, 1972):

$$\frac{1}{k_e}=\frac{\sum_{i=1}^{n}M_i}{\sum_{i=1}^{n}M_ik_i}\cos^2(\alpha)+\frac{\sum_{i=1}^{n}\frac{M_i}{k_i}}{\sum_{i=1}^{n}M_i}\sin^2(\alpha)=\frac{1}{k_{\parallel}}\cos^2(\alpha)+\frac{1}{k_{\perp}}\sin^2(\alpha) \tag{30}$$

Then, in order to calculate the flow in any direction α, it is enough to know the value of hydraulic conductivity for all the layered structure in the direction parallel and perpendicular to the surface constituting layer boundary.
As expected, the form of relation (30) is identical to the function representing hydraulic conductivity in orthotropic soils, where two-dimensional anisotropy occurs. (Batu, 1998; Cheng, 2000; Snow, 1969; Wieczysty, 1982).

$$\frac{1}{k_e} = \frac{1}{k_{xx}}\cos^2(\alpha) + \frac{1}{k_{yy}}\sin^2(\alpha) \tag{31}$$

In equation (31), k_{xx} and k_{yy} correspond to the values of hydraulic conductivity in the principal directions of hydraulic conductivity tensor and $k_{\parallel} = k_{xx}$, while $k_{\perp} = k_{yy}$.
If based on equation (30) or (31) one determines the values of $\sqrt{k_e}$ for various α values, then plotting them in a circle diagram (a graph in polar coordinates) will produce an ellipse. It is referred to as hydraulic conductivity ellipse.

2.4 Angle between hydraulic gradient and filtration velocity

Angle γ between hydraulic gradient (grad$H$) and filtration velocity $\mathbf{v}$ can be determined from the ratio of the dot products of vectors grad$H$ and $\mathbf{v}$ to the product of their lengths. In the case of a 2D flow, and adopting a Cartesian coordinate system whose axes are directed along the principal axes of hydraulic conductivity tensor, angle γ is (Fig. 4)

$$\gamma = \arccos\left(\frac{\mathrm{grad}H \bullet \mathbf{v}}{|\mathrm{grad}H||\mathbf{v}|}\right) = \arccos\left[\frac{\frac{\partial H}{\partial x}k_{xx}\frac{\partial H}{\partial x} + \frac{\partial H}{\partial y}k_{yy}\frac{\partial H}{\partial y}}{\sqrt{\left(\frac{\partial H}{\partial x}\right)^2 + \left(\frac{\partial H}{\partial y}\right)^2}\sqrt{k_{xx}^2\left(\frac{\partial H}{\partial x}\right)^2 + k_{yy}^2\left(\frac{\partial H}{\partial y}\right)^2}}\right] \tag{32}$$

where k_{xx} and k_{yy} denote the values of hydraulic conductivity tensor k:

$$k = \begin{pmatrix} k_{xx} & 0 \\ 0 & k_{yy} \end{pmatrix} \tag{33}$$

Remembering that

$$\tan(\alpha) = \frac{k_{yy}\frac{\partial H}{\partial y}}{k_{xx}\frac{\partial H}{\partial x}} \tag{34}$$

one obtains

$$\gamma = \arccos\left(\frac{k_{yy}\cos^2(\alpha) + k_{xx}\sin^2(\alpha)}{\sqrt{k_{yy}^2\cos^2(\alpha) + k_{xx}^2\sin^2(\alpha)}}\right) = \arccos\left(\frac{\cos^2(\alpha) + \lambda\sin^2(\alpha)}{\sqrt{\cos^2(\alpha) + \lambda^2\sin^2(\alpha)}}\right). \tag{35}$$

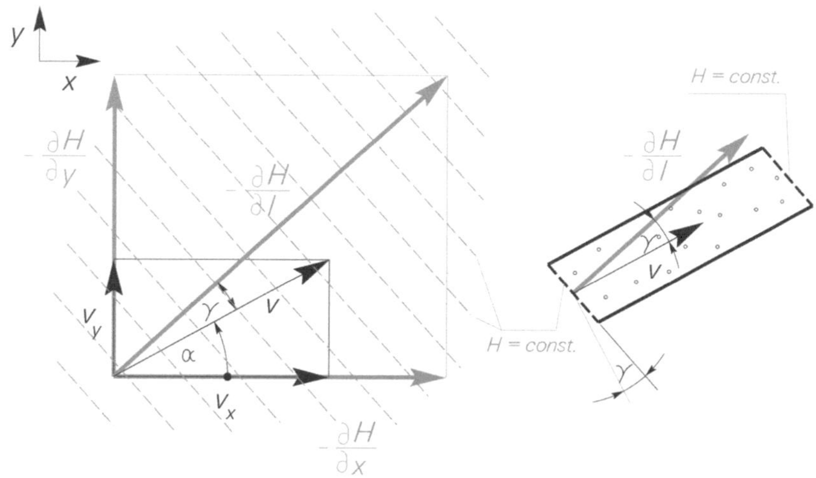

Fig. 4. Direction differences between hydraulic gradient and filtration velocity.

where α denotes the angle between velocity vector and the principal axis of hydraulic conductivity tensor, along which value k reaches the maximum values.
From the above relation, one can determine the value of angle α, for which angle γ has the maximum value. In order to do this, one must determine the derivative $\frac{d\gamma}{d\alpha}$ and compare it to zero. The angle determined in this way is:

$$\alpha = \pm \arctan \sqrt{\frac{k_{yy}}{k_{xx}}} = \pm \arctan \sqrt{\frac{1}{\lambda}} \tag{36}$$

Assuming that $k_{xx} > k_{yy}$, $\lambda = \frac{k_{xx}}{k_{yy}}$ denotes anisotropy factor. For layered media composed of homogeneous isotropic and parallel layers, value k_{xx} corresponds to $k_{\parallel}$, and value k_{yy} – to $k_{\perp}$.

3. Problems with determining the hydraulic conductivity of anisotropic media.

Defining water permeability of anisotropic media requires determining values of hydraulic conductivity in the direction of principal axes of hydraulic conductivity (Renard, 2001; Mozely et al., 1996). For media characterized by two-dimensional orthotropy, including layered media, these will be two values - one in parallel direction, and the other - perpendicular to layering. In accordance with equation (30), determining hydraulic conductivity in these directions is possible based on tests of hydraulic conductivity k_e for any two known angles α. Then one obtains two equations with two unknowns $k_{\parallel}$ and $k_{\perp}$. When determining these values, one can later calculate equivalent hydraulic conductivity k_e in any direction α. However, if there are any difficulties defining angles α, for which water permeability measurements are performed, $k_{\parallel}$ and $k_{\perp}$ can be determined by performing a larger number of measurements for various angles (at least three different angles), and then matching the obtained results with the ellipse equation (Cheng, 2000).

This is the theory. However, determining hydraulic conductivity of layered media in permeameters involves additional problems. Analysing relation (35), one can observe that the angle between grad$H$ and **v** is equal to zero, i.e. both directions coincide only when angle α is equal to 0, 90, 180 and 270 degrees, then if the flow occurs in the direction parallel or perpendicular to layering. When performing measurements along these directions, one does not make additional errors resulting from anisotropy. However, if such a measurement is not possible, (e.g. samples are obtained from hole coring and the hole axis is diagonal to layering), determining the value of hydraulic conductivity in the direction diagonal to layering entails an error. It results from the fact that during measurements inside permeameters the appropriate geometric shapes of the sample, the proper size and the difference between filtration velocity **v** and grad$H$ are not maintained.

Geometric dimensions should allow for maintaining the angle between filtration direction and the plane through which water flows to the sample, and the lateral surfaces of the sample. The inflow and outflow surfaces should be perpendicular to the gradient of hydraulic head, and the lateral surfaces - parallel to streamlines composed of straight line segments (Fig. 3).

Meeting the above conditions is very difficult technically, moreover it would require prior knowledge of filtration parameters of particular layers of the medium. The knowledge of these parameters would enable theoretical calculation of hydraulic conductivity without a need to analyse all the layer structure. If we assume however that we do not know these parameters, maintaining appropriate shapes of the sample is not possible. In such a case, as an approximation of the abovementioned theoretical solution, reflecting a flow diagonal to layering, one can adopt a flow through a medium sample whose geometric dimensions and flow conditions are consistent with the idea presented in Fig. 5. One should remember here that the cross-section of the sample in the direction perpendicular to filtration direction should be a rectangle.

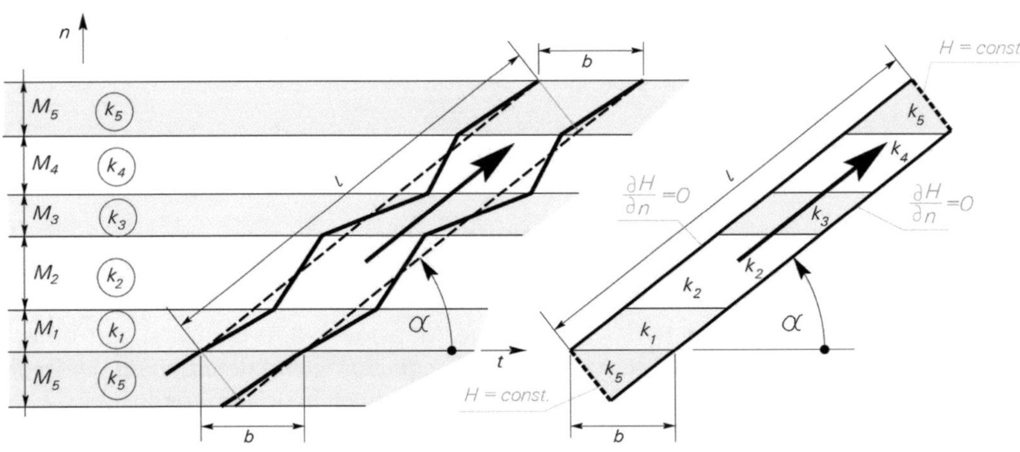

Fig. 5. Diagram of the pattern of adopting a geometric shape for a layered medium sample for permeameter analysis.

Unfortunately, when performing permeameter tests on samples with the shape presented in Fig. 5, one cannot avoid errors in determination of hydraulic conductivity.

4. Model calculations

In order to estimate error values connected with determining hydraulic conductivity of layered media tested in a permeameter, water flow through an imaginary soil medium composed of layers with known properties was calculated. The calculations consisted of a numerical solution of a flow through a specific arrangement of layers in a permeameter and then defining equivalent hydraulic conductivity. The hydraulic conductivity defined in this way was then compared to theoretically calculated hydraulic conductivity of a layered structure corresponding to the layer pattern in the permeameter.

4.1 Characteristics of the medium

Model calculations were performed on a soil sample composed of five homogeneous and isotropic layers with constant thickness (Fig. 3). The thickness of particular layers M_i, and hydraulic conductivity k_i are presented in Tab. 1.

Layer No.	M_i	k_i	$k_{\parallel}$	$k_{\perp}$	λ
	cm	$m \cdot s^{-1}$	$m \cdot s^{-1}$	$m \cdot s^{-1}$	-
1	21	$7{,}0 \cdot 10^{-4}$	$2{,}06 \cdot 10^{-4}$	$1{,}98 \cdot 10^{-5}$	10,37
2	20	$1{,}0 \cdot 10^{-5}$			
3	19	$6{,}0 \cdot 10^{-6}$			
4	20	$8{,}0 \cdot 10^{-5}$			
5	31	$2{,}0 \cdot 10^{-4}$			

Table 1. Basic properties of a layered medium.

For such a layer pattern, formula (30) was used to calculate theoretical values of hydraulic conductivity $k_{e(t)}$ and the angle between vectors grad$H$ and **v** for different values of angle α between filtration velocity and the direction of principal axis t. Based on the obtained results, hydraulic conductivity ellipse ($\sqrt{k_e}$), and the values of angles between grad$H$ and filtration velocity **v** (angles γ) depending on filtration direction, were drawn (Figs. 6 and 7). Table 1 also contains calculated hydraulic conductivity values for all the layered structure in the direction parallel and perpendicular to layer boundaries, as well as the anisotropy factor.

4.2 Numerical solution of permeameter flow

The problem of permeameter flow was solved based on Laplace's equation, assuming the steady flow of an incompressible fluid through an undeformed medium. On the surfaces forming the filtration area boundary, boundary conditions were adopted according to Fig. 5., while on the boundary of layers with different water permeability, $H_i = H_{i+1}$ and $v_{ni} = v_{ni+1}$ were adopted, with H_i, H_{i+1} denoting the hydraulic heads, and v_{ni} and v_{ni+1} - filtration velocity components normal to the boundary of ith and $i+1^{th}$ layers. All the filtration area was divided into 110x14 blocks. The calculations were performed with finite difference method in Excel spreadsheet, using Gauss-Seidel iteration. The calculation results for different angles α are shown in Tab. 2.

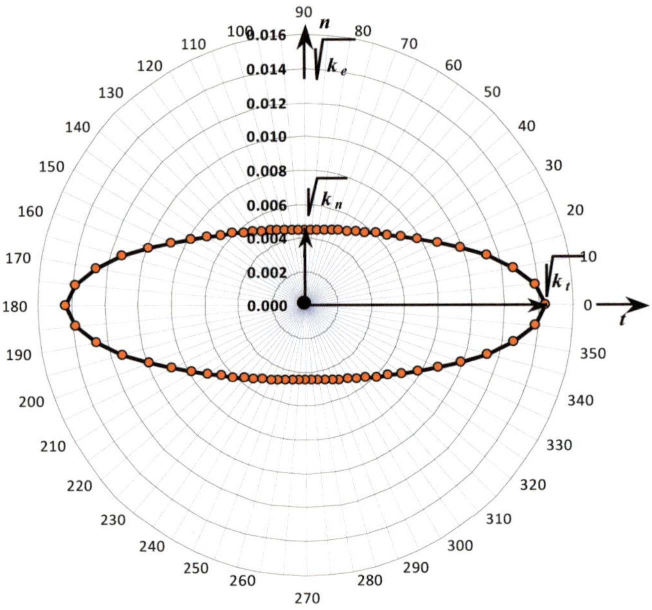

Fig. 6. Hydraulic conductivity ellipse.

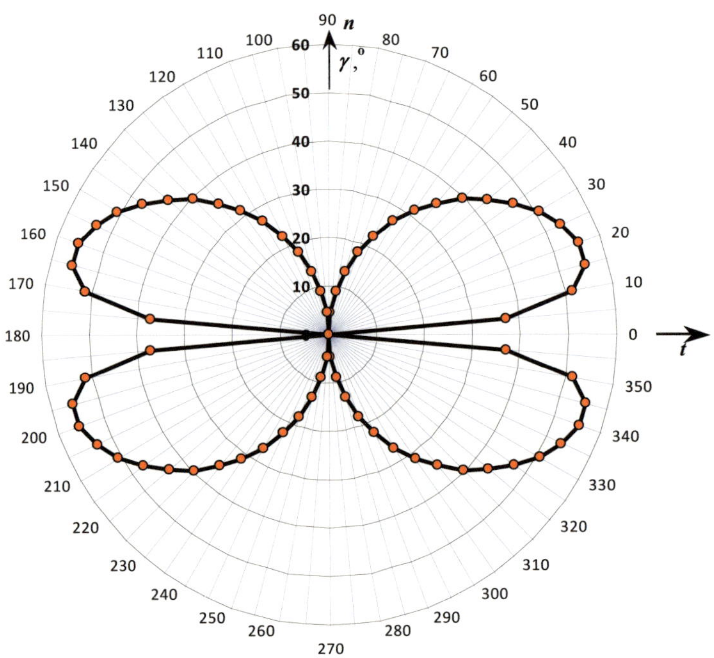

Fig. 7. Angle between hydraulic gradient and filtration velocity depending on filtration direction.

No.	α	γ	$k_{e(t)}$	$k_{e(num)}$	$\frac{\Delta k}{k_{e(t)}}$	$\frac{\Delta Q}{Q_{sr}}$
	°	°	m·s-1	m·s-1	%	%
1	0	0,00	$2{,}06\cdot10^{-4}$	$2{,}06\cdot10^{-4}$	0,00	0,00
2	45	39,49	$3{,}61\cdot10^{-5}$	$2{,}31\cdot10^{-5}$	36,15	-0,03
3	50	35,37	$3{,}16\cdot10^{-5}$	$2{,}15\cdot10^{-5}$	32,01	3,82
4	55	31,14	$2{,}82\cdot10^{-5}$	$2{,}13\cdot10^{-5}$	24,39	3,04
5	60	26,81	$2{,}56\cdot10^{-5}$	$2{,}11\cdot10^{-5}$	17,40	2,43
6	65	22,43	$2{,}36\cdot10^{-5}$	$2{,}09\cdot10^{-5}$	11,45	1,68
7	70	17,99	$2{,}22\cdot10^{-5}$	$2{,}07\cdot10^{-5}$	6,64	1,09
8	75	13,52	$2{,}11\cdot10^{-5}$	$2{,}04\cdot10^{-5}$	3,18	0,61
9	80	9,03	$2{,}04\cdot10^{-5}$	$2{,}01\cdot10^{-5}$	1,08	0,27
10	85	4,52	$2{,}00\cdot10^{-5}$	$1{,}99\cdot10^{-5}$	0,17	0,06
11	90	0,00	$1{,}98\cdot10^{-5}$	$1{,}98\cdot10^{-5}$	0,09	0,00

Table 2. Results of numerical calculations of angle γ between hydraulic gradient and filtration velocity and equivalent hydraulic conductivity $k_{e(num)}$.

It contains theoretically and numerically defined values of equivalent hydraulic conductivity ($k_{e(t)}$ and $k_{e(num)}$), theoretical values of angles γ between grad$H$ and **v**, error in numerical determination of conductivity in relation to the theoretical value $\frac{\Delta k}{k_{e(t)}} = \frac{k_{e(t)} - k_{e(num)}}{k_{e(t)}} 100\%$, as well as error values $\frac{\Delta Q}{Q_{sr}} = \frac{Q_w - Q_d}{0{,}5(Q_w + Q_d)} \cdot 100\%$ resulting from a numerical comparison of the quantity of water flowing into (Q_d) and out of (Q_w) the analysed sample. This error was regarded as representative for estimating the accuracy of numerical calculation results.

4.3 Error of hydraulic conductivity determination

In order to estimate the error in determining hydraulic conductivity of a layered medium, it was assumed that the analyses had been performed in a permeameter. The tests were carried out on a layered medium presented in Fig. 5 for two flow directions in relation to the principal axis of permeability tensor. One direction corresponded to angle α of 45°, while the other - 80°. The values of equivalent hydraulic conductivity determined correctly through research should be consistent with Tab. 2 and amount to $2{,}31\cdot10^{-5}$ m·s^{-1} and $2{,}01\cdot10^{-5}$ m·s^{-1} respectively. Using them later to calculate the hydraulic conductivity in the direction of the principal axis of permeability tensor from formula (30), one obtains $k_{\parallel} = 2{,}75\cdot10^{-5}$ m·s^{-1} and $k_{\perp} = 1{,}99\cdot10^{-5}$ m·s^{-1}. The values calculated in this way were compared with the theoretical properties of a layered medium presented in Tab.1. On can see from the comparison that the values of $k_{\perp}$ are only slightly different (the theoretical value is 1,98 10^{-5} m s^{-1}), while in the case of $k_{\parallel}$, the difference is very distinct, reaching 7,5 times. One should also emphasize that such a big difference between these results is not caused by the low accuracy of numerical flow calculations. This is proved by a very small difference in the discharge of water flowing into and out of the sample, no more than 0,3% (Tab. 2).

The error in hydraulic conductivity determination in a permeameter may be also caused by the order of layering. In order to estimate its value, numerical calculations were performed for the adopted flow through a layered medium composed of the same layers but arranged in a different order. One should emphasize that according to formula (30), the theoretical value of equivalent hydraulic conductivity k_t does not depend on the layering order. The calculation results for a flow at the angle $\alpha = 45^{\circ}$ and for four layering variants are shown in Tab. 3.

Variant No	Layering order	$k_{e(t)}$	$k_{e(num)}$	$\frac{\Delta k}{k_{e(t)}}$	$\frac{\Delta Q}{Q_{sr}}$
		$m \cdot s^{-1}$	$m \cdot s^{-1}$	%	%
1	5-1-2-3-4-5	$3{,}61 \cdot 10^{-5}$	$2{,}31 \cdot 10^{-5}$	36,15	-0,027
2	5-3-2-4-1-5		$2{,}16 \cdot 10^{-5}$	40,14	-0,058
3	5-1-4-2-3-5		$2{,}27 \cdot 10^{-5}$	37,07	0,033
4	5-2-4-3-1-5		$2{,}55 \cdot 10^{-5}$	29,54	0,012

Tab 3. Calculation results of hydraulic conductivity of a layered medium for various layering orders.

One can see that depending on the considered variant, different results are obtained, and anisotropy-related determination error is contained within wide limits. In the analysed examples it oscillates between 30 and over 40% in relation to the theoretical value. In order to illustrate the causes of differences in hydraulic conductivity, calculations of streamline pattern were also performed. For the layering in variants 2 and 4, streamline patterns are shown in Figs. 6 and 7.

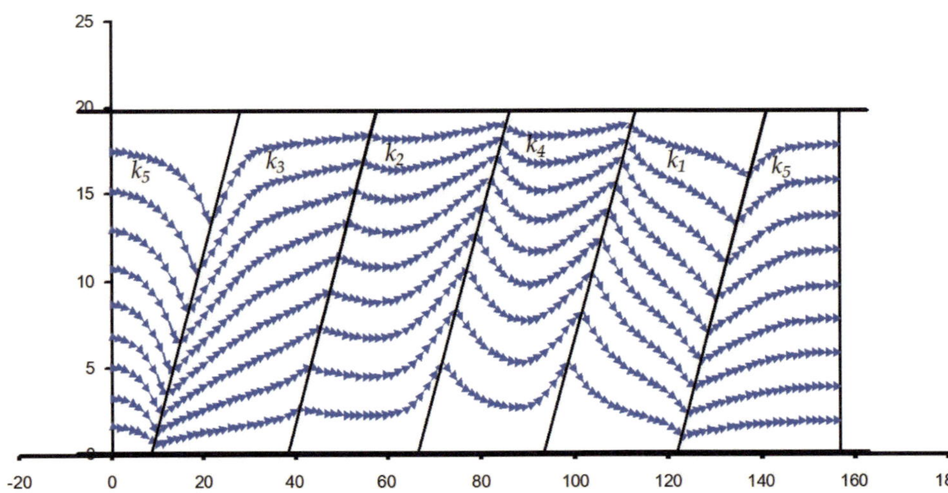

Fig. 6. Streamlines in a sample for layering variant 2.

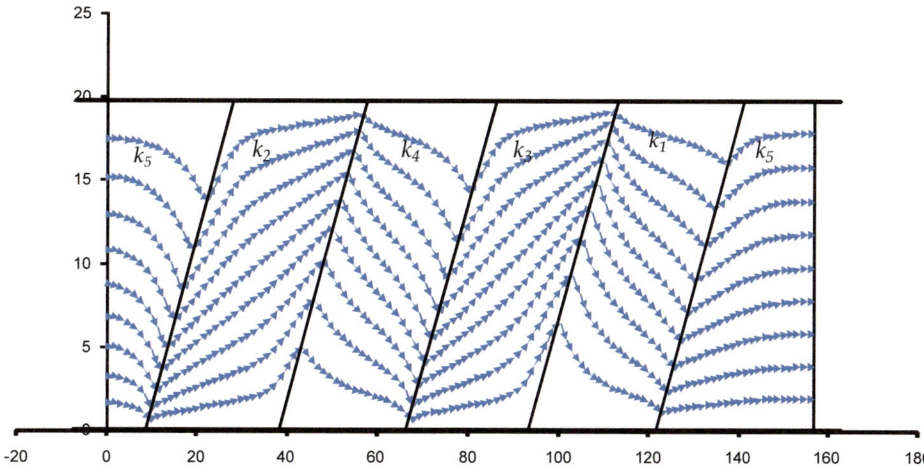

Fig. 7. Streamlines in a sample for layering variant 4.

A distinct difference in streamline patterns is noticeable. Certainly the streamline pattern is basically different from the theoretical streamlines shown in Fig. 3, which are polygonal chains composed of straight line segments.

5. Conclusion

The presented theoretical discussion has demonstrated that flows in layered media composed of homogeneous and anisotropic layers cannot be modelled as flows in a homogeneous medium. However, it is possible in the case of a structure composed of parallel layers characterised by homogeneity and orthotropy in two dimensions. A flow in such a structure can be modelled as a flow in a single anisotropic layer. In such a case one does not have to know the permeability parameters and the thickness of particular layers. In order to calculate the flow, it is then enough to know the hydraulic conductivity in the direction parallel and perpendicular to layering. These values can be determined theoretically by analysing a flow through such a layered structure in any two directions to the boundary surfaces of the layers. However, the calculation results point to a possibility of vital errors occurring while determining the permeability of layered media in a permeameter. This regards those tests where flow occurs in the direction diagonal to layering. The occurrence of major determination errors is connected with a failure to maintain the appropriate directions of grad$H$ and $\mathbf{v}$ and the proper geometric shape of the sample in a permeameter. The value of those errors depends on the value of anisotropy factor λ and the angular difference between grad$H$ and $\mathbf{v}$. In the performed model calculations of a flow through a layered medium with anisotropy factor $\lambda = 10{,}37$, the obtained value of hydraulic conductivity in the direction perpendicular to layer boundary was very close to the theoretical one, while in the parallel direction it was 7,5 times underrated. Clearly then, such analyses result in very serious underrating of anisotropy factor. The conducted calculations also reveal a considerable influence of layering order on the determination results of hydraulic conductivity in a permeameter. The differences obtained from model calculations oscillated between c. 30 to 40 % in relation to the theoretical value.

To sum up, one should clearly state that the correct determination results of hydraulic conductivity of layered media can be obtained by forcing a flow in the direction parallel and perpendicular to layering. Any departures from this rule may lead to very large determination errors.

6. References

Batu, V. (1998). *Aquifer hydraulics: A comprehensive quide to hydrogeologic data analysis,* John Wiley & Sons, ISBN 0-471-18502-7, New York

Bear, J. (1972). *Dynamics of Fluids in Porous Media.* Dover Publications. ISBN 0-486-65675-6.

Cheng, A. (2000). *Multilayered aquifer systems: Fundamentals and applications,* Marcel Dekker Inc., ISBN 0-8247-9875-9, New York, Basel

Mozely, P.S. & Davis, J.M. (1996). Relationship between oriented calcite concretions and permeability correlation structure in alluvial aquifer, Sierra Ladrones Formation, New Mexiko. *Journal of Sedimentary Research,* Vol. 66, No. 1 (January 1996), pp. (11-16), ISSN: 1527-1404

Renard Ph.; Genty A.; Stauffer F. (2001). Laboratory determination of full permeability tensor, *Journal of Geophysical Research,* Vol. 106, No. B11, (November 2001) pp. 26,443-26,452, ISSN 0148-0227

Rogoż M. (2007). *Dynamika wód podziemnych.* GIG, ISBN 978-83-87610-97-5, Katowice

Snow D. T. (1969). Anisotropic permeability of fractured media, *Water Resour., Res.,* 5(6), 1273-1289, ISSN 0043-1397

Part 3

Laboratory Hydraulic Conductivity Assessment

7

Unsaturated Hydraulic Conductivity for Evaporation in Heterogeneous Soils

Dongmin Sun[1] and Jianting Zhu[2]
[1]*University of Houston-Clear Lake, Environmental Science Program,*
[2]*Desert Research Institute, Nevada System of Higher Education,*
U.S.A.

1. Introduction

Simulations of vadose zone moisture flow and contaminant transport typically use closed-form soil hydraulic properties (i.e., unsaturated hydraulic conductivity and soil water retention characteristics). Understanding large-scale vadose zone hydrological processes requires a solid approach to characterizing the large degree of heterogeneity of hydraulic properties in the field (e.g., Dagan & Bresler, 1983; Bresler & Dagan, 1983; Vereecken et al., 2007). As a result, the impact of soil heterogeneity on vadose zone flow and transport has been the focus of considerable research in recent decades (e.g., Hopmans & Stricker, 1989; Butters & Jury, 1989; Ellsworth & Jury, 1991; Destouni, 1992; Russo, 1993, 1998; Mallants et al., 1996; Hendrayanto et al., 2000; Avanidou & Paleologos, 2002; Hristopulos, 2003; Jhorar et al., 2004; Das & Hassanizadeh, 2005; Kozak & Ahuja. 2005; Kozak et al., 2005; Neuweiler & Cirpka 2005; Ward et al., 2006; Lu et al., 2007; Coppola, et al., 2009). Local scale soil hydraulic properties have been studied extensively (e.g., Gardner, 1958; Brooks & Corey, 1964; Laliberte, 1969; Farrell & Larson, 1972; Campbell, 1974; Mualem, 1976; Clapp & Hornberger, 1978; van Genuchten, 1980; Libardi et al., 1980; van Genuchten & Nielson, 1985; Hutson & Cass, 1987; Russo, 1988; Bumb et al., 1992; Setiawan & Nakano, 1993; Rossi & Nimmo, 1994; Kosugi, 1994; Zhang & van Genuchten, 1994; Leij et al., 1997). However, connecting heterogeneous properties and processes at different scales remains a major scientific challenge in hydrology (Dagan, 1989; Gelhar, 1993; Renard & de Marsily, 1996; Sposito, 1998; Grayson & Bloschl, 2000; Kasteel et al., 2000; Cushman et al., 2002; Farmer, 2002; Zhang, 2002; Williams & Ahuja, 2003; Pachepsky et al., 2003; Zhang et al., 2004; Vereecken et al., 2007). One way to connect soil hydrologic processes at different scales is to employ hydraulic property upscaling. The upscaling algorithms seek to aggregate a mesh of hydraulic properties defined at the small (support) scale into a coarser mesh with "effective" hydraulic properties that can be used in large-scale (e.g., landscape-scale, watershed-scale, basin-scale) hydro-climate models. The main goal of using effective hydraulic properties is to capture particular flow and transport processes in a heterogeneous soil, through conceptualization of heterogeneous formation as an equivalent homogeneous formation. In this way, the heterogeneous system is replaced by an equivalent homogeneous medium (e.g., Rubin, 2003; Zhu & Mohanty, 2003a,b; Zhu & Mohanty, 2004; Zhu et al., 2007; Zhu, 2008; Zhu & Sun, 2009; Zhu & Sun, 2010). Hydraulic parameters that define the equivalent homogeneous medium are known as effective parameters.

However, upscaling studies have revealed significant challenges and problems in representing soil hydrologic processes and parameters at different scales (e.g., Bresler & Dagan, 1983; Milly & Eagleson, 1987; Kim & Stricker, 1996; Smith & Diekkruger, 1996; Kim et al., 1997; Harter & Hopmans, 2004). For example, upscaled effective hydraulic properties derived from stochastic analysis, which account for local-scale hydraulic property heterogeneities perform well for deep and unbounded unsaturated zones where gravity-dominated flow is the main process and where the mean hydraulic gradient is approximately constant (e.g., Zhang et al., 1998). The gravitational flow regime enables the use of relatively simple approaches in stochastic analysis of subsurface flow (e.g., Tartakovsky et al., 1999; Harter & Zhang, 1999; Russo, 2003; Severino et al., 2003; Severino & Santini, 2005; Russo, 2005; Russo & Fiori, 2009). In a recent study, Zhang (2010) investigated the effective hydraulic conductivity of unsaturated media through numerical experiments of gravity-induced flow with multidimensional heterogeneity. Under this flow scenario, the use of average unit hydraulic gradient assumption implies that pressure head is constant throughout the profile. It should be pointed out that the unit-gradient assumption is applicable for a limited range of infiltration conditions where the pressure gradient is close to zero; and thus this assumption may be of limited applicability for near-surface processes, such as vadose zone and atmosphere interactions. On the other hand, many studies on hydraulic property upscaling that focus on near-ground surface interactions mostly deal with steady state flux exchanges (e.g., Zhu & Mohanty, 2002a,b; 2003a; Zhu et al., 2006) and results indicate that upscaling behaviors are distinctly different for infiltration and evaporation and an effective hydraulic property is usually more difficult to define for evaporation. In a study of effective hydraulic parameters for transient hydrological processes, Zhu and Mohanty (2006) combined the one-dimensional local process and the Miller-Miller (Miller & Miller, 1956) media concept, thereby illustrating that effective hydraulic parameters depend on the time frame being considered. Zhu & Sun (2009) investigated the use of effective soil hydraulic properties (expressed in terms of hydraulic parameters) applicable to near surface large-scale transient infiltration problems in a landscape with horizontally heterogeneous soil hydraulic properties. These studies show that methods to aggregate and upscale local hydraulic parameters are critical to improve the understanding of near surface large-scale hydrologic processes.

Two widely used upscaling approaches in vadose zone flows include homogenization theory and Monte Carlo type of simulations based on the stream tube approximation. In the homogenization theory, an upscaled flow equation is developed based on a separation of length scales in the medium for the limit at which the typical length scale of heterogeneities became negligible compared to the size of the medium (e.g., Sviercoski et al., 2009; Neuweiler & Eichel, 2006; Neuweiler & Cirpka, 2005; Lewandowska & Laurent, 2001). The second approach is the stream tube approach in which the heterogeneous field is conceptualized as a series of vertically homogeneous and horizontally independent stream tubes or parallel columns (Dagan & Bresler, 1983; Bresler & Dagan, 1983; Govindaraju et al., 1992; Rubin & Or, 1993; Chen et al., 1994a,b; Toride & Leij, 1996a, b; Kim et al., 1997; Wildenschild & Jensen, 1999; Zhu & Mohanty, 2002b). Both approaches have challenges and limitations in dealing with heterogeneity and upscaling of hydraulic properties.

First, the homogenization theory uses an approach based on a separation of length scales in the medium and aims at deriving an upscaled flow equation in a heterogeneous medium for the limit at which the typical length scale of heterogeneities became negligible compared with the size of the medium (e.g., Hornung, 1997). In other words, a clear separation of scales is required to derive upscaled model based on the homogenization theory. The flow system is sorted for the different orders and solved for each order separately, which requires that system dimensionless numbers, as well as parameter ratios, be of a fixed order of this length scale ratio. Different orders of dimensionless numbers and different parameter ratios could lead to different upscaled flow models (e.g., Van Duijn et al., 2002 and Lewandowska et al., 2004, respectively). The homogenization approach is mostly applicable to a certain flow and parameter regime. Previous studies using homogenization theory usually considered two flow regimes. The first flow regime was quantified by small Bond number, meaning forces due to pressure gradients are dominant at the small scale. The second flow regime is quantified by large Bond number, meaning that forces due to pressure gradients and gravity contribute equally at the small scale. Most notably, Bond number is assumed to be a fixed order in relation to the ratio between the small and the large scale. In addition, due to its requirement of separation of scales, the homogenization theory is difficult to use for formation without clear scale separation such as in the situation when there are only a finite number of layers in the soil formations.

Second, the stream tube approach is often used in upscaling vadose zone hydrological processes in which many studies conceptualize the heterogeneous field as a series of vertically homogeneous and horizontally independent stream tubes or parallel columns. In the study of Severino et al. (2003), the effective hydraulic conductivity was obtained by an ensemble average over all the stream tubes of a local analytical solution of Richards equation that regards the hydraulic parameters as horizontally correlated random space functions. Leij et al. (2007) simulated and aggregated unsaturated zone flows using the stream tube approach in which the heterogeneous field is conceptualized as a series of vertically homogeneous and horizontally independent stream tubes or parallel columns. Leij et al. (2007) focused on aggregating a posteriori unsaturated flow processes and illustrated that a priori aggregation (effective hydraulic properties) would overestimate the large-scale average infiltration by more than 40%, if the effective water retention curve was obtained from the aggregated suction head and the water content and the arithmetic mean of the saturated hydraulic conductivity. More recently, Coppola et al. (2009) studied the effects of using unimodal and bimodal interpretative models of hydraulic properties on the ensemble hydrological behavior of stream tubes by comparing predictions to mean water contents measured over time at several field scale sites. Zhu & Sun (2009) examined how the effective hydraulic parameters are sensitive to the time frame of hydrologic processes, by using the stream tube concept to study the effective hydraulic parameters for transient infiltration. Ahuja et al. (2010) who also used the steam tube approach, explored effective parameter sets to describe field-average infiltration and redistribution under different rainfall conditions and investigated whether an effective field saturated hydraulic conductivity and correlated hydraulic parameters derived from matching early-stage average ponded infiltration could give reasonable results for infiltration under lower rainfall rates as well as for soil water redistribution. These results showed that there were no unique effective average properties that gave the best results for both infiltration and redistribution, even for the same initial pressure-head condition. It should be emphasized that when the stream tube approach is used, no interactions among these tubes are considered. The stream tube approach is most appropriate to model flows where the effective cross-sectional diameter of a "tube" or "column" is larger than its length (e.g., Protopapas and Bras, 1991; Leij et al., 2006; Leij et al., 2007).

In this chapter, we seek to provide some practical guidelines of how the commonly used simple averaging schemes (arithmetic, geometric, or harmonic mean) perform in simulating large scale evaporation in a heterogeneous landscape. As discussed earlier, previous studies on hydraulic property upscaling focusing on steady state flux exchanges illustrated that an effective hydraulic property is usually more difficult to define for evaporation. This chapter mainly focuses on upscaling hydraulic properties of large scale transient evaporation dynamics based on the stream tube approach. Specifically, we examine large scale hydraulic parameters in two practical aspects: (1) if the three simple averaging schemes (i.e., arithmetic, geometric and harmonic means) of hydraulic parameters are appropriate in representing large scale evaporation processes, and (2) how the applicability of these simple averaging schemes depends on the time scale of evaporation processes in heterogeneous soils. Multiple realizations of local evaporation processes are carried out using HYDRUS-1D computational code (Simunek et al, 1998). The three averaging schemes of soil hydraulic parameters are used to simulate the cumulative flux exchange, which is then compared to the large scale average cumulative evaporation. The relative error between the cumulative evaporation based on simple averaging schemes and the average cumulative evaporation is used to judge the applicability of the simple averaging schemes in predicting the large scale evaporation from the heterogeneous soils. The sensitivity of the relative errors to the time frame of evaporation processes is also discussed.

2. Methods

2.1 Hydraulic properties and hydraulic parameters

The hydraulic properties are characterized by the soil water retention curve which defines the water content (θ) as a function of the suction head (h), and the hydraulic conductivity function which establishes the relationship between the hydraulic conductivity (K) and the water content or the suction head. Some of the more commonly used models describing these functional relationships include: the Gardner-Russo model (Gardner, 1958; Russo, 1988), the Brooks-Corey model (Brooks & Corey, 1964), and the van Genuchten (1980) model.

The unsaturated hydraulic conductivity (K)-suction head (h) and the suction head (h)-water content (θ) are represented by the Gardner-Russo model (Gardner, 1958; Russo, 1988),

$$Se(h) = [e^{-0.5\alpha h}(1 + 0.5\alpha h)]^{-2/(l+2)} \quad (1)$$

$$K(h) = K_S e^{-\alpha h} \quad (2)$$

where $Se = (\theta-\theta_r)/(\theta_s-\theta_r)$ is the effective degree of saturation, θ is the volumetric water content, θ_r is the residual volumetric water content, θ_s is the saturated volumetric water content, h is the suction head (positive for unsaturated soils), K is the hydraulic conductivity, K_S is the saturated hydraulic conductivity, α is related to pore-size distributions, *l* is a parameter which accounts for the dependence of the tortuosity, and the correlation factors on the water content estimated to be about 0.5 as an average for many soils.

Brooks & Corey (1964) established the constitutive relationship between K and h and between Se and h using the following empirical equations from the analysis of a large soil database,

$$Se(h) = (\alpha h)^{-\lambda} \qquad (\alpha h > 1) \tag{3a}$$

$$Se(h) = 1 \qquad (\alpha h \leq 1) \tag{3b}$$

$$K(h) = K_S(\alpha h)^{-\beta} \qquad (\alpha h > 1) \tag{4a}$$

$$K(h) = K_S \qquad (\alpha h \leq 1) \tag{4b}$$

where λ is a pore-size distribution parameter affecting the slope of the retention function, and $\beta = \lambda(l+2) + 2$.

The model developed by van Genuchten (1980) is an S-shaped function. The function was combined with Mualem's hydraulic conductivity function (Mualem, 1976) to predict the unsaturated hydraulic conductivity. Van Genuchten's equations for the soil water retention curve and the hydraulic conductivity can be expressed as follows,

$$Se(h) = [1 + (\alpha h)^n]^{-m} \tag{5}$$

$$K(h) = K_S Se^l \left\{1 - \left(1 - Se^{1/m}\right)^m\right\}^2, \quad m = 1 - 1/m \tag{6}$$

where m and n are empirical parameters.

In this chapter, we use van Genuchten model since it closely fits measured water-retention data for many types of soils (Leij et al., 1997). Other hydraulic property models can also be similarly used. While results of simulated hydrologic processes using other hydraulic property models may differ quantitatively, they demonstrate similar trends.

2.2 Field-measured and re-generated hydraulic parameter data

Heterogeneity in hydraulic properties (as expressed in terms of hydraulic parameters) largely determines the variability in the water content and flux. The hydraulic parameters we used in this study are an 84-point set of van Genuchten parameters that were derived from field measurements at the Corn Creek Fan Complex (see Fig. 1) at the Desert National Wildlife Refuge, north of Las Vegas, Nevada, U.S.A. (Young et al., 2005). One main purpose of the study by Young et al. (2005) was to characterize the hydraulic properties of surface materials that exist in distinct geomorphic surfaces at the study site. Field work first identified distinct geologic units on both the proximal and distal portions of the Corn Creek Fan, where each unit had distinct morphologic or geologic surface features. After digitizing the geologic unit identification into a geographic information system database, hydraulic and physical properties were obtained from 84 locations in a large area of >100 km^2. Hydraulic properties, including soil water retention and hydraulic conductivity functions, were estimated at sites underneath plant canopies and at intercanopy locations. The locations were chosen based on the geologic unit mosaic of the Corn Creek Fan to cover the various geologic units present on the site, assuming that each unit has distinct hydraulic property characteristics. The field work hence was not designed to obtain a detailed spatial structure for the study site. We use the data set, which has a strong correlation between Ks and α, along with synthetic hydraulic parameter data sets having different levels of parameter correlation, to investigate the effects of hydraulic parameter correlations, not the spatial autocorrelations of the parameters. The soil hydraulic properties of the surface soil at each field site were determined using tension infiltrometry. Hydraulic and physical properties were obtained from 84 locations resulting in 84 samples. Full van Genuchten hydraulic parameters were estimated using the tension infiltrometer, resulting in a hydraulic parameter set of 84 points (called Field Set in the subsequent analysis). Additional details of the field test methodology and procedures are given by Young et al. (2005). Table 1 lists the basic statistics for the Field Set.

In practice, van Genuchten parameter n can be determined with greater certainty than the other van Genuchten parameters (e.g., Schaap & Leij, 1998). Hills et al. (1992) also demonstrated that random variability of α is more important than that of other van Genuchten parameters. Spatial variability in α has a larger impact on the ensemble behavior of soil hydrologic processes than that in other van Genuchten parameters (Zhu & Mohanty, 2002b). Therefore, it is reasonable to treat n as deterministic to examine the influence of more important hydraulic parameters. Following these findings, we treated n as a deterministic parameter using the mean value from the Field Set in this study. The variabilities of other van Genuchten parameters, θ_r and θ_s, are also relatively insignificant in comparison to K_S and α due to the fact that we are mainly concerned with the cumulative evaporation, not the moisture content.

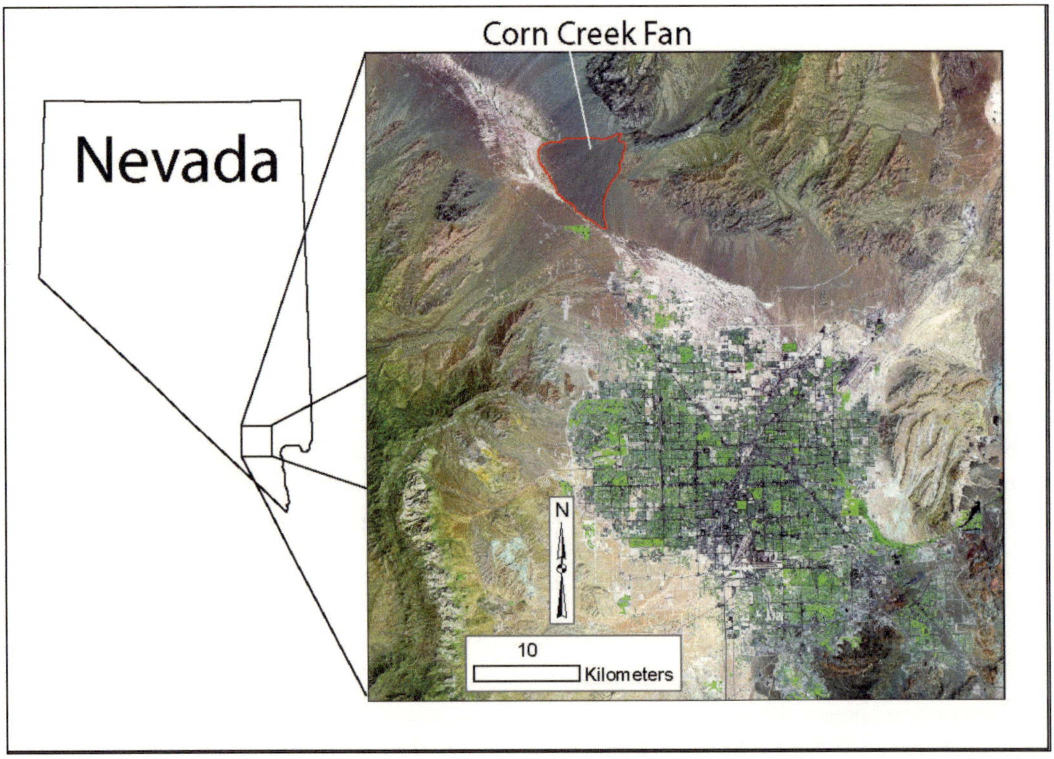

Fig. 1. Location of the Corn Creek Fan Complex, north of Las Vegas, Nevada, U.S.A., where the variability of the hydraulic parameters were characterized.

Since the field measurement-derived hydraulic parameter data set that we use for the subsequent analysis in the upscaling context indicates a fairly strong correlation between the two parameters K_S and α and has only 84 measured points, we seek to investigate the influence of hydraulic parameter correlation, the variances of hydraulic parameters on appropriateness of averaging schemes by synthetically re-generating random hydraulic parameter fields for those two parameters. In doing so, we generate two additional data sets for the K_S and α: 1) an 84-point set that has the same mean and variance with the Field Set, but has zero correlation between the K_S and α (called Set 1), 2) an 84–point set that has the same mean but has two times bigger variance in comparison to the Field Set and also has zero correlation between the K_S and α (called Set 2). Specifically, Set 1 is designed to examine the importance of parameter correlation, since it only differs in correlation level between the K_S and α in comparison to the Field Set. Set 2 can be used to investigate the influence of parameter variance since it has larger variance than the Field Set. The basic statistics of the hydraulic parameter data sets that will be used in the subsequent analyses is listed in Table 1.

Data set	$<K_S>$(cm/min)	$<\alpha>$(1/cm)	CV(K_S)	CV(α)	CC(K_S, α)
Field Set	0.123	0.092	0.837	0.287	0.74
Set 1	0.123	0.092	0.837	0.287	0.0
Set 2	0.123	0.092	1.18	0.41	0.0

Table 1. Basic statistics of field hydraulic parameter set and regenerated sets. $<K_S>$ and $<\alpha>$ represent means of K_S and α respectively; CV(K_S) and CV(α) are the variances of K_S and a respectively; CC(K_S, α) is the coefficient of correlation between K_S and α.

2.3 Aggregation of local scale evaporation processes and relative error calculation

For the local scale evaporation process, we use HYDRUS-1D package (Simunek et al, 1998) to simulate the one-dimensional flow subjected to the head-type conditions on the land surface (1000cm) and on the bottom (1cm or 10cm). The HYDRUS-1D modular program uses fully implicit, Galerkin-type linear finite element solutions of the governing equation for a variably-saturated porous medium (Richards, 1931). Initial suction is assumed to be constant in the profile representing initial wetness of soils. A large suction head of 1000 cm is used as top boundary condition to induce the evaporation. Multiple realizations (84) of HYDRUS-1D simulations are performed and the average cumulative evaporation is calculated. Another simulation is performed to calculate the cumulative evaporation using the simple averaging schemes for the K_S and α. Three simple averaging schemes are used for K_S and α, as described in the following.

1. Arithmetic mean

$$AM(K_S) = < K_S > \tag{7}$$

$$AM(\alpha) = < \alpha > \tag{8}$$

2. Geometric mean

$$GM(K_S) = \exp[<\ln K_S>] \tag{9}$$

$$GM(\alpha) = \exp[<\ln\alpha>] \tag{10}$$

3. Harmonic mean

$$HM(K_S) = [<1/K_S>]^{-1} \tag{11}$$

$$HM(\alpha) = [<1/\alpha>]^{-1} \tag{12}$$

In the above expressions, "<>" denote simple average (i.e., arithmetic mean).
Relative error of using the simple averaging schemes to simulate average evaporation in a large scale heterogeneous landscape is then defined as

$$e(t) = [E_{SA}(t) - <E(t)>]/<E(t)> \tag{13}$$

where $E_{SA}(t)$ denotes the cumulative evaporation calculated using the simple averaging schemes of hydraulic parameters, $<E(t)>$ is the average cumulative evaporation for the heterogeneous landscape. A close to zero e(t) indicates that the simple averaging hydraulic parameters represent the large scale average evaporation well. Positive e(t) simply means the simple averaging hydraulic parameters over-predict the large scale average evaporation, while negative e(t) signals the under-prediction of large scale evaporation by using the simple averaging hydraulic parameters.

3. Discussion

Evolution of relative errors by using the simple averaging schemes of hydraulic parameters to simulate large scale cumulative evaporation as functions of time when the initial suction head is 1 cm is shown in Fig. 2 to Fig. 4 for Field Set, Set 1 and Set 2, respectively. The initial suction head of only 1 cm indicates a wet initial condition close to saturation prior to the evaporation. The geometric mean of hydraulic parameters leads to the smallest relative errors, illustrating geometric mean is the most optimal averaging scheme. The relative errors tend to be larger at the beginning, and decrease as time evolves. The arithmetic mean over-predicts the average evaporation, while the harmonic mean under-predicts the average evaporation. As expected, a larger variance would result in larger relative errors, as evidenced from Fig. 4 in comparison to Fig. 2 and Fig. 3.
In Fig. 5 to Fig. 7, we plot the results of relative errors under otherwise same conditions as Fig. 2 to Fig. 4, but for the initial suction head of 10 cm. This initial condition indicates a drier condition as the water content decreases dramatically as the suction head increases. In general, drier initial condition leads to relatively smaller errors, but signifies an increasing difficulty in selecting a consistent simple averaging scheme that can be used in predicting the average evaporation. The most appropriate simple averaging scheme changes with correlation between the hydraulic parameters and the variances of hydraulic parameters. Under initial drier condition, the simple averaging schemes work better when the hydraulic

parameters are correlated (see Fig. 5). Under drier condition with uncorrelated hydraulic parameters, the three simple averaging schemes all under-predict average evaporation (see Fig. 6 and Fig. 7).

In general, the results indicate that the effective hydraulic parameters are mostly constant (i.e., do not change much with time) except during the initial stage of evaporation, when the errors of using simple averaging schemes vary more significantly with time. In later stage, the errors tend to be more uniform with time. This suggests that the average evaporation behavior at the initial stage in a heterogeneous soil is more difficult to represent using an equivalent homogeneous medium.

4. Conclusions

In this chapter, we examined how the time frame of hydrologic processes affects the performance of averaging schemes and how the hydraulic parameter correlation and variability impact the performance of simple averaging schemes. The average cumulative evaporation in the heterogeneous soils was quantified through multiple realizations of local scale evaporation processes. The suitability of using the simple averaging schemes to represent the heterogeneous evaporation processes was quantified by the difference between the cumulative evaporation based on the simple averaging schemes and the average cumulative evaporation.

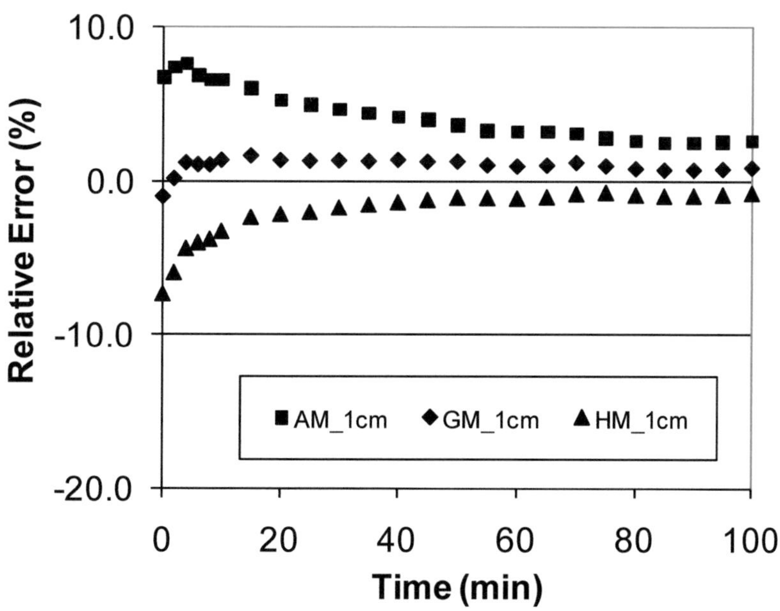

Fig. 2. Evolution of the relative errors for the cumulative evaporation when the surface suction head is 1000 cm. The initial suction head is 1 cm. Results based on Field Set. AM=Arithmetic mean, GM=Geometric mean, HM=Harmonic mean.

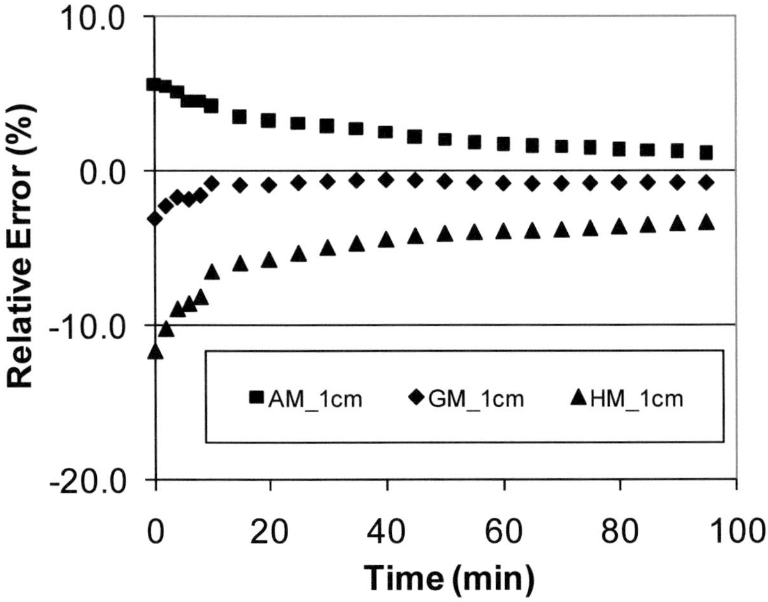

Fig. 3. Evolution of the relative errors for the cumulative evaporation when the surface suction head is 1000 cm. The initial suction head is 1 cm. Results based on Set 1. AM=Arithmetic mean, GM=Geometric mean, HM=Harmonic mean.

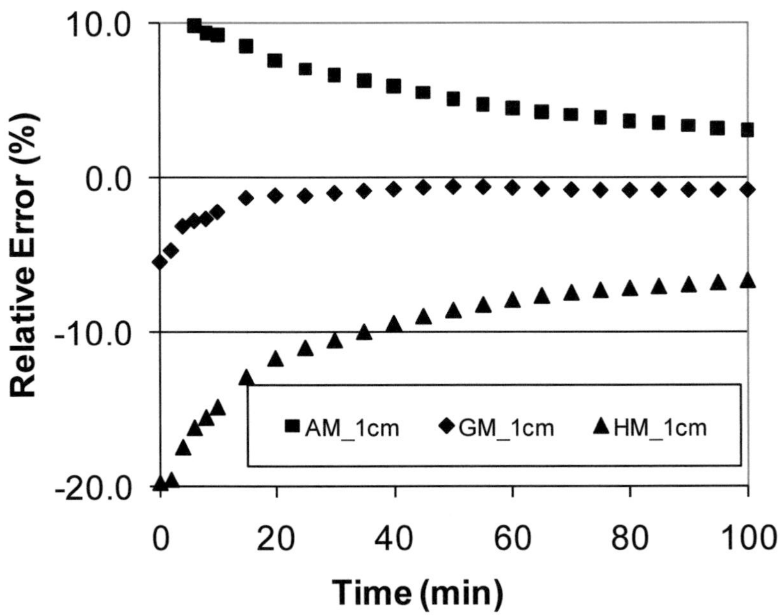

Fig. 4. Evolution of the relative errors for the cumulative evaporation when the surface suction head is 1000 cm. The initial suction head is 1 cm. Results based on Set 2. AM=Arithmetic mean, GM=Geometric mean, HM=Harmonic mean.

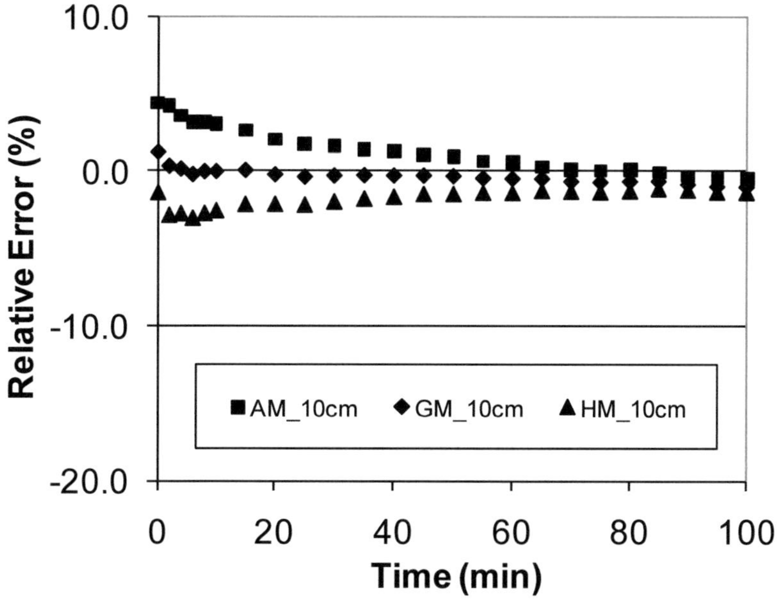

Fig. 5. Evolution of the relative errors for the cumulative evaporation when the surface suction head is 1000 cm. The initial suction head is 10 cm. Results based on Field Set. AM=Arithmetic mean, GM=Geometric mean, HM=Harmonic mean.

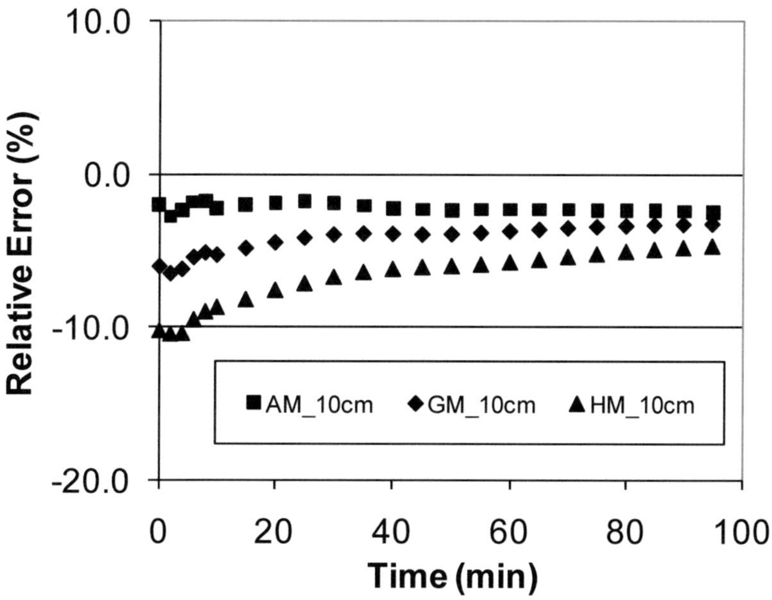

Fig. 6. Evolution of the relative errors for the cumulative evaporation when the surface suction head is 1000 cm. The initial suction head is 10 cm. Results based on Set 1. AM=Arithmetic mean, GM=Geometric mean, HM=Harmonic mean.

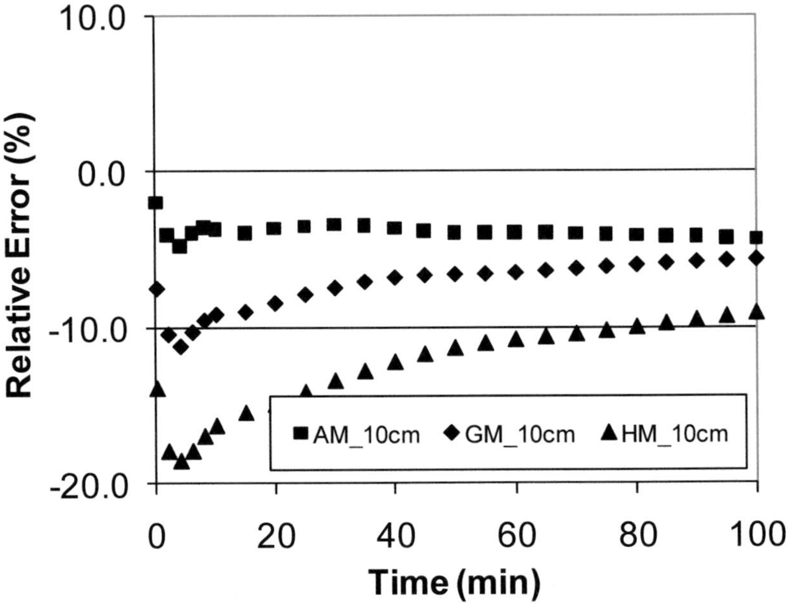

Fig. 7. Evolution of the relative errors for the cumulative evaporation when the surface suction head is 1000 cm. The initial suction head is 10 cm. Results based on Set 2. AM=Arithmetic mean, GM=Geometric mean, HM=Harmonic mean.

In general, all averaging schemes produce larger deviation from averaged evaporative fluxes at the beginning of evaporation. Relative Errors based on using the simple averaging schemes of hydraulic parameters are generally larger for initial wetter conditions due to larger pressure gradient near the surface. The appropriateness of different averaging schemes is sensitive to the correlation between the hydraulic parameters and the variances of hydraulic parameters. At most time frames, average large scale evaporation behavior is better captured (i.e., relative errors of cumulative evaporation are smaller) when the geometric mean is used as the effective parameters. As expected, larger hydraulic parameter variances would introduce larger relative errors when using the simple averaging schemes of hydraulic parameters of the heterogeneous soils.

5. References

Ahuja, L. R.; Ma, L. & Green, T.R. (2010). Effective soil Properties of heterogeneous areas for modeling infiltration and redistribution. *Soil Sci. Soc. Am. J.*, 74, 1469-1482, doi:10.2136/sssaj2010.0073.

Avanidou, T. & Paleologos, E.K. (2002). Infiltration in stratified, heterogeneous soils: relative importance of parameters and model variations. *Water Resour. Res.*, 38(11), 1232, doi:10.1029/2001WR000725.

Bresler, E. & Dagan, G. (1983). Unsaturated flow in spatially variable fields, 2. Application of water flow models to various fields. *Water Resour. Res.*, 19, 421-428.

Brooks, R.H. & Corey, A.T. (1964). *Hydraulic Properties of Porous Media*, Colorado State Univ., Hydrology Paper No. 3, 27 pp, Fort Collins, CO., U.S.

Bumb, A.C.; Murphy, C.L. & Everett. L.G. (1992). A comparison of three functional forms for representing soil moisture characteristics. *Ground Water*, 3, 177-185.

Butters, G.L. & Jury, W.A. (1989). Field scale transport of bromide in an unsaturated soil. 2. Dispersion modeling. *Water Resour. Res.*, 25, 1583–1589.

Campbell, G.S. (1974). A simple method for determining unsaturated conductivity from moisture retention data. *Soil Sci.*, 117, 311-314.

Chen, Z.; Govindaraju, R.S. & Kavvas, M.L. (1994a). Spatial averaging of unsaturated flow equations under infiltration conditions over areally heterogeneous fields, 1, Development of models. *Water Resour. Res.*, 30(2), 523-533.

Chen, Z.; Govindaraju, R. S. & Kavvas. M.L. (1994b). Spatial averaging of unsaturated flow equations under infiltration conditions over areally heterogeneous fields, 2, Numerical simulations. *Water Resour. Res.*, 30(2), 535-548.

Clapp, B.E. & Hornberger, G.M. (1978). Empirical equations for some soil hydraulic properties. *Water Resour. Res.*, 14, 601-604.

Coppola, A.; Basile, A., Comegna, A. & Lamaddalena. N. (2009). Monte Carlo analysis of field water flow comparing uni- and bimodal effective hydraulic parameters for structured soil. *J. Contam. Hydrol.*, 104, 153–165, doi:10.1016/j.jconhyd.2008.09.007.

Cushman, J. H.; Bennethum, L.S. & Hu, B. X. (2002). A primer on upscaling tools for porous media. *Adv. Water Resour.*, 25, 1043–1067.

Dagan, G. & Bresler, E. (1979). Solute dispersion in unsaturated heterogeneous soil at field scale: I. Theory. *Soil Sci. Soc. Am. J.*, 43, 461-467.

Dagan, G. & Bresler. E. (1983). Unsaturated flow in spatially variable fields, 1. Derivation of models of infiltration and redistribution. *Water Resour. Res.*, 19, 413–420.

Das, D.B. & Hassanizadeh, S.M. (2005). *Upscaling Multiphase Flow in Porous Media: From Pore to Core and Beyond*, Springer, Berlin.

Destouni, G. (1992). Prediction uncertainty in solute flux through heterogeneous soils. *Water Resour. Res.*, 28, 793–801.

Ellsworth, T.R. & Jury. W.A. (1991). A three-dimensional field study of solute transport through unsaturated, layered porous media. 2. Characterization of vertical dispersion. *Water Resour. Res.*, 27, 967–981.

Farmer, C.L. (2002). Upscaling: A review. *Int. J. Numerical Meth. Fluids*, 40, 63–78.

Farrell, D.W. & Larson, W.E. (1972). Modeling the pore structure of porous media. *Water Resour. Res.*, 3, 699-706.

Gardner, W.R. (1958). Some steady state solutions of unsaturated moisture flow equations with applications to evaporation from a water table. *Soil Sci.*, 85(4), 228-232.

Gelhar, L.W. (1993). *Stochastic Subsurface Hydrology*, Prentice Hall, Upper Saddle River, NJ.

Govindaraju, R.S.; Or, D., Kavvas, M.L., Rolston, D.E. & Biggay, J. (1992). Error analyses of simplified unsaturated flow models under large uncertainty in hydraulic properties. *Water Resour. Res.*, 28(11), 2913-2924.

Grayson, R. & Bloschl, G. (2000). *Spatial Patterns in Catchment Hydrology*, Cambridge Univ. Press, Cambridge, UK.

Harter, T. & Hopmans, J.W. (2004). Role of vadose zone flow processes in regional scale hydrology: Review, opportunities and challenges. *Unsaturated Zone Modeling:*

Processes, Applications, and Challenges, eds. Feddes, R.A.; de Rooij,G.H. & van Dam, J.C., Kluwer, 179–208.

Harter, T. & Zhang, D. (1999). Water flow and solute spreading in heterogeneous soils with spatially variable water content. *Water Resour. Res.*, 35, 415–426.

Hendrayanto, K.; Kosugi K. & Mizuyama, T. (2000). Scaling hydraulic properties of forest soils. *Hydrol. Process.*, 14, 521–538.

Hills, R.G.; Hudson; D.B. & Wierenga, P.J. (1992). Spatial variability at the Las Cruces trench site. *Indirect Methods for Estimating the Hydraulic Properties of Unsaturated Soils*, eds. van Genuchten, M. Th.; Leij, F.J. & Lund, L.J.), University of California, Riverside, CA., pp.529-538.

Hopmans, J.W. & Stricker, J.N.M. (1989). Stochastic analysis of soil water regime in a watershed. *J. Hydrol.*, 105, 557–584.

Hornung, U. (1997). *Homogenization and Porous Media*, Springer, New York.

Hristopulos, D.T. (2003). Renormalization group methods in subsurface hydrology: Overview and applications in hydraulic conductivity upscaling. *Adv. Water Resour.*, 26, 1279–1308.

Hutson, J.L. & Cass, A. (1987). A retentivity function for use in soil-water simulation model. *J. Soil Sci.*, 38, 105-113.

Jhorar, R.K.; van Dam, J.C., Bastiaanssen, W.G.M. & Feddes, R.A. (2004). Calibration of effective soil hydraulic parameters of heterogeneous soil profiles. *J. Hydrol.*, 285, 233–247.

Kasteel, R.; Vogel, H. & Roth, K. (2000). From local scale hydraulic properties to effective transport properties in soils. *Eur. J. Soil Sci.*, 51, 81–91.

Kim, C.P. & Stricker, J.N.M. (1996). Influence of spatially variable soil hydraulic properties and rainfall intensity on the water budget. *Water Resour. Res.*, 32(6), 1699-1712.

Kim, C.P.; Stricker, J.N.M. & Feddes. R.A. (1997). Impact of soil heterogeneity on the water budget of the unsaturated zone. *Water Resour. Res.*, 33(5), 991-999.

Kosugi, K. (1994). Three-parameter lognormal distribution model for soil water retention. *Water Resour. Res.*, 30, 891-901.

Kozak, J.A. & Ahuja, L.R. (2005). Scaling of infiltration and redistribution of water across soil textural classes. *Soil Sci. Soc. Am. J.*, 69, 816–827.

Kozak, J.A.; Ahuja, L.R. & Green, T.R. (2005). Scaling and estimation of evaporation and transpiration of water across soil textures. *Vadose Zone J.*, 4, 418–427.

Laliberte, G.E. (1969). A mathematical function for describing capillary pressure-desaturation data. *Bull. Int. Assoc. Sci. Hydrol.*, 14, 131-149.

Leij, F.J.; Russell, W.B. & Lesch, S.M. (1997). Closed-form expressions for water retention and conductivity data. *Ground Water*, 35(5), 848-858.

Leij, F.J.; Sciortino, A. & Warrick, A.W. (2006). Infiltration in two parallel soil columns. *Water Resour. Res.*, 42, W12408. doi:10.1029/2006WR00500.

Leij, F.J.; Sciortino, A., Haverkamp, R. & Ugalde, J.M.S. (2007). Aggregation of vertical flow in the vadose zone with auto- and cross-correlated hydraulic properties. *J. Hydrol.*, 338, 96- 112.

Lewandowska, J. & Lautent, J.-P. (2001). Homogenization modeling and parametric study of moisture transfer in an unsaturated heterogeneous porous media. *Transp. Porous Media*, 45, 321-345.

Lewandowska, J.; Szymkiewicz, A., Burzynski, K. & Vauclin, M. (2004). Modeling of unsaturated water flow in double-porosity soils by the homogenization approach. *Adv. Water Resour.*, 27, 283-296.

Libardi, P.L.; Reichardt, K., Nielsen, D.R. & Biggar, J.W. (1980). Simple field methods for estimating soil hydraulic conductivity. *Soil Sci. Soc. Am. J.*, 44, 3-7.

Lu, Z.; Zhang, D. & Robinson, B.A. (2007). Explicit analytical solutions for one-dimensional steady state flow in layered, heterogeneous unsaturated soils under random boundary conditions. *Water Resour. Res.*, 43, W09413, doi:10.1029/2005WR004795.

Mallants, D.; Jacques, D., Vanclooster, M., Diels, J. & Feyen, J. (1996). Stochastic approach to simulate water flow in a macroporous soil. *Geoderma*, 70(2–4), 299-324.

Miller, E.E. & Miller, R.D. (1956). Physical theory of capillary flow phenomena. J. *Appl. Phys.*, 27, 324–332.

Milly, P.C.D. & Eagleson, P.S. (1987). Effects of spatial variability on annual average water balance. *Water Resour. Res.*, 23(11), 2135-2143.

Mualem, Y. (1976). A new model for predicting the hydraulic conductivity of unsaturated porous media. *Water Resour. Res.*, 12, 513–522.

Neuweiler, I. & Cirpka, O.A. (2005). Homogenization of Richards equation in permeability fields with different connectivities. *Water Resour. Res.*, 41, W02009, doi:10.1029/2004WR003329.

Neuweiler, I. & Eichel. H. (2006). Effective parameter functions for the Richards equation in layered porous media. *Vadose Zone J.*, 5, 963-977.

Pachepsky, Y.; Radcliffe, D. & Selim, H.M. (2003). *Scaling Methods in Soil Physics*, CRC Press, Boca Raton, FL.

Protopapas, A.L. & Bras. R. L. (1991). The one-dimensional approximation for infiltration in heterogeneous soils. *Water Resour. Res.*, 27, 1019–1027.

Renard, P. & de Marsily, G. (1996). Calculating equivalent permeability: A review. *Adv. Water Resour.*, 20, 253–278.

Richards, L.A. (1931). Capillary conduction of liquids through porous media. *Physics*, 1, 318-333.

Rossi, C. & Nimmo, J.R. (1994). Modeling of soil water retention from saturation to oven dryness. *Water Resour. Res.*, 30, 701-708.

Rubin, Y. (2003). *Applied Stochastic Hydrogeology*, Oxford Univ. Press, New York.

Rubin, Y. & Or, D. (1993). Stochastic modeling of unsaturated flow in heterogeneous soils with water uptake by plant roots: The parallel columns model. *Water Resour. Res.*, 29, 619–631.

Russo, D. (1988). Determining soil hydraulic properties by parameter estimation: On the selection of a model for the hydraulic properties. *Water Resour. Res.*, 24(3), 453-459.

Russo, D. (1993). Stochastic modeling of macrodispersion for solute transport in a heterogeneous unsaturated porous formation. *Water Resour. Res.*, 29, 383–397.

Russo, D. (1998). Stochastic analysis of flow and transport in unsaturated heterogeneous porous formation: Effects of variability in water saturation. *Water Resour. Res.*, 34, 569–581.

Russo, D. (2003). Upscaled conductivity in gravity-dominated flow through variably saturated heterogeneous formations. *Water Resour Res.*, 39, 7–11.

Russo, D. (2005). Stochastic analysis of solute mass flux in gravity-dominated flow through bimodal heterogeneous unsaturated formations. *Vadose Zone J.*, 4, 939–953, doi:10.2136/vzj2004.0183.

Russo, D. & Fiori, A. (2009). Stochastic analysis of transport in a combined heterogeneous vadose zone–groundwater flow system. *Water Resour. Res.*, 45, W03426, doi:10.1029/2008WR007157, 2009.

Schaap, M.G. & Leij, F.J. (1998). Database-related accuracy and uncertainty of pedotransfer functions. *Soil Sci.*, 163, 765-779.

Setiawan, B.I. & Nakano, M. (1993). On the determination of unsaturated hydraulic conductivity from soil moisture profiles and from water retention curves. *Soil Sci.*, 156, 389-395.

Severino, G. & Santini, A. (2005). On the effective hydraulic conductivity in mean vertical unsaturated steady flows. *Adv. Water Resour.*, 28, 964-974.

Severino, G.; Santini, A. & Sommella, A. (2003). Determining the soil hydraulic conductivity by means of a field scale internal drainage. *J. Hydrol.*, 273, 234–248.

Simunek, J.; Šejna, M. & van Genuchten, M.Th. (1998). *The HYDRUS-1D Software Package for Simulating One-dimensional Movement of Water, Heat, and Multiple Solutes in Variably Saturated Media*, Version 2.0, IGWMC - TPS - 70, International Ground Water Modeling Center, Colorado School of Mines, Golden, Colorado, 186 pp.

Smith, R.E. & Diekkruger, B. (1996). Effective soil water characteristics and ensemble soil water profile in heterogeneous soils. *Water Resour. Res.*, 32, 1993–2002.

Sposito, G. (1998). *Scale Dependence and Scale Invariance in Hydrology*, Cambridge Univ. Press, Cambridge, UK.

Sviercoski, R.F.; Warrick, A.W. & Winter. C.L. (2009). Two-scale analytical homogenization of Richards' equation for flows through block inclusions. *Water Resour. Res.*, 45, W05403, doi:10.1029/2006WR005598.

Tartakovsky, D. M.; Neuman, S.P. & Lu, Z. (1999). Conditional stochastic averaging of steady state unsaturated flow by means of Kirchhoff transformation. *Water Resour Res.*, 35, 731–45.

Toride, N. & Leij, F.J. (1996a). Convective-dispersive stream tube model for field-scale solute transport: I. Moment analysis. *Soil Sci. Soc. Am. J.*, 60, 342–352.

Toride, N. & Leij, F.J. (1996b). Convective-dispersive stream tube model for field-scale solute transport: II. Examples and calibration. *Soil Sci. Soc. Am. J.*, 60, 352–361.

Van Duijn, C.J.; Mikelic, A. & Pop. I.S. (2002). Effective equations for two-phase flow with trapping on the micro scale. *SIAM J. Appl. Math.*, 62, 1531–1568.

van Genuchten, M.Th. (1980). A closed-form equation for predicting the hydraulic conductivity of unsaturated soils. *Soil Sci. Soc. Am. J.*, 44, 892-898.

van Genuchten, M.Th. & Nielsen. D.R. (1985). On describing and predicting the hydraulic conductivity of unsaturated soils. *Annales Geophysicae*, 3, 615-628.

Vereecken, H.; Kasteel, R., Vanderborgh, J. & Harter, T. (2007). Upscaling hydraulic properties and soil water flow processes in heterogeneous soils: A review. *Vadose Zone J.*, 6, 1-28.

Ward, A.L.; Zhang, Z.F. & Gee, G.W. (2006). Upscaling unsaturated hydraulic parameters for flow through heterogeneous anisotropic sediments. *Adv. Water Resour.*, 29, 268–280.

Wildenschild, D & Jensen. K.H. (1999). Numerical modeling of observed effective flow behavior in unsaturated heterogeneous sands. *Water Resour. Res.*, 35, 29-42.

Williams, R.D. & Ahuja. R.L. (2003). Scaling and estimating the soil water characteristic using a one-parameter model, *Scaling Methods in Soil Physics*, p. 35–48, CRC Press, London.

Young, M.H.; Mcdonald, E.V., Caldwell, T.G., Benner, S.G. & Meadows, D.G. (2004). Hydraulic properties of a desert soil chronosequence in the Mojave Desert. *Vadose Zone J.*, 3, 956-963.

Zhang, D. (2002). *Stochastic Methods for Flow in Porous Media: Coping with Uncertainties*, Academic Press, San Diego.

Zhang, R. & van Genuchten, M.Th. (1994). New models for unsaturated soil hydraulic properties. *Soil Sci.*, 158, 77-85.

Zhang, D.; Wallstrom, T.C. & Winter, C.L. (1998). Stochastic analysis of steady-state unsaturated flow in heterogeneous media: Comparison of the Brooks-Corey and Gardner-Russo models. *Water Resour. Res.*, 34(6), 1437-1449.

Zhang, Z.F. (2010). Effective hydraulic conductivity of unsaturated isotropic soils with multidimensional heterogeneity. *Soil Sci. Soc. Am. J.*, 74, 734-743, doi:10.2136/sssaj2009.0405.

Zhang, Z.F.; Ward, A.L. & Gee, G.W. (2004). A combined parameter scaling and inverse technique to upscale the unsaturated hydraulic parameters for heterogeneous soils. *Water Resour. Res.*, 40, W08306. doi:10.1029/2003WR002925.

Zhu, J. (2008). Equivalent parallel and perpendicular unsaturated hydraulic conductivities: Arithmetic mean or harmonic mean?. *Soil Sci. Soc. Am. J.*, 72, 1226-1233, doi:10.2136/sssaj2007.0337.

Zhu, J. & Mohanty, B.P. (2002a). Upscaling of hydraulic properties for steady state evaporation and infiltration. *Water Resour. Res.*, 38(9), 1178, doi:10.1029/2001WR000704.

Zhu, J. & Mohanty, B.P. (2002b). Spatial averaging of van Genuchten hydraulic parameters for steady-state flow in heterogeneous soils: A numerical study. *Vadose Zone J.*, 1, 261-272.

Zhu, J. & Mohanty, B.P. (2003a). Effective hydraulic parameters for steady state flows in heterogeneous soils. *Water Resour. Res.*, 39(8), 1227, doi:10.1029/2002WR001831.

Zhu, J. & Mohanty, B.P. (2003b). Upscaling of hydraulic properties in heterogeneous soils. *Scaling Methods in Soil Physics*, eds., Pachepsky, Y.; Radcliffe, D.E. & Selim, H.M., CRC Press, 97- 117.

Zhu, J. & Mohanty, B.P. (2004). Soil hydraulic parameter upscaling for steady-state flow with root water uptake. *Vadose Zone J.*, 3, 1464–1470.

Zhu, J. & Mohanty, B.P. (2006). Effective scaling factor for transient infiltration in heterogeneous soils. *J. Hydrol.*, 319(3), 96-108.

Zhu, J.; Mohanty, B.P. & Das, N.N. (2006). On the effective averaging schemes of hydraulic properties at the landscape scale. *Vadose Zone J.*, 5, 308-316.

Zhu, J. & Sun, D. (2009). Effective soil hydraulic parameters for transient flows in heterogeneous soils. *Vadose Zone J.*, 8, 301-309, doi:10.2136/vzj2008.0004.

Zhu, J. & Sun, D. (2010). Capillary pressure-dependent anisotropy of unsaturated soils. *Can. J. Soil Sci.*, 90(2), 319-329.

Zhu, J.; Young, M.H. & van Genuchten, M.Th. (2007). Upscaling schemes and relationships for the Gardner and van Genuchten hydraulic functions for heterogeneous soils. *Vadose Zone J.*, 6, 186-195.

8

Determination of Hydraulic Conductivity of Undisturbed Soil Column: a Measurement Accomplished with the Gamma Ray Transmission Technique

Anderson Camargo Moreira[1*], Otávio Portezan Filho[2], Fábio Henrique de Moraes Cavalcante[3] and Carlos Roberto Appoloni[2]

[1]*Programa de Pós Graduação em Ciências e Engenharia de Materiais/EMC, Universidade Federal de Santa Catarina - UFSC,*

[2]*Depto. de Física, Universidade Estadual de Londrina - UEL,*

[3]*Centro de Física Nuclear, Instituto Superior de Engenharia/DEEA/ISEL, Universidade de Lisboa,*

[1,2]*Brasil*

[3]*Portugal*

1. Introduction

The determination of hydraulic conductivity is an important parameter on the description of water transportation in complex porous systems, such as soil. Darcy Buckingham equation describes the water movement in unsaturated porous media. This parameter is given in terms of moisture content expressed by their moisture content or its potential energy. The soil hydraulic conductivity is a property of each pore system since it depends on the shape and geometrical arrangement of its particles. The study of hydraulic conductivity for unsaturated soils can be used to analysis the irrigation, drainage and nutrient transport, also for development of an agricultural crop. The description of the water flow in soil is important for design irrigation projects, erosion control and even control of movement of toxic waste from the surface to deeper horizons.

The first experimental study on the movement of water in the soil was done by Henry Darcy in 1856, using soil columns of saturated sand (Darcy, 1856). In 1907, Edgar Buckinghan described the water flow in unsaturated porous media modifying the equation obtained by Darcy (Buckingham, 1907). In 1931, Lorenzo Adolph Richards combined the equations of Darcy and Buckingham with the equation of continuity to establish an overall relationship, which is yet used (Richards, 1931). Klute describes several methods for estimating the hydraulic conductivity and diffusivity for unsaturated soils *in natura* and in laboratory (Klute, 1972), including the Instantaneous Profile Method, (Sisson et al., 1980).

The Gamma Ray Transmission (GRT) technique in physics of soil was first applied to density measurements, according to Reginato & Van Bavel, 1964, and later for determination of soil moisture content. One of the earliest work using the GRT technique was carried out by Reichardt, 1965. The GRT allows nondestructive measurements of density and soil

moisture content, study of soil compaction caused by agricultural machinery and also the monitoring of changes in soil profile density (Portezan Filho, 1997).
In order to do descriptive measurements of hydraulic conductivity of soil, crop growing conditions should be complied with *in situ* analysis (Chong et al., 1981, Libardi et al., 1980, Hillel et al., 1972). However, laboratory analysis can recreate the environment field simulating rainfall or irrigation procedures with controlled water infiltration in undisturbed soil columns. The undisturbed columns preserve the soil structure *in natura* and can be moved to laboratory for analysis. Studies of disturbed soil columns for some purposes are found in literature (Ying et al., 2010, Al-Houri et al., 2009), however, analysis of undisturbed soil columns are relatively novel concept (Coppola et al., 2004).
This chapter introduces a non conventional technique for determining the hydraulic conductivity in soils [K (θ)], the of Gamma Rays Transmission method, which together with the methods of Sisson et al., 1980 was adopted to analyze the conductivity of undisturbed soil columns.

2. Methodology

2.1 Gamma RaysTransmission

Each one of these processes produces a complete or partial transfer of energy from a photon to an electron of the sample, contributing to definition of the mass attenuation coefficient [μ ($cm^2.g^{-1}$)] of the material. The μ coefficient will define the rate of radiative absorption by matter. It is an intrinsic property of each material with density [ρ ($g.cm^{-3}$)] and atomic number (Z) defined. In the case of materials with several density and chemical composition, the volume fraction crossed by the radiation beam has a particular mass attenuation coefficient. The exponential absorption which characterizes the passage of a monoenergetic electromagnetic radiation through a homogeneous material is given by the Beer-Lambert Law (Siegbahn, 1965) given by:

$$I = I_0 e^{-\mu\rho x} \quad (1)$$

A photon detector converts photons into electrical pulses. These pulses are measurable by the detection system, each counted pulse is called "count". The intensity of the radiation from a source "I_0" is the ratio between the number of counts and the exposure time (*s*). The intensity "I" of the beam that was attenuated in by the object with a thickness x (cm) is quantified in the same way, ie:

$$I = \frac{N}{t} \quad (2)$$

were N is the total number of counts.
The radiation beam in the technique GRT has a thickness defined by the use of collimators, assuming a cylindrical shape with a diameter defined. Thus, the determination of the mass attenuation coefficient can be done point to point in the sample, allowing analyzing a density profile of the material.

2.2 Sisson method

Working in laboratory with saturated porous material, Darcy found a proportional relationship between density of water flow and hydraulic gradient.

The constant of proportionality was called "saturated hydraulic conductivity". With the experimental results for the water flow in unsaturated soils columns, Buckingham obtained a similar equation. Richards called it the Darcy-Buckingham equation. For the water flow in a soil sample with rigid structure, the flow can be written as:

$$\vec{q} = -K(\theta) gradH \tag{3}$$

where the $\vec{q}$ vector is the density of water flow in soil (m.s^{-1}), *grad H* is the vector gradient of hydraulic potential (cm.cm^{-1}) and K (θ) is the hydraulic conductivity of the sample (m.s^{-1}) of soil, function of soil moisture content with rigid porous media. The Sisson method assumes that the one-dimensional water flow in porous media, in z-direction from the surface of soil to down, can be described by Richards equation:

$$\frac{\partial\theta(z,t)}{\partial t} = \frac{\partial}{\partial z}\left\{K(\theta)\frac{\partial H(z,t)}{\partial z}\right\} \tag{4}$$

and the hydraulic potential gradient is unitary:

$$\frac{\partial H(z,t)}{\partial z} = \frac{\partial(h+z)}{\partial z} = -1 \tag{5}$$

During the process of water redistribution in the saturated soil, Sisson considers the hydraulic conductivity K (θ) an exponential function of soil moisture θ, according to the equation below:

$$K(\theta) = K_s e^{[\gamma(\theta-\theta_s)]} \tag{6}$$

where K_S and θ_S are the values of saturated hydraulic conductivity and moisture respectively. The equation (6) can be described as follows:

$$\frac{\partial\theta}{\partial t} = -\frac{\partial K(\theta)}{\partial z} = -\frac{dK(\theta)}{d\theta}\left(\frac{\partial\theta}{\partial z}\right) \tag{7}$$

According to Jones & Wagenet, 1984, for the solution of this equation Sisson, used the techniques of resolution of ordinary differential equations presented by Lax, 1972. The algorithm of Lax allows the solution of equation (7). The solution for z = *L*, is given by:

$$\theta_s - \theta = \left(\frac{1}{\gamma}\right)\ln t + \left(\frac{1}{\gamma}\right)\ln\left(\frac{\gamma K_s}{L}\right) \tag{8}$$

where θ_S is the moisture content saturation calculated by experimental data moisture θ of the water redistribution for every depth level z = L, through the model θ versus lnt for a short time (0.0007 day, practically zero). By some algebraic manipulations of the equation (8), it is obtained:

$$\ln\left(\frac{L}{t}\right) = \ln(\gamma K_s) + \gamma(\theta - \theta_s) \tag{9}$$

This is the equation presented by Sisson et al., 1980. Performing a linear regression lnt versus (θ - θ_S) for each depth level L, the values of the constants K_S (saturated hydraulic conductivity) and γ can be determined.

3. Materials and methods

3.1 Nuclear apparatus

The equipment adopted at this work was a 2"x2" diameter NaI (Tl) scintillation detector, 2 mm diameter source and 5 mm diameter detector lead collimators, standard gamma spectrometry electronic constituted by pre-amplifier, amplifier and high voltage supply (HV). The radiation source was an Americium-241 (^{241}Am) source with 59.6 keV energy and 100 mCi activity. The nuclear setup is coupled to a sample holder system that movement vertically the soil column. Figure 1 shows the gamma ray transmission experimental set-up.

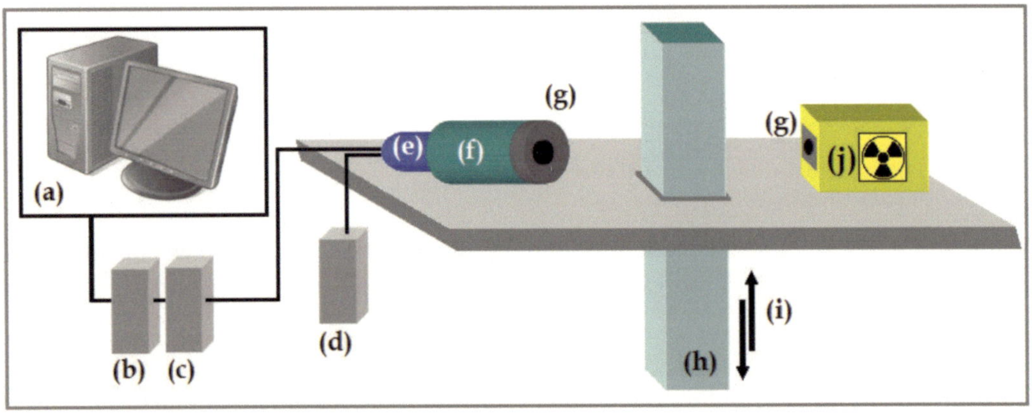

Fig. 1. Nuclear setup constituted by (a) data acquisition station, (b) signal amplifier and (c) pre-amplifier, (d) high voltage supplier, (e) photomultiplier, (f) NaI detector, (g) Pb collimators, (h) soil column, (i) sample holder with vertically movement and (j) ^{241}Am gamma ray source.

3.2 Soil column

Undisturbed soil columns were taken from the campus of the State University of Londrina-UEL, Brazil, in a region destined to agronomic studies. The soil of the place is categorized as dystrophic dark red latossol. For the soil columns extract, some equipment are required like shovels, hoes and garden trowels, gauzes and paraffin as well stove and metal container for melting the paraffin. In a trench with about 60 cm deep, the column was molded into one wall of the trench with the garden trowels. With 40 cm, 10 cm wide and 5 cm thick, the column was wrapped with a layer of gases that absorbed the previously melted paraffin. When dry, the layer of gas and paraffin formed a shield that contained the undisturbed soil for its transport to the laboratory LFNA, Applied Nuclear Physics Laboratory, located in the Physics Department/UEL.

Six columns of undisturbed soil were extracted from the trench, but only one column resisted the transporting to the laboratory or no presented cracks in the interface between shield and soil. About 5kg of soil samples were also collected from the trench for determination of initial moisture content of the soil.

3.3 Soil column gammagraphy

To use the Sisson model the gammagraphy of the column should be accomplished. Gammagraphy is a scan of the column with the gamma beam used for the estimation of any parameters of the equation 9. As shown in Figure 2, the scanning is performed at the measuring points pre-determined over the soil column.

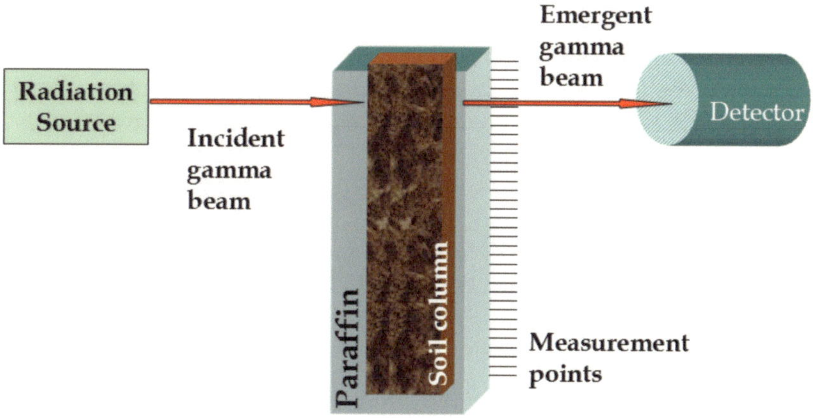

Fig. 2. Process of gammagraphy of the soil column.

The gammagraphy should be performed in the column with soil and in the empty column, ie, only in the paraffin that surrounds the soil. For the column with soil, Figure 3 shows the thickness crossed by the gamma beam. The air and paraffin shield attenuation are inserted in the initial beam intensity I_o.

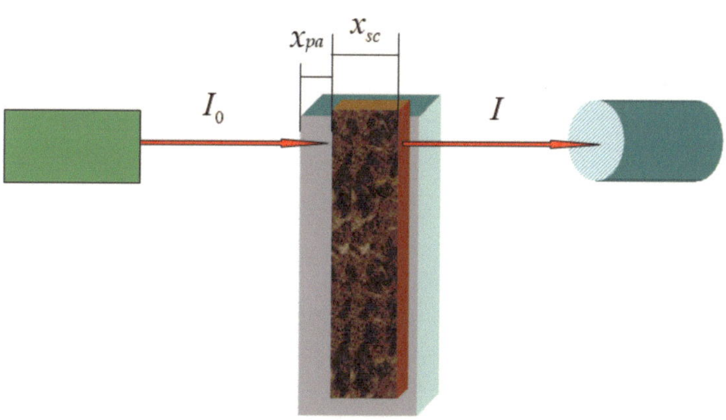

Fig. 3. Gammagraphy setup of column with soil.

The undisturbed soil column is constituted by soil, initial moisture content and porosity, so the x_{sc} thickens is given by:

$$x_{sc} = x_s + x_w + x_{pa} \tag{10}$$

where x_{sc} is the thickness of the soil column, x_s is the equivalent thickness of the soil, x_w is the equivalent thickness of the water and x_p is the equivalent thickness of the pores. Considering the equation 10, equation 1 can be written as follows:

$$I = I_0\, exp-\left(\mu_s \rho_s x_s + \mu_w \rho_w x_w + \mu_p \rho_p x_{pa}\right) \tag{11}$$

Denoting by θ_s, θ_w and θ_p, the volumetric fraction of soil, water and the pore within the sample, respectively, is given that:

$$\theta_s + \theta_w + \theta_a = 1 \tag{12}$$

Considering the space of the soil, water and pore in terms of "volume", $V_s = V\theta_s$, $V_w = V\theta_w$, $V_p = V\theta_p$, where V is the total sample volume, and also the equivalent thickness as $x_s = x\theta_s$, $x_w = x\theta_0$, $x_a = x\theta_a$ (θ_0 is the initial moisture content), equation 11 can be written as:

$$I = I_0\, exp-\left(\mu_s \rho_s x\theta_s + \mu_w \rho_w x\theta_0 + \mu_p \rho_p x\theta_p\right) \tag{13}$$

According to Portezan Filho, 1997 $\mu_p\rho_p x\theta_p$ is very small in relation to the other terms of the exponential and can be unvalued. Thus, equation 13 becomes:

$$I = I_0\, exp-\left(\mu_s \rho_s x\theta_s + \mu_w \rho_w x\theta_0\right) \tag{14}$$

The soil bulk density ρ is denoted by $\rho=\rho_s\theta_s$, and considering the water density as $\rho_w = 1$ g.cm^{-3}, equation 14 becomes:

$$I = I_0\, exp-\left(\mu_s x\rho + \mu_w x\theta_0\right) \tag{15}$$

For the gammagraphy of the paraffin shield, first it is considered the column with dry soil. As can be seen in the Figure 3, x_{pa} is the paraffin thickness, I_0 and I (intensity of the gamma beam emergent from the soil column without paraffin shield) become I_{0s} (average of the measurement points) and I_s (intensity of the gamma beam emergent from the soil column with paraffin shield), respectively, thus the following equation can be written as:

$$I = \overline{I_{0s}}\, exp-\left(x_{pa}\mu_p\rho_p + x\mu_s\rho + x\mu_w\theta_0\right) \tag{16}$$

Considering the empty column, ie, only the paraffin that shields the soil, as shown in Figure 4, the equation 16 can be written as:

$$I = \overline{I_{0pa}}\, exp-\left(x_{pa}\mu_p\rho_p\right) \tag{17}$$

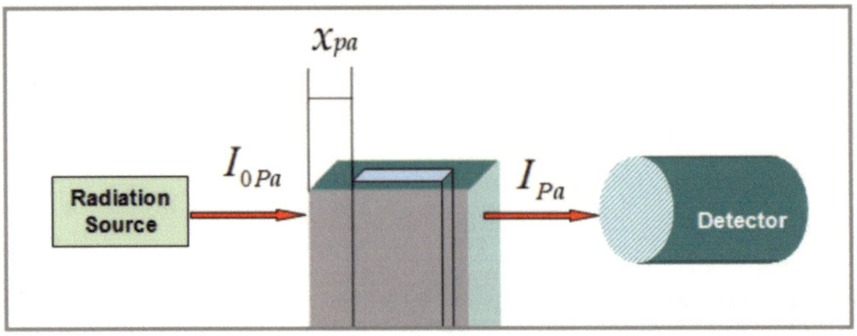

Fig. 4. Gammagraphy of the paraffin that shields the empty the soil column.

Dividing the equation 16 by 17, and with some algebraic manipulations, the result is given by:

$$\mu_s \rho = \frac{1}{x_{sc}} ln\left(\frac{I_{pa}}{I_s} \frac{\overline{I_{0s}}}{\overline{I_{0pa}}} \right) - \mu_w \theta_0 \tag{18}$$

where I_{0s} and I_{0pa} are the average intensity of the gamma beam incident on the soil column and empty column (cont.s^{-1}), respectively, I_s and I_{pa} are the average intensity of the gamma beam incident on the soil column and empty column (cont.s^{-1}), x_{sc} is the thickness of the soil column (cm), θ_0 is the initial moisture content (cm^3.cm^{-3}) and μ_w is the water mass attenuation coefficient (cm^2.g^{-1}). This equation allows the determination of mass attenuation coefficient of the soil μ_s. If soil density ρ (g.cm^{-3}) is not known, this parameter should be determined by the gravimetric method presented in the preceding item.

For the measurements of moisture content profiles $\theta(z,t)$ in function of the time and depth level, after water infiltration in soil column, equation 16 can be described as:

$$I = \overline{I_{0s}}\, exp - \left(x_{pa} \mu_p \rho_p + x \mu_s \rho + x \mu_w \theta \right) \tag{19}$$

where I is the intensity of the gamma beam emergent from the soil with water (cont.s^{-1}) and θ is the moisture content (cm^3.cm^{-3}).

Inserting the equation 16 in the equation 19, the new equation is given by:

$$I = I_s\, exp\left(x_{pa} \mu_p \rho_p + x \mu_s \rho + x \mu_w \theta_0 \right) exp - \left(x_{pa} \mu_p \rho_p + x \mu_s \rho + x \mu_w \theta \right) \tag{20}$$

With some algebraic manipulations, equation 20 becomes:

$$I = I_s\, exp - \left\{ x \mu_w \left(\theta - \theta_0 \right) \right\} \tag{21}$$

Inserting the exponential term of the equation 17 in the equation 16, the result is given by:

$$I_s = \frac{\overline{I_{0s}} I_{pa}}{\overline{I_{0pa}}} exp - \left\{ x \left(\mu_s \rho_s - \mu_w \theta_0 \right) \right\} \tag{22}$$

Inserting the I_s from the equation 22 in the equation 21:

$$I = \frac{\overline{I_{0s}} I_{pa}}{\overline{I_{0pa}}} exp - \left\{ x \left(\mu_s \rho_s - \mu_w \theta_0 \right) \right\} exp - \left\{ x \left(\mu_w \theta - \mu_w \theta_0 \right) \right\} \tag{23}$$

With some algebraic manipulations, equation 23 becomes:

$$\frac{\overline{I_{0pa}} I}{\overline{I_{0s}} I_{pa}} = exp - \left\{ x \left(\mu_s \rho_s + \mu_w \theta \right) \right\} \tag{24}$$

Applying the neperian logarithm (ln) on each side of the equation 23 and inserting the exponential term on the equation 24, and isolating $\theta(z,t)$, the result is:

$$\theta(z,t) = \frac{1}{x \mu_w} \left[ln\left(\frac{I_{pa} \overline{I_{0s}}}{I_s \overline{I_{0pa}}} \right) - x \mu_s \rho_s \right] \tag{25}$$

where μ_w is the water mass attenuation coefficient ($cm^2.g^{-1}$), x is the thickness of the soil column (cm), I_P and I_S are the intensity of the gamma beam emergent from the empty column and soil column ($cont.s^{-1}$), respectively, I_{0P} and I_{0S} are the average intensity of the gamma beam incident on the empty column and soil column ($cont.s^{-1}$), respectively. With the processes of infiltration and redistribution of the water in soil, the values of moisture content $\theta(z,t)$ at each depth level of the soil column were calculated using the equation 25. With the data of moisture content variation at each depth level, the hydraulic conductivity K (θ) was determined applying the model of Sisson.

3.4 Soil density and humidity

In order to determine the θ_0 and ρ_s parameters, the gravimetric method was adopted. For these determinations beakers, electric kiln and digital balance are required. The soil sample that was collected from the trench was sieved and packed in the beakers with know volume. The mass of the beaker with soil was measured after and before the drying process in the kiln (24 hours at 105° C). The difference in mass before and after drying is the mass of moisture content of the sample from the trench, which is also considered as being the moisture content present in the soil column.
To measure the initial moisture content of soil sample (θ_0), we have the following equation:

$$\theta_0 = \frac{m_{H_2O}}{V_B} \tag{26}$$

where, water density is 1 ($g.cm^{-3}$), θ_0 is the initial moisture content ($cm^3.cm^{-3}$), m_{H2O} is the water mass (g) and V_B is the water volume (cm^3).
Soil density [ρ_s ($g.cm^{-3}$)] was determinate with the following equation:

$$\rho_s = \frac{m_s}{V_{AC}} \tag{27}$$

where m_s is the soil mass (g) and V_{AC} is the sample holder volume.

3.5 Mass attenuation coefficients

The determination of the water mass attenuation coefficient [μ ($cm^2.g^{-1}$)] has dual importance in the process of analysis of soil hydraulic conductivity, besides being an important factor present in the equations that predict the conductivity, is also used as a parameter in checking the effectiveness of the transmission system, including collimation, stability of electronic and deviations inherent of the equipment.
In this procedure a rectangular acrylic box is used as a sample holder. The box is irradiated empty at several measurement points, as can be seen in the Figure 5. The transmission is performed in the empty box to measure I_0 of the equation 1. The I_0 is given as the intensity of the gamma beam radiation emergent from the sample holder, instead the intensity of the beam incident on the sample.

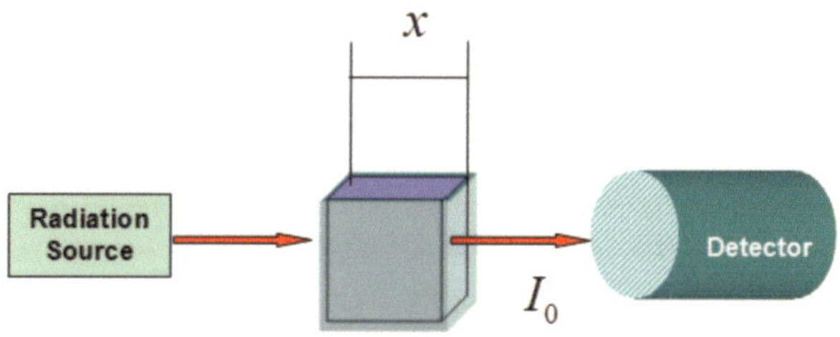

Fig. 5. Sample holder setup for μ_w and μ_s determinations.

After that, the box is filled with water and the transmission is performed again, where the equation 1 takes the following form:

$$I = I_0 e^{-\mu_w x} \quad (28)$$

The same procedure is accomplished with the sample holder filled with soil:

$$I = I_0 e^{-\mu_s \rho_s x} \quad (29)$$

where ρ_s is given by the equation 27 and x is the internal width of the sample holder.

4. Preliminary measurements

4.1 Soil density and mass attenuation coefficients

Following the script of section 3.4, the soil density (ρ_s) of the trench and its initial moisture content (θ_0) was determined. The values are ρ_s = 1.48 g.cm^{-3} and θ_0= 0.223 cm^3.cm^{-3}.
To determine the mass attenuation coefficients of water and soil, was used the equations 28 and 29. Was used the exposure time t = 300s for both. Six measures points were adopted in the acrylic box for the three situations, empty box, box filled with water and soil. The determined values, an average of six measures, were μ_s = 0.4114 ± 0.0045 cm^2.g^{-1} and μ_w = 0.02016 ± 0.00002 cm^2.g^{-1}.
Table 1 presents values of mass attenuation coefficients of water found in literature, for the same gamma energy used in this work, and also the value determined with the WinXCOM program. This software determines the mass attenuation coefficients for materials by their chemical composition and energy of the radioactive source (Gerward et al., 2001, Moreira et al., 2006).

authors	μ_w (cm^2.g^{-1})
Current work	0.02016 ± 0.00002
Oliveira, 1991	0.0200 ± 0.0002
Ferraz, 1974	0.0200 ± 0.0002
Appoloni & Rios, 1994	0.0267 ± 0.0001
WinXCOM (Gerward et al., 2001)	0.0207

Table 1. Values of the water mass attenuation coefficient.

The good agreement among the μ_w values shown in Table 1 indicates that the gamma ray transmission system is able to accomplish the measurements.

4.2 Undeformed soil column gammagraphy

With the soil column placed on the measurement table (Figure 1), the process of gammagraphy of soil column has begun and the column is still dry. The column was measured at 12 points (space between two points of 2 cm), starting at z = 2 cm to z = 24 cm depth levels and exposure time was 60 s per point. This procedure is also known as densitometry, because analyze the homogeneity of the density profile within the soil column in relation to the depth. The presence of voids that have great influence on the process of hydraulic conductivity can be detected.

The densitometry process is described in section 3.3. and the equation used was the 18. The gammagraphy of the paraffin shield of the column was performed after the process of infiltration and redistribution of water in soil. At this stage, the column needs to be open and have the soil removed.

Figure 6 presents the graphic ρ (g.cm^{-3}) versus z (cm). The density homogeneity along the column can be noted. The lowest value of density was $\rho = 1.347 \pm 0.039$ g.cm^{-3} at depth level z= 10 cm and the highest was $\rho = 1.654 \pm 0.048$ g.cm^{-3} at depth level z=4 cm.

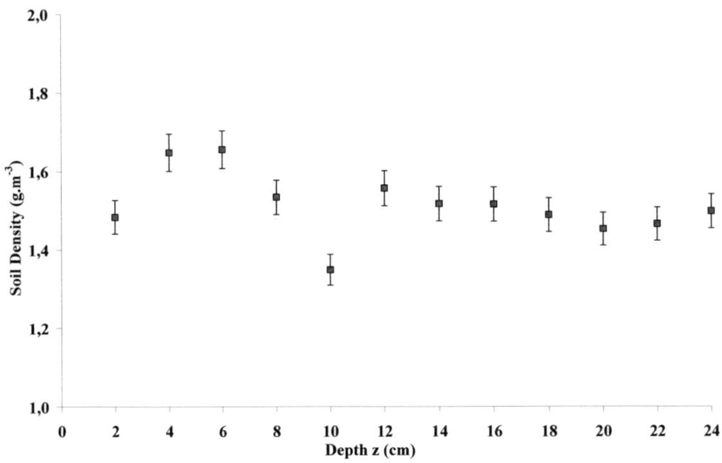

Fig. 6. Soil column densitometry.

4.3 Infiltration and redistribution of water in soil column

With the densitometry performed, the soil column was prepared for the infiltration of water into undisturbed soil. The top of the column was cut and a thin layer of soil was removed. Five layers of filter paper and a polymeric foam were placed on the surface of soil. This way, the direct contact between soil and water was avoided and a homogeneous infiltration with a low water flow was performed. Water was infiltrated keeping a water layer of approximately 0.5 cm on the top of the column. The gamma beam was kept in depth level z = 32 cm and the gamma beam intensity was constantly measured with t = 60 s. The infiltration was completed when the gamma beam intensity was changed. This change indicated the water reached the point z = 32 cm. When the infiltration finished, the soil column was sealed to avoid water evaporation and the redistribution measurements started. The redistribution measurements analyze the movement of water in the soil column and determine the function of hydraulic conductivity K (θ) with the methodology adopted. The distance between measurement points was 2 cm, from the depth z = 2 cm to z = 20 cm. The determination of moisture content at each point was done with the equation 25. Several data were needed to follow the variation of moisture content in the depth levels. The redistribution process was accomplished in 22 days. In the beginning, 9 measurements were made with interval of 20 min, and subsequently increased to 40 min.

Figures 7, 8, 9, 10 and 11 presents the graphics of moisture content $\theta(z,t)$ at each point versus time for the depth levels z = 8, 12, 16, 18 and 20 cm, respectively, the curve fitting coefficient (regression **r**) to the experimental data are also presented. Analyzing the graphs, we notice a decrease in moisture content with the increase of the time, characteristic of the process of draining water. Theres a accentuated decrease of moisture content in the first two days of redistribution and the results from the four day no present significant changes.

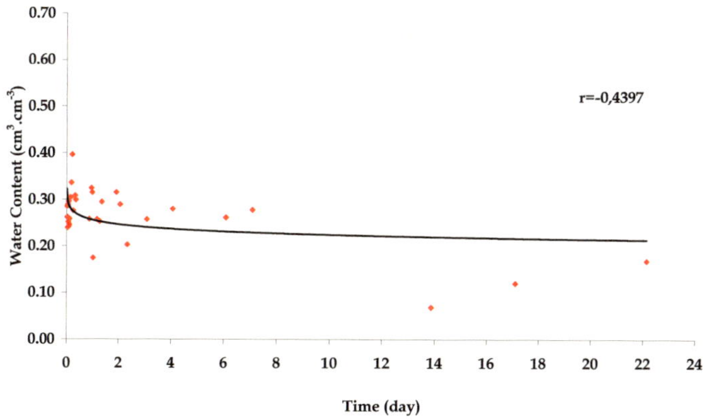

Fig. 7. Water redistribution z = 8cm.

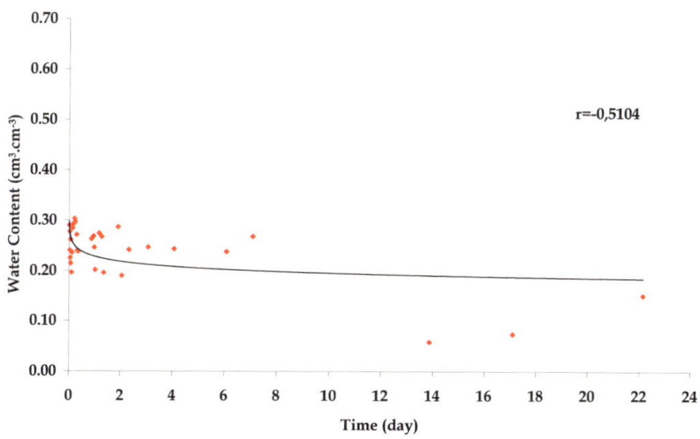

Fig. 8. Water redistribution z = 12cm.

Figures 12 and 13 present the results of moisture content versus time for two depths (14 and 28 cm) of another a soil column that was discarded. Increase of moisture content is verified in the course of time, quite different from the behavior presented in the previous figures. In this other column particularly, the soil was dried and cracks were created in the interface soil/paraffin, preferential pathways for water were found. The determination of the hydraulic conductivity of that soil column was not accomplished.

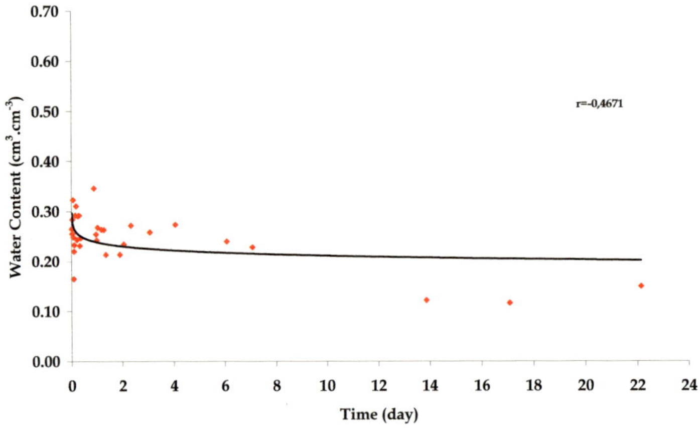

Fig. 9. Water redistribution z = 16 cm.

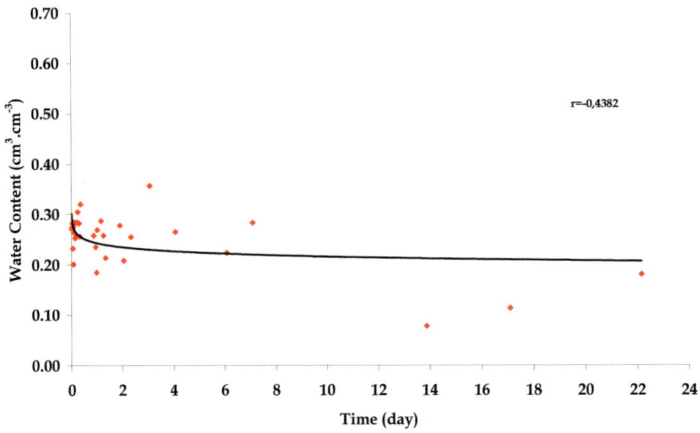

Fig. 10. Water redistribution z = 18 cm.

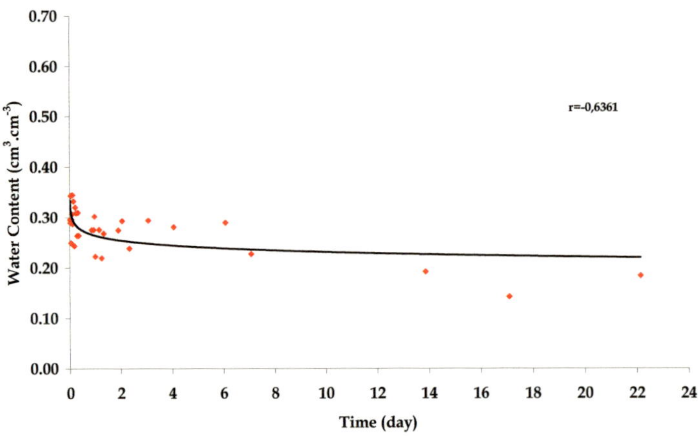

Fig. 11. Water redistribution z = 20 cm.

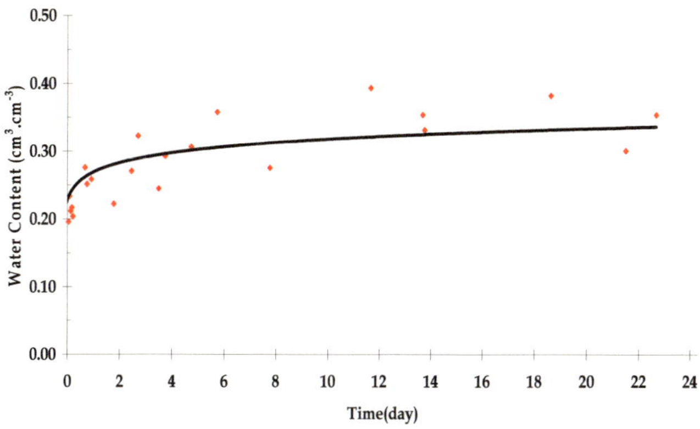

Fig. 12. Water redistribution z = 14 cm from a discarded soil column.

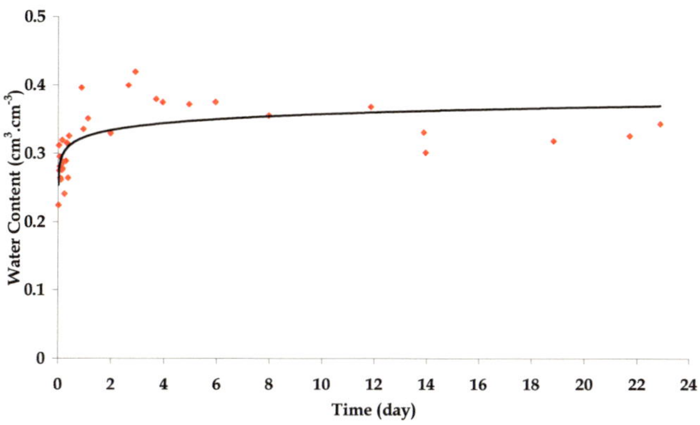

Fig. 13. Water redistribution z = 14 cm from a discarded soil column.

4.4 Empty column gammagraphy

When the redistribution measurements finished the soil column was opened. The soil was removed from the column and only the shield (paraffin and gauzes) was preserved. The gammagraphy of empty column was performed following the procedures of section 3.3. The gammagraphy data were used to calculate the soil linear attenuation coefficient and also the soil moisture content (equations 18 and 25) in the process of water infiltration and redistribution.

4.5 Hydraulic conductivity determination

With the values of moisture content [$\theta(z, t)$] in function of time for each depth level and also the data of redistribution, the values of hydraulic conductivity [K(θ)] by the method of Sisson et al., 1980 was determined.

The experimental values of moisture content are normalized with the linear regression θ= a ln t + b (according to equation 8), for a very short time (t = 0.0007 day). The normalized values of moisture content (which will be used in the method of Sisson), called moisture model or theoretical values of moisture content redistribution are shown in Table 2. In the first line of the table the values of saturation moisture can be found.

Using the data of moisture content redistribution with the model, the value of parameters γ, Ks was determined. With these parameters calculated for each depth level, it is possible to obtain the hydraulic conductivity function K (θ). There is a small correlationship between moisture content with densities measured in each depth level. The depth level z = 10 cm, for example, has the lowest density value, as shown in Figure 6. indicating a high porous region compared to the other depths.
The curve fitting coefficient r and the values a and b of the model $\theta = a \ln t + b$ to the experimental data are shown in Table 3.
The worst curve fittings to the experimental data were the depths z = 10 cm and 14 cm, where r = -0.2875 and -0.2721 respectively. In depths z=2 cm and 20 cm the values of r = -0.6730 and -0.6361, respectively, are the best results of the model application of the experimental data.
Table 4 presents the values of K_s and γ, obtained by the method of Sisson with the equation (6), for all depth levels of the soil column (z = 2 cm to 20 cm). There's an increase of the conductivities K_s in relationship to the depth. With these two parameters was possible to determine the hydraulic conductivity function K (θ) at each depth level, as can be seen in Table 5. Figure 14 shows the behavior of lnK [K ($cm.h^{-1}$)] in function of θ ($cm^3.cm^{-3}$) for the depth levels of z = 2 to z=20 cm.

Time (day)	z (cm)									
	2	4	6	8	10	12	14	16	18	20
0.0007	0.426	0.250	0.288	0.354	0.453	0.331	0.392	0.324	0.329	0.367
0.2	0.290	0.185	0.196	0.277	0.400	0.251	0.315	0.257	0.262	0.287
0.3	0.281	0.180	0.189	0.272	0.396	0.245	0.310	0.252	0.257	0.281
0.4	0.274	0.177	0.185	0.268	0.394	0.241	0.306	0.249	0.254	0.277
0.5	0.269	0.174	0.181	0.265	0.392	0.238	0.303	0.246	0.251	0.274
0.7	0.261	0.170	0.176	0.260	0.388	0.233	0.299	0.243	0.247	0.269
1.0	0.252	0.166	0.170	0.255	0.385	0.228	0.294	0.238	0.243	0.264
1.5	0.242	0.161	0.163	0.250	0.381	0.222	0.288	0.234	0.239	0.258
2.0	0.236	0.158	0.159	0.246	0.379	0.218	0.284	0.230	0.235	0.254
2.5	0.230	0.156	0.155	0.243	0.377	0.215	0.281	0.228	0.233	0.251
3.0	0.226	0.153	0.152	0.240	0.375	0.212	0.279	0.225	0.230	0.248
3.5	0.222	0.152	0.150	0.238	0.373	0.210	0.277	0.224	0.229	0.246
4.0	0.219	0.150	0.147	0.237	0.372	0.208	0.275	0.222	0.227	0.244
4.5	0.216	0.149	0.146	0.235	0.371	0.207	0.273	0.221	0.226	0.243
6.0	0.214	0.148	0.144	0.234	0.370	0.205	0.272	0.219	0.224	0.241
7.0	0.209	0.145	0.141	0.231	0.368	0.203	0.270	0.217	0.222	0.239
8.0	0.206	0.144	0.138	0.229	0.367	0.200	0.267	0.215	0.220	0.236
8.5	0.201	0.141	0.135	0.226	0.365	0.198	0.265	0.213	0.218	0.234
10.0	0.197	0.140	0.133	0.224	0.364	0.195	0.263	0.211	0.216	0.231
11.5	0.194	0.138	0.130	0.222	0.362	0.193	0.261	0.210	0.215	0.229
13.0	0.191	0.137	0.128	0.221	0.361	0.192	0.259	0.208	0.213	0.228
15.0	0.187	0.135	0.126	0.219	0.360	0.190	0.257	0.207	0.211	0.225
18.0	0.183	0.133	0.123	0.216	0.358	0.187	0.255	0.204	0.209	0.223
20.0	0,181	0,132	0,121	0,215	0,357	0,186	0,253	0,203	0,208	0,221

Table 2. Moisture content ($cm^3.cm^{-3}$) in each depth level.

terms	z (cm)									
	2	4	6	8	10	12	14	16	18	20
a	-0.024	-0.012	-0.016	-0.014	-0.009	-0.014	-0.013	-0.012	-0.012	-0.014
b	0.252	0.166	0.170	0.255	0.385	0.228	0.294	0.238	0.243	0.264
r	-0.673	-0.415	-0.484	-0.440	-0.287	-0.510	-0.272	-0.467	-0.438	-0.636

Table 3. Curve fitting correlation **r**.

z (cm)									
2	4	6	8	10	12	14	16	18	20
Hydraulic Conductivity (cm.h^{-1}) K_s									
2.863	2.771	5.823	6.489	5.632	10.166	11.323	11.275	12.679	17.038
γ coefficient									
41.915	86.605	61.820	73.969	106.525	70.824	74.184	85.142	85.179	70.430

Table 4. K_s and γ values.

Due to the exponential model for K (θ), small variations for water moisture θ provide great differences between the values of hydraulic conductivity.

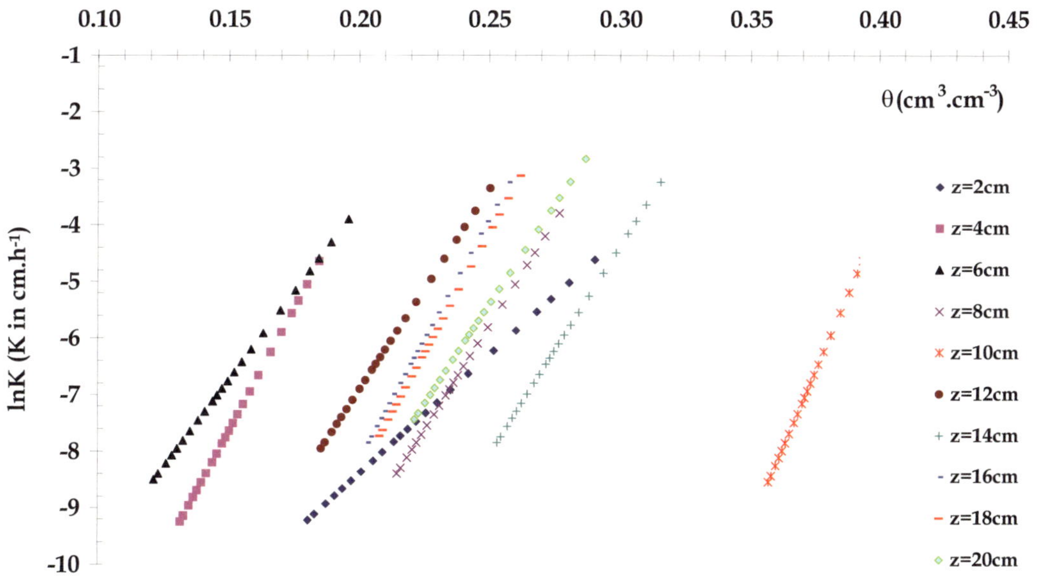

Fig. 14. ln hydraulic conductivity versus moisture content.

Time	z (cm)									
(day)	2	4	6	8	10	12	14	16	18	20
0.2	0.00994	0.00962	0.02022	0.02253	0.01956	0.03530	0.03932	0.03915	0.04402	0.05916
0.3	0.00663	0.00641	0.01348	0.01502	0.01304	0.02353	0.02621	0.02610	0.02935	0.03944
0.4	0.00497	0.00481	0.01011	0.01127	0.00978	0.01765	0.01966	0.01958	0.02201	0.02958
0.5	0.00398	0.00385	0.00809	0.00901	0.00782	0.01412	0.01573	0.01566	0.01761	0.02366
0.7	0.00284	0.00275	0.00578	0.00644	0.00559	0.01009	0.01123	0.01119	0.01258	0.01690
1.0	0.00199	0.00192	0.00404	0.00451	0.00391	0.00706	0.00786	0.00783	0.00880	0.01183
1.5	0.00133	0.00128	0.00270	0.00300	0.00261	0.00471	0.00524	0.00522	0.00587	0.00789
2.0	0.00099	0.00096	0.00202	0.00225	0.00196	0.00353	0.00393	0.00392	0.00440	0.00592
2.5	0.00080	0.00077	0.00162	0.00180	0.00156	0.00282	0.00315	0.00313	0.00352	0.00473
3.0	0.00066	0.00064	0.00135	0.00150	0.00130	0.00235	0.00262	0.00261	0.00293	0.00394
3.5	0.00057	0.00055	0.00116	0.00129	0.00112	0.00202	0.00225	0.00224	0.00252	0.00338
4.0	0.00050	0.00048	0.00101	0.00113	0.00098	0.00176	0.00197	0.00196	0.00220	0.00296
4.5	0.00044	0.00043	0.00090	0.00100	0.00087	0.00157	0.00175	0.00174	0.00196	0.00263
6.0	0.00040	0.00038	0.00081	0.00090	0.00078	0.00141	0.00157	0.00157	0.00176	0.00237
7.0	0.00033	0.00032	0.00067	0.00075	0.00065	0.00118	0.00131	0.00131	0.00147	0.00197
8.0	0.00028	0.00027	0.00058	0.00064	0.00056	0.00101	0.00112	0.00112	0.00126	0.00169
8.5	0.00023	0.00023	0.00048	0.00053	0.00046	0.00083	0.00093	0.00092	0.00104	0.00139
10.0	0.00020	0.00019	0.00040	0.00045	0.00039	0.00071	0.00079	0.00078	0.00088	0.00118
11.5	0.00017	0.00017	0.00035	0.00039	0.00034	0.00061	0.00068	0.00068	0.00077	0.00103
13.0	0.00015	0.00015	0.00031	0.00035	0.00030	0.00054	0.00060	0.00060	0.00068	0.00091
15.0	0.00013	0.00013	0.00027	0.00030	0.00026	0.00047	0.00052	0.00052	0.00059	0.00079
18.0	0.00011	0.00011	0.00022	0.00025	0.00022	0.00039	0.00044	0.00044	0.00049	0.00066
20.0	0.00010	0.00010	0.00020	0.00023	0.00020	0.00035	0.00039	0.00039	0.00044	0.00059

Table 5. Hydraulic conductivity K (θ) ($cm.h^{-1}$) values.

5. Conclusion

With the methodology employed was possible to analyze the temporal and spatial evolution of moisture content θ $(z,\ t)$ in the undisturbed soil column. The results of hydraulic conductivity using the method of Sisson indicate a decrease of K (θ) in relation to the depth. With the gamma ray transmission technique was also possible to detect a variation on the soil compaction inside the column, by the densitometry data.

The combination of the two methods applied in this study, Sisson and Gamma Ray Transmission, was effective to determine the hydraulic conductivity of undisturbed soil column. However, the combination can be used for application in infiltration of water in other porous media to evaluate the temporal evolution of the water flow.

Another important tool of the GRT, a non-destructive technique, is to analyze the porous structural variability of samples along its length. The gammagraphy, specially the densitometry presented in section 3.3, with some approximations, can also be applied to study internal structures of a large range of samples.

6. References

Al-Houri, Z.M.; Barber, M.E.; Yonge, D.R.; Ullman, J.L.; Beutel, M.W. (2009). Impacts of frozen soils on the performance of infiltration treatment facilities. *Cold Regions Science and Technology,* Vol. 59, 51–57

Appoloni, C.R. & Rios, E.A. (1994), Mass attenuation coefficients of brazilian soils in the range 10-1450 keV. *Applied Radiation Isotopes,* Vol. 45, 287-291

Buckingham, E. (1907) *Studies on the movement of soil moisture.* Bureau Soils Bulletin, 38, Washington DC, USA

Chong, S.K.; Green, R.E.; Ahuja, L.R. (1981). Simple in situ determination of hydraulic conductivity by power function descriptions of drainage. *Water Resources Research,* Vol. 17, 1109–1114

Coppola, A.; Santinib, A.; Botti, P.; Vacca, S.; Comegna, V.; Severino, G. (2004) Methodological approach for evaluating the response of soil hydrological behavior to irrigation with treated municipal wastewater. *Journal of Hydrology,* Vol. 292, 114–134

Darcy, H. (1856). *Les fontaines publique de la ville de Dijon,* Dalmont, Paris

Ferraz, E.S.B. (1974). *Determinação simultânea de densidade e umidade de solos por atenuação de raios gama do 137Cs e 241Am.* PhD. Thesis, ESALQ/USP, Piracicaba, Brazil

Gerward, N.; Guilbert, K.B.; Jensen and H. Levring. (2001), X ray absorption in matter. Reengineering XCOM. *Radiation Physics and Chemistry,* Vol. 60, 23-24

Hillel, D.; Krentos, V.D.; Stylianou, Y. (1972). Procedure and test of an internal drainage method for measuring soil hydraulic characteristics in situ. *Soil Science.* Vol. 114, 395–400

Jones, A.J. & Wagenet, R.J. (1984). In situ estimation of hydraulic conductivity using simplified methods. *Water Resource Research,* Vol. 20, 1620-1626

Klute, A. (1972). The determination of the hydraulic conductivity and diffusivity of unsaturated soils. *Soil Science,* Vol. 113, 264-276

Lax, P.D. The formation and decay of shock waves. (1972). *American Mathematical Monthly,* Vol. 79, 227-241, apud Sisson, J.B.; Ferguson, A.H.; Van Genuchten, M.TH. (1980). Simple method for predicting drainage from field plots. *Soil Science Society American Journal,* Vol. 44, 1147-1152

Libardi, P. L.; Reichardt, K.; Nielsen, D. R.; Biggar, J. W. (1980).Simple field methods for estimating soil hydraulic conductivity. *Soil Science Society American Journal,* V.44, 3-7.

Ma, Y.; Feng, S.; Su, D.; Gao, G.; Huo, Z. (2010). Modeling water infiltration in a large layered soil column with a modified Green–Ampt model and HYDRUS-1D, *Computers and Electronics in Agriculture,* 71S S40 - S47

Moreira, A.C., Appoloni, C.R. (2006). Mass attenuation coefficient of the Earth, Moon and Mars samples over 1 keV-100 GeV energy range. Applied Radiation and Isotopes. Vol. 64, 1065-1073

Oliveira, J.C.M. (1991). *Determinação de parâmetros do solo durante a infiltração horizontal e redistribuição da água por atenuação de raios gama e tensiometria.* PhD Thesis CENAQ/USP, Piracicaba, Brazil,

Portezan Filho, O. (1997). *Análise crítica da determinação da condutividade hidráulica do solo utilizando atenuação de radiação gama monoenergética.* PhD Thesis, IPEN/USP, São Paulo, Brazil

Reichardt, K. (1965) *Uso da radiação gama na determinação da umidade e densidade do solo.* PhD Thesis, Escola Superior de Agricultura Luis de Queiros. Piracicaba, Brazil

Reginato, R.J. & Van Bavel, C.H.M. (1964) Soil water measurement with gamma attenuation. *Soil Science Society American Journal,* Vol. 28, 721-724

Richards, L.A. (1931). Capillary conduction of liquid through porous media. *Physics*, Vol. 1 , 318-333

Siegbahn, K. *Alpha-, Beta- and Gamma-Ray Spectroscopy*. (1965) North-Holland Publishing Company, Amsterdam, New York, Oxford, 37-76

Sisson, J.B.; Ferguson, A.H.; Van Genuchten, M.TH. (1980). Simple method for predicting drainage from field plots. *Soil Science Society American Journal*, Vol. 44, 1147-1152

9

Hydraulic Conductivity of Semi-Quasi Stable Soils: Effects of Particulate Mobility

Oagile Dikinya
Department of Environmental Science, University of Botswana, Botswana

1. Introduction

Particulate mobility within the intra-soil aggregates and soil pores has profound effects on the stability of soils and consequently hydraulic conductivity. Soil structure or aggregation essentially describes the way the soil constituents (sand, silt, clay, organic matter) are arranged and the size and shape of pores between them (Geeves *et al.* 1996). On the basis of size, soil aggregates can be distinguished as macro-aggregates (> 250 μm) and micro-aggregates (< 250 μm) and with further breakdown, they can release finer particles (< 20 μm, Oades and Waters 1991). However, aggregates < 2 μm (predominately clay floccules) are held together by forces derived from the interaction of clay particles. The loss of macro-porosity is attributable to detachment and subsequent transport of aggregates (Roth *et al.* (1991; Sutherland *et al.* 1996) and further dis-aggregation can affect soil water retention (Neufeldt *et al.* 1999). Aggregates with low stability fracture easily and the breakdown into smaller sizes is a major cause reduced permeability through pore clogging as a result of loss of aggregation and porosity of surface soils. Once a crust is formed the water and solute transport properties change, for example increased rate of water infiltration with removal of crust by application of gypsum (Ramirez *et al.*,1999). This surface crusts or sealing is often attributable to slaking and dispersion of clay minerals.
Dispersion is most commonly associated with sodicity and is highly sensitive to both the exchangeable sodium percentage (ESP) and total electrolyte concentration. Soils with a high ESP are more susceptible to dispersion due to its greater ability to develop diffuse double layers. The exchangeable sodium percentage (ESP) can be defined as:

$$ESP = \frac{(100 * ExchangeableNa)}{CEC} \tag{1}$$

where, CEC is the cation exchange capacity.
Similarly, Sodium adsorption ratio (SAR) can also be used to define sodicity, particularly when using soil solutions.

$$SAR = \frac{\left[Na^{+}\right]}{\left(0.5\left(\left[Ca^{2+}\right]+\left[Mg^{2+}\right]\right)\right)^{0.5}} \tag{2}$$

Where, all [concentrations are expressed as mmol(+)/L].
In general, high levels of SAR decrease the stability of soil structures and microstructures usually become more unstable, deflocculated and dislodged as the ESP increases with more

production of fine or mobile particles. For example, in particulate facilitated transport, mobile colloids must be present in large concentrations and must be transported over significant distances (Kretzschmar *et al.* 1999). In most environments, mobilization is favoured by high pH, high SAR and low ionic strength resulting in severe decreases in permeability. Permeability or hydraulic conductivity is estimated using Darcy's law for one-dimensional vertical flow (Klute and Dirksen 1986):

$$q = K * \partial H / \partial z \tag{3}$$

where q is the flux density, K is the hydraulic conductivity, $\partial H / \partial z)$ is gradient of hydraulic head H ,and z is the gravitational head.

Despite the considerable amount of research carried out, the processes by which the structural breakdown and pore clogging by particulate transport occur in soils are far from satisfactorily understood and also limited by suitable modelling procedures. For instance McDowel-Boyer *et al.*(1986) and Harvey and Garabedian (1991) by applying filtration theory, attributed poor estimates of transport with natural porous media to; (i) the wider pore size and particle size distributions, complex pore geometry and rough matrix surface or these materials and (ii) a wide particle size distribution mobile surface colloids. Using a cake filtration model, Mays and Hunt (2005) suggested that the increased head loss results from the formation of deposits with a decline in hydraulic conductivity. Reduced flow capacity of the soil matrix can be induced by the invasion and geochemical transformation due to plugged pore channels (Song and Elimelech, 1995) thus minimising erosion of base soils with reduction in filter permeability (Lee *et al.*, 2002).

Thus research was undertaken in an attempt to throw light on the dynamics of the processes of mobilisation during the flow of NaCl solutions through two soil materials of markedly different structural cohesiveness. The experiments were carried using saturated soil columns to determine hydraulic conductivity and particle size distribution as the criterion for the assessment of the dynamic flow processes.

2. Materials and methods

2.1 Soil samples and their characteristics

Two different soil samples were used as porous media; sample A and B with distinct physico-chemical characteristics (Table 1) analysed using standard methods (Klute 1986;

Soil Property	Sample A	Sample B
Sand (%)	83.3	89.0
Silt (%)	6.6	1.9
Clay (%)	10.1	9.1
Texture	Sandy loam	Loamy sand
Electrical conductivity (μS/cm)	42	380
pH(water)	6.1	8.7
CEC (mmol$_c$/g)	3.4	8.1
Organic Carbon (%)	0.03	0.55
Clay mineralogy	Kaolinite,vermiculite	Kaolinte, smectite
Bulk density (g/cm^3)	1.61	1.58

CEC-cation exchange capacity

Table 1. Physico-chemical characteristics of soil samples

Klute 1986b). Sample A provides a classic example of a relatively fragile agricultural soil which is highly susceptible to structural breakdown and permeability problems while sample B is a completely disrupted and reconstituted material (sand and clay) and presents similar structural problems.

2.2 Particle size distribution measurements

The particle size distributions (Table 2) for the various sized fractions of these porous media were determined using the sedimentation or grain size mechanical wet sieving method (Day 1965) for the particle size range from 45-2000 μm while a laser light scattering technique (Mastersizer Microplus Ver.2.18, c/o Malvern Instruments Ltd, 1995) was used for the range < 45 μm. A mixture of air dried soil (<2 mm) with water was boiled and 6 % hydrogen peroxide solution was added to remove organic matter followed by a calgon/NaOH mixture to disperse the soil. Suspensions were collected for various fraction sizes according to the sedimentation theory or Stokes's law (Stokes, 1891). For the sand fractions suspensions were separated using sieves of various diameters (1000, 500, 250, 125, 45 μm). Suspensions passing through 45 μm together with effluents from subsequent leaching experiments were analyzed for particle size distribution using a Malvern Mastersizer analyser following leaching experiments. Suspensions were dispersed in the Mastersizer's ultrasonic bath unit (equipped with a small angle light scattering apparatus, Helium-neon laser; λ = of 633 nm, as a light source) for about 25 minutes. Suspension concentrations were adjusted until an obstruction of a least 0.2 % is reached for best results using refractive indices of 1.59 and 1.33, respectively for clay and deionised water with a particle density of 2.6 g/cm^3. Both values assumed to be representative for the soil material used for analysis.

	Sample A	**Sample B**
Size fractions (μm)	*% particle fraction (g/g)*	
1000 - 2000 μm	12.37	1.70
500 - 1000 μm	26.78	12.81
250 - 500 μm	21.30	57.46
125 - 250 μm	15.92	13.67
45 - 125 μm	2.41	2.40
< 45 μm	4.52	1.00

Table 2. Particle size fractions for samples A and B

2.3 Hydraulic conductivity measurements

2.3.1 Constant head hydraulic conductivity measurements

A glass Marriotte bottle, filled with deionised water was used to maintain a constant head, with ambient temperature during the experiment varying from 20 to 22 °C. The flow was from the bottom to the top of the columns to prevent air entry. The hydraulic head across the soil samples (ΔH) was kept at an average of 30 cm and 40 cm for samples A and B, respectively. ΔH was measured as equivalent to the vertical distance from upper water level (on the Marriotte bottle) to the bottom at the leachate collection point. The flow was continuous during leaching in order to maintain saturation throughout the experiments. In the case of flow interruptions, columns were kept saturated by tightly closing the system to avoid air entry. Hydraulic condtivity (HC) was initially measured using the deionised water followed by 1mmol NaCL to subsequently measure pore size distribution before and after

leaching with 1mmol NaCL. Any turbid percolate observed during the experiment was collected and stored in plastic bottles for further analysis of effluent particle size distribution using the Malvern Mastersizer technique, and for mineralogical characterization. Using a column and constant head experimental set up, equation 3 can be re-written as $K = (4VL) / (\pi d^2 \Delta t \Delta H)$:where V (cm^3) is the volume of water collected during time interval Δt (minutes) , L (cm) is the length of column soil sample, d (cm) is the inner diameter of the column and ΔH (cm) is the change in hydraulic head across the soil sample.
Pore-size distributions of the soil columns were estimated from water-retention versus water-potential curves determined using sintered glass funnels and the Haines method over the low pressure ranges (measured at 5 cm suction interval) and pressure-plate, ceramic-membrane apparatus for the higher pressure range (10 to 800 KPa). The column samples were pre-wetted with 0.1*M* $CaCl_2$ solution to avoid swelling and subsequently leached with deionised water followed by 1 mmol/L NaCL solution adjusted at pH 8.5. Since the soils were predominately kaolinitic in nature and therefore have low swelling potential. Similarly, the water-retention data was transformed to pore-size distribution using the form of Kelvin equation:

$$\Delta P = (2\sigma / rp) \tag{4}$$

where, ΔP is the pressure difference (Pa) across an air-water interface, σ is the surface tension of water(Jm-2) and *rp* is the radius of a circular capillary tube (m).

2.3.2 Pressure leaching experiments

Pressure leaching simulations and/or experiments were carried out using soil columns as shown schematically in Figure1. Soil samples (< 2 mm) were initially mixed with 0.9–1 mm acid-washed sand to provide a rigid skeletal structure to facilitate flow and colloidal mobility. The proportions of mixtures were 50% soil–50% sand and 30% soil–70% sand, for samples B and A, respectively. The mixtures were uniformly wet packed into acrylic soil columns (400 mm long, 25 mm diameter, 2 mm wall thickness, cross-sectional area 4.91 cm2, volume 196 cm3) and saturated with the desired electrolyte concentrations. Holes were bored at various column depths (Xi), for horizontal insertion of pressures sensors (Pi) at 50 mm (X1) and 250 mm (X2), along the column from the inlet (Fig. 1). Miniature pressure transducer with tensiometers T5-7/5 model (UMS, 2000) each with 1.5 m cable length and 5 cm shaft length and 20 mm diameter were connected to the system for online pressure monitoring and measurements of saturated hydraulic conductivity with time.
A peristalitic pump (designed with a system maximum discharge pressure of 300 kPa at 0 - 40° C), was used to pump solutions of ionic strengths of 1 mmol/L NaCl into the soil columns in a vertical upward flow. During leaching, pressures were measured. For continuous online pressure measurements, an *Agilent 34970A* model data logger-PC system (Agilent Technologies 2003) was used as a data acquisition system. To ensure equilibration of the system and to enable particle-hydrodynamic settling, the columns were saturated with 1 mmol/L NaCl solution overnight (for at least 12 hours). Prior to experimental runs, a steady state was attained by measuring outflow rates until a constant value was achieved. Hydraulic conductivity was then estimated from equation 3 (Darcian flow under hydrodynamic dispersion). Further, to simulate or quantify the effect of ionic strength, the columns were also leached with 10, 2.5 and 1 mmol/L NaCL solution to measure the extent of hydraulic conductivity decrease in response to the electrical conductivity.

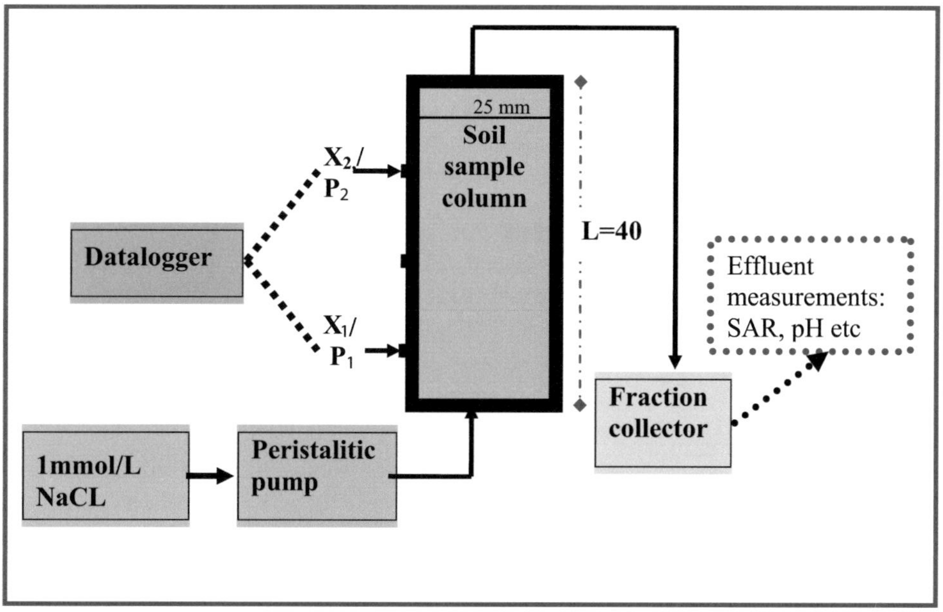

Fig. 1. Schematic diagram for pressure leaching experiment for hydraulic conductivity measurements ($X_{1(inlet)}$= 50 mm and $X_{2(outlet)}$= 250 mm from bottom of column and P1 and P2 are respective pressures points)

3. Results and discussion

3.1 Hydraulic properties associated with dispersion of aggregates

The pore-size distributions were obtained from water retention data (θ vs h) in Figs. 2a and 3a for sample A and B, respectively; where θ is volumetric water content and h is suction or pressure head. The pore radius r of samples were generating by differentiating *dθ/dh* from the measured water contents and pressure head or suction measurements. The pore-size distributions for the columns packed with samples are shown in Figs. 2b and 3b for sample A and B, respectively. The pore size distribution for the sample A soil columns were broader with the peak occurring at approximately 8 μm while those for the sample B were much narrower with a peak at approximately 12 μm.

The small shift in the narrow distribution for Sample A following leaching would be consistent with small dispersion and relocation of particles (see Figs 4 and 5). The broader distribution for the Sample B residue became substantially narrower following leaching (Figs 2 and 3). This change would explain the dramatic decrease of hydraulic conductivity by 4 orders of magnitude (See section 3.2). Similar observations were made by Leij *et al.* (2002) who observed that over the total number of pores decrease, the mean pore size decreased from an initial 49.4 to 28 μm after disturbance. This dispersion-dependent pore size distribution changes in samples were manifested by decrease of porosities: 0.370 to 0.354 cm^3/cm^3 for sample A and 0.431 to 0.417 cm^3/cm^3 for sample B following leaching with 1 mmol/L NaCl. i.e a decrease of porosity by 9% and 3% for sample A and B, respectively. This accounted for the subsequent decreases of actual hydraulic conductivity from 6.1 to 0.02 cm/h and from 1.5 to 0.14 cm/h for sample A and B, respectively (see section 3.2)

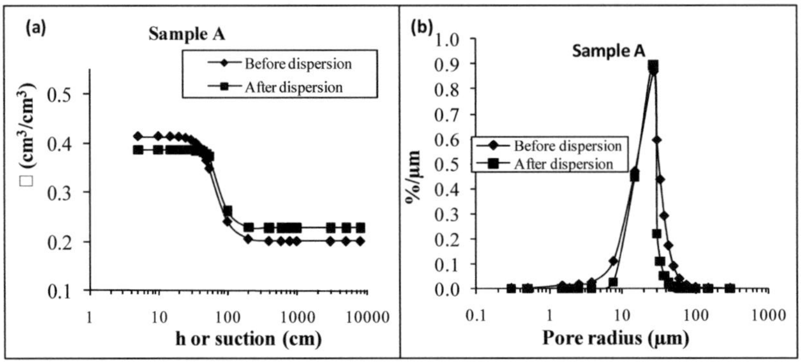

Fig. 2. Soil water retention (a) and pore-size distribution (b) for sample A

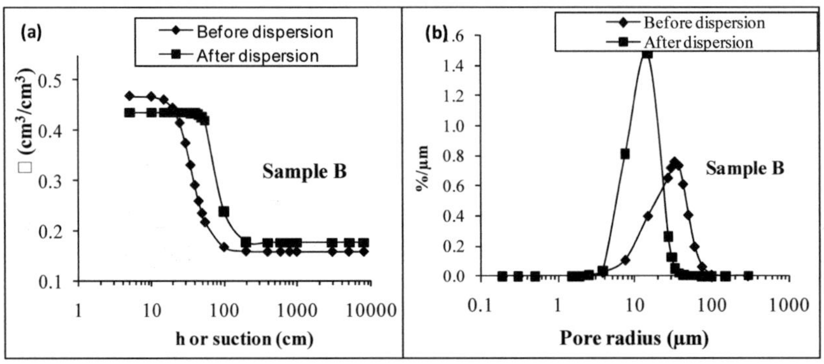

Fig. 3. Soil water retention (a) and pore-size distribution (b) for sample B

Similarly, using measured particle size distribution (Fig 4), soil structure was modelled at the pore scale (Dikinya *et al.*, 2007) to explain the different response of the two samples to the experimental conditions (see Fig 5). The size of the pores was determined as a function of deposited clay particles. The modal pore size of sample B as indicated by the constant water retention curve was 45 µm and was not affected by the leaching process. In the case of the sample B, the mode changed from 75 to 45 µm. This reduction of pore size corresponds to an increase of capillary forces that is related to the measured shift of the water retention curve of the two samples.

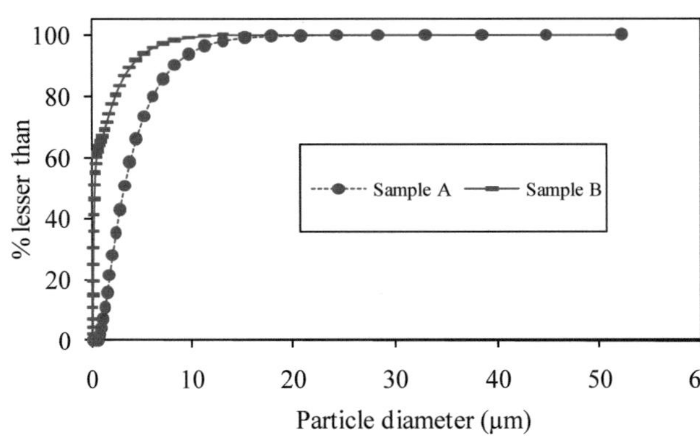

Fig. 4. Particle size distribution of effluents

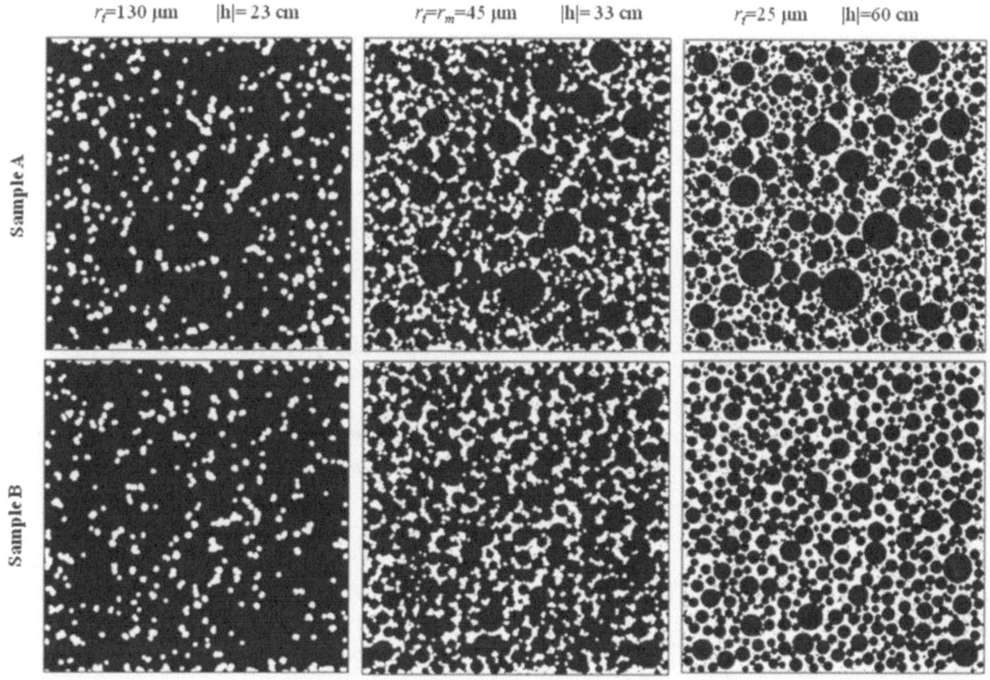

Fig. 5. Size-dependent pore connectivity for soil samples A and B. All particles and pores smaller than the threshold value r_t , are shown in black and pores of radius $r \geq r_t$ in white.

Source: [Dikinya *et al.* 2007].

Comparison of the modelled pore size distribution with the measured water retention curves for the four cases; (i) Sample B before the release process, (ii) Sample A before the release, (iii) Sample B after the re-deposition and (iv) Sample A after the re-deposition, have revealed that measured water retention curves were very similar for the cases i), iii) and iv). The air entry values of these three curves were 2-3 times the value observed for sample B before the release process (ii). The same phenomenon was simulated with the pore model: the mode of the modelled pore size distribution was similar (about 45μm) for (i), (iii) and (iv) and about 75 to 95 μm for ii). The ratio of the capillary forces associated with these values of the modes is somehow equal to 1.67 to 2.11 and is similar to the measured ratio of the air entry values.

3.2 Hydraulic conductivity under constant head using Mariotte bottle device

The hydraulic conductivity (HC) was found to progressively decrease with time. This decrease was attributed to decreases of pore radii associated with increased instability of soil intra-aggregates during leaching. Sample B was substantially more prone to structural disintegration than sample A with actual hydraulic conductivity decreases from 6.1 to 0.02 cm/h and from 1.5 to 0.14 cm/h, respectively as manifested by a relative marked measured shift of the water retention curve. In sample B there was appreciable decrease of the HC to 8.5 % of the initial value. This was attributed to localized pore clogging (similar to a surface seal) affecting hydraulic conductivity, but not the microscopically measured pore size distribution or water retention. This decrease could be explained with the modelled destruction of large pores due to particle re-deposition.

3.3 Hydraulic conductivity associated with particle mobility and pressure build-up

The dynamics of the process of structural disintegration in soil columns were further evaluated by simultaneously measuring changes in pressure gradients along the columns and sodium adsorption ratio (SAR) during pressure leaching with solutions of 10, 2.5, 1 mmol/L NaCl. Steady increases in pressure gradient (ΔP/ΔL) and corresponding decreases in RHC with time were observed for both soils and follow similar trends at all column depths indicating continuous particle accumulation in filter pores. The most severe increases in ΔP/ΔL and decreases in RHC always occurred near the inlet to the columns and the decline gradually decreased along the column. An increase of ΔP/ΔL and decrease in RHC with decreasing ionic strengths was also observed for both soils. The decreases in RHC and increases in ΔP/ΔL were clearly influenced by the size as well as the concentration of migrating particles in the porous medium. The finer mobile particles in the mining residue were clearly more readily self-filtered at the lower electrolyte concentration than the larger sample B particles, producing more rapid increases in ΔP/ΔL and decreases in RHC. This more effective particulate movement and more rapid plugging is undoubtedly due to increased development of the diffuse double layer, swelling and dispersion within the soil matrix at these concentrations.

A more dramatic decrease of saturated hydraulic conductivity was noticeable for the less cohesive sample A than for the aggregated sample B. Due to differences in the clay mineralogy and its treatment history involving greater disruption, sample A is far more sensitive to particle mobilisation and pore clogging than sample B. Sample A is therefore more likely to encounter constriction and retention thus allowing smaller sized particles to become entrapped behind the coarser fraction, forming a filtration zone. The more clogging occurs the greater is the decrease of hydraulic conductivity and the smaller is the amount of particles being released. This particularly enhanced at the lower electrolyte concentrations. Further, RHC of both soil samples has been shown to decrease with time and with increasing SAR with the reduction being substantially greater for the less cohesive sample B. This was probably due to dilution of high-sodicity soil irrigation water that can cause induce swelling, aggregate slaking and particle clay dispersion (Bagarello *et al.* 2006).

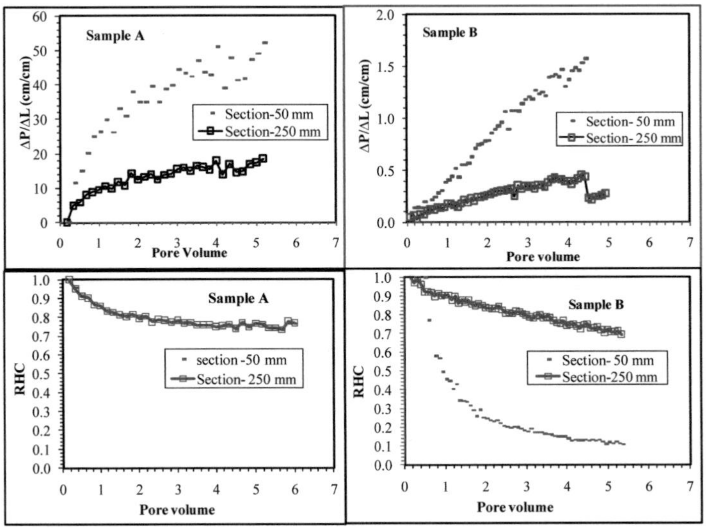

Fig. 6. Pressure build up (upper curves) and hydraulic conductivity (lower curves) attributable to structural disintegration and particle migration during leaching experiments.

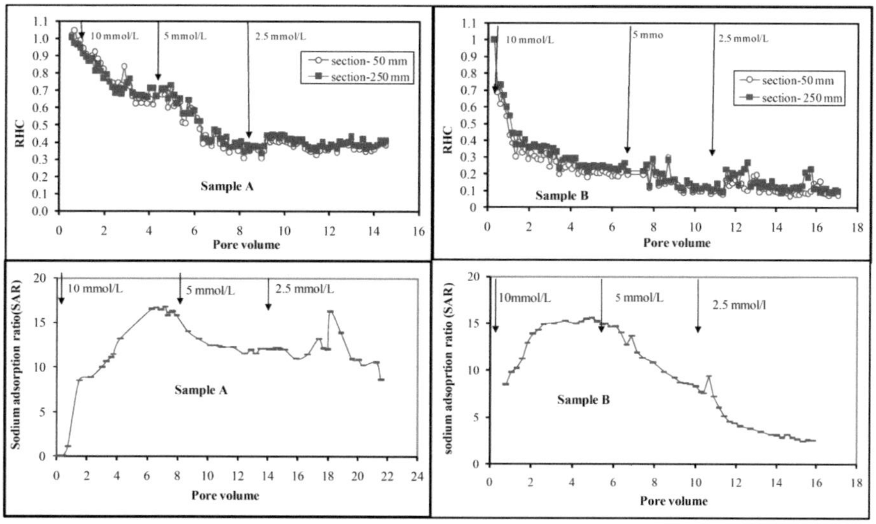

Fig. 7. Effect of ionic strength on relative hydraulic conductivity (RHC) and its associated sodium adsorption ratio (SAR) during leaching experiments.

4. Conclusions

The dynamics of the process of structural disintegration in soil columns were evaluated by simultaneously measuring changes in pressure gradients along the columns and concentrations during pressure leaching. The decreases in RHC and increases in $\Delta P/\Delta L$ were clearly influenced by the size as well as the concentration of migrating particles in the porous medium. The finer mobile particles in Sample B were clearly more readily self-filtered at the lower electrolyte concentration than the larger sample A, producing more rapid increases in $\Delta P/\Delta L$ and decreases in RHC. This more rapid decrease in RHC particularly at the lowest concentration (1 mmol/L) was consistent with measured SAR. The effects of clay mineralogy are evident with the kaolinite- smectite sample B having a more marked decrease in RHC with increasing ionic strength compared with the kaolinitic sample A clays.

5. References

Bagarello V, Iovino M, PalazzoloE, Panno M, Reynolds WD (2006) Field and laboratory approaches for determining sodicity effects on saturated soil hydraulic conductivity. *Geoderma 130, 1–13.*

Day PR (1965) Particle fractionation and particle-size analysis. In 'Methods of soil analysis part I'. Agronomy 9 (Ed. CA Black) pp 545–567, (American. Society of Agronomy Madison, Wisconsin).

Dikinya O, Lehmann P, Hinz C and Aylmore LG (2007) Using a pore scale model to quantify the effect of particle re-arrangement on pore structure and hydraulic properties. *Hydrological Processes,* 21, 989–997.

Greene RS, Nettleton WD, Chartres CJ, Leys JF, Cunningham RB (1998) Runoff and micromorphological properties of a grazed haplargid near Cobar, NSW, Australia. *Australian Journal of Soil Research 36, 87–108.*

Geeves G, Cresswell H, Murphy B, Chartres C (1996) Productivity and sustainability from managing soil structure. *Extension brochure. (CSIRO Division of Soils and NSW Department of Land and water conversation: Canberra)*

Harvey RW, Garabedian SP (1991) Use of colloid filtration theory in modelling movement of bacteria through a contaminated sandy aquifer. *Environmental Science Technolology 25, 178–185*

Klute A (1986b) Water retention: Laboratory methods. In "Methods of soil analysis. Part 1–Physical and Mineralogical Methods. Second Edition". Agronomy Monograph No 9: American Society of Agronomy/ Soil Science Society of America (Ed. A Klute), Madison, Wisconsin.

Klute A (1986) Methods of soil analysis. Part 1–Physical and Mineralogical Methods. Second Edition". Agronomy Monograph No 9: American Society of Agronomy/ Soil Science Society of America (Ed. A Klute), Madison, Wisconsin.

Kretzschmar R., Borkovec M, Grolimund D, Elimelech M (1999) Mobile Subsurface Colloids and their role in contaminant transport. *Advances in Agronomy 65, 1 – 96.*

Leij FL, Ghezzehei TA, Or D (2002) Analytical models for soil pore-size distribution after tillage. *Soil Science Society of America Journal 66, 1104–1114.*

Lee IM, Park YJ, Reddi LN (2002) Particle transport characteristics and filtration of granitic residual soils from the Korean peninsula. *Canadian Geotechnical Journal 39, 472–482.*

Lehmann P, Wyss P, Flisch A, Lehmann E, Vontobel P, Kaester A, Beckmann F, Frey O, Gygi A, Flühler H (2006) Mapping the structure of porous media using tomography with X-ray, thermal neutrons and synchrotron radiation, *Vadoze Zone Journal 5, 80–97; doi:10.2136.*

Malvern Instruments Ltd (1995). "Malven Mastersizer Microplus version ver 2.18". (Engigma Business Park: Gronwood Road, UK).

Mays DC, Hunt JR. (2005) Hydrodynamic Aspects of Particle clogging in Porous Media. *Environmental Science and Technology 39, 577 –584*

McDowell-Boyer LM, Hunt JR, Sitar N (1986) Particle transport through porous media. *Water Resources Research 22, 1901–1921.*

Neufeldt H, Ayarza MA, Resck DVS, Zech W (1999) Distribution of water-stable aggregates and aggregating agents in Cerrado. *Geoderma 93, 85–99.*

Oades JM, Waters (1991) Aggregate hierarchy in soils. *Australian Journal Soil Research 29, 815–828.*

Roth CH, de Castro Filho C, de Mediros GB (1991). Analise de fatores fiscos e quimicos relacionados com a agregacao de um latossolo xoxo distrofico. *R Bras. Ci. Solo 15, 241–248.*

Song L, Elimelech M (1995) Particle deposition onto a permeable surface in laminar flow. *Journal of colloid and interface science 173, 165–180.*Stokes GG (1891) Mathematical and Physical Papers III, *Cambridge Press.*

Sutherland RA, Watung RL, El-Swaify (1996) Splash transport of organic carbon and associated concentrations and mass enrichment ratios for Oxisol, Hawaii. *Earth surface processes landforms 21, 1145–1162.*

Part 4

Implications of Hydraulic Conductivity on Land Management and Policy Development

10

Saturated Hydraulic Conductivity and Land Use Change, New Insights to the Payments for Ecosystem Services Programs: a Case Study from a Tropical Montane Cloud Forest Watershed in Eastern Central Mexico

Alberto Gómez-Tagle (Jr.) Ch.[1], Daniel Geissert[2], Octavio M. Perez-Maqueo[2], Beatriz E. Marin-Castro[2] and M. Beatriz Rendon-Lopez[1]
[1]*INIRENA, Universidad Michoacana de San Nicolás de Hidalgo, Morelia, Michoacán,*
[2]*Instituto de Ecología, A.C., Xalapa, Veracruz,*
Mexico

1. Introduction

Water infiltration into soil is a complex process that in field conditions varies for every precipitation event (Wit, 2001) due mainly to its dependence of antecedent soil moisture (Cerdà, 1995; Lassen & Lull, 1951). Some authors use saturated hydraulic conductivity (K_s) as a descriptor of the infiltration process (Wit, 2001; Ziegler, et al., 2004). This hydrophysical variable allows field based comparison between sites with different initial moisture contents and soil characteristics. Several researches report differences in infiltration and K_s, associated to vegetation patches (Cerdà & Doerr, 2005), land use change patterns (Buytaert, et al., 2005; Tobón, et al., 2004; Ziegler, et al., 2004) and vegetation recovery (Li & Shao, 2006; Zimmermann & Elsenbeer, 2008). This trend has been reported in many different ecosystems and vegetation types ranging from tropical rain forests (Zimmermann & Elsenbeer, 2008) to semiarid and Mediterranean shrublands (Cerdà & Doerr, 2005; Li & Shao, 2006). Differences can be marginal or up to several orders of magnitude (Li & Shao, 2006). This allowed the possibility of using land use and plant cover as an indicative variable of the infiltration process.

Payment for ecosystem services (PES) schemes include in most cases a "consumer" that pays the "provider" for maintaining the ecosystem functions that generate the ecosystem services in question. Perhaps, two of the most common examples of payment for ecosystem services are the carbon sequestration programs (Ordoñez, et al., 2008) and the hydrologic service initiatives (Naranjo & Murgueitio, 2006).

In Mexico, the Federal Government has recently developed a strategy of payment for ecosystem services (PES) which encompass biodiversity, carbon sequestration and hydrologic ecosystem services. The Mexican program in 2008 had the largest budget worldwide for such an initiative (60 millions US dollars) (CONAFOR, 2008). While an important step in the incorporation of economics to conservation, some authors pointed out that the Federal government initiative is based on unverified relationships between land use/cover and water flow in the soil and hydrologic response of watersheds (Gómez-Tagle, 2009; Pérez-Maqueo, et al., 2005).

Initiatives at the local scale have also been established to compliment Federal programs. The first local PES initiative in Mexico was developed in 2003 in the municipality of Coatepec. The city of Coatepec (population 73,500) is in the state of Veracruz and receives over 98 percent of its' water from surface flow from the Gavilanes river. Therefore the PES municipal program is focused on the preservation of cloud forests in the Gavilanes watershed through monetary incentives. This program makes the assumption that mature cloud forest cover maintains year round stream water flow in the headwaters, and that this type of forest favors water infiltration into soil allowing both ground water recharge and water storage in the soil. Nevertheless, implementation of PES in the area follows a binary approach with two levels; forested, namely land with tree coverage, and unforested or land without tree coverage. By definition within the local PES initiative, the first level is considered appropriate for economic compensation while the latter is not.

Recent research (Gómez-Tagle and Geissert, unpublished) indicate that this watershed includes many different land use types/coverages, some of them are related to diminished infiltration and hydraulic conductivity (Karlsen, 2010; Marín-Castro, 2010), thus infiltration capacity may vary significantly from one to another.

In this chapter we examine the relationships between land use/cover and key hydrophysical variables in order to strengthen and aid policy making related to PES initiatives in the area. Specifically, the analysis addresses the following questions: 1) What is the relationship between land use/cover and saturated hydraulic conductivity? and 2) Which transitional/successional stages are linked to higher infiltration rates and are prone to be included in PES initiatives?

2. Methods

2.1 Study area

The Río Gavilanes watershed is a 33.2 km^2 exoreic catchment in the headwaters of La Antigua River basin. The watershed is located between 97°06´09.46" and 96° 59´52.67" W and 19°31´46.38" and 19°27´33.2" N on the windward slope of Cofre de Perote strato-volcano in the state of Veracruz in eastern central Mexico (Figure 1). The elevation of the watershed ranges from 1180 to 2960 m above sea level. Environmental conditions result from three different climatic subtypes according to Köppen's system modified by García (2004): subtropical humid in the low portion, temperate humid in the middle, and cool temperate humid in the high portion. Mean annual temperature for the three portions is 19.3, 14.3 and 9.3 °C respectively, while total annual rainfall is 1800.4 (low), 3036.9 (middle) and 1724.4 (high) mm. The hottest month is May and the coldest is January (Figure 2). Precipitation shows a different trend; the low and high portions of the watershed receive similar amounts of rainfall during the year while the middle portion receives a significant higher amount of precipitation (Figure 2). The annual potential evapotranspiration estimated by means of the modified Thornthwaite (1948) method is 1455.0 (low), 850.3 (middle) and 588.9 (high) mm and the annual estimated water budget is 345.5 (low), 2186.7 (middle) and 1135.4 (high) mm. The low portion has seven months (November-May) of negative water budget, while

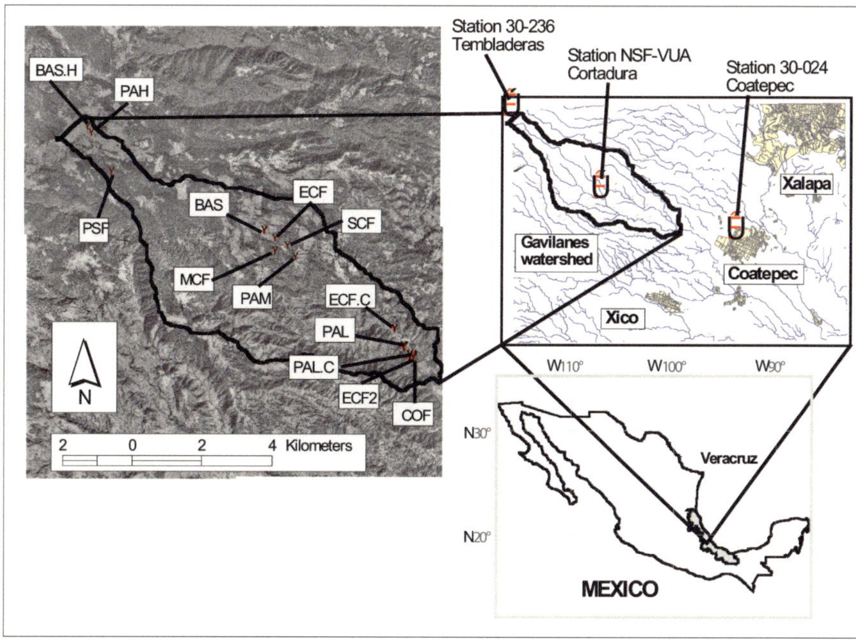

Fig. 1. Study area and sampling sites; *Bacharis-Pteridium* shrub middle (BAS), *Bacharis* shrub high (BASH), coffee plantation low (COF), early cloud forest regeneration middle (ECF), early cloud forest regeneration low (ECF2, early cloud forest regeneration from coffee plantation low (ECFC), mature cloud forest middle (MCF), pasture high (PAH), pasture low (PAL), well managed pasture low (PALC), pasture middle (PAM), pine-spruce forest (PSF), secondary cloud forest (SCF).

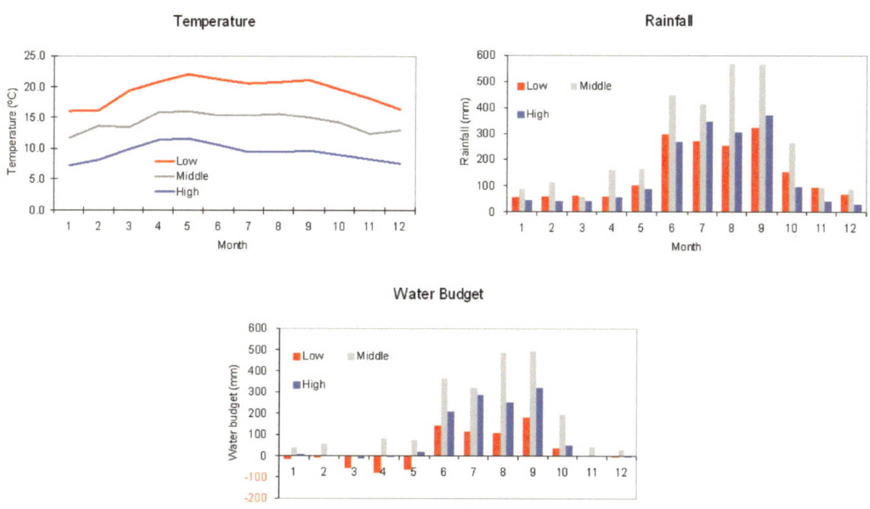

Fig. 2. Climatic variables in the study area; mean monthly temperature, mean monthly rainfall and monthly water budget. (Data from weather stations 30-236 Tembladeras and 30-024 Coatepec. The low and high portion climatic data from García, (2004). Middle portion data from NSF-UVA Cortadura station Holwerda et al. (2010).

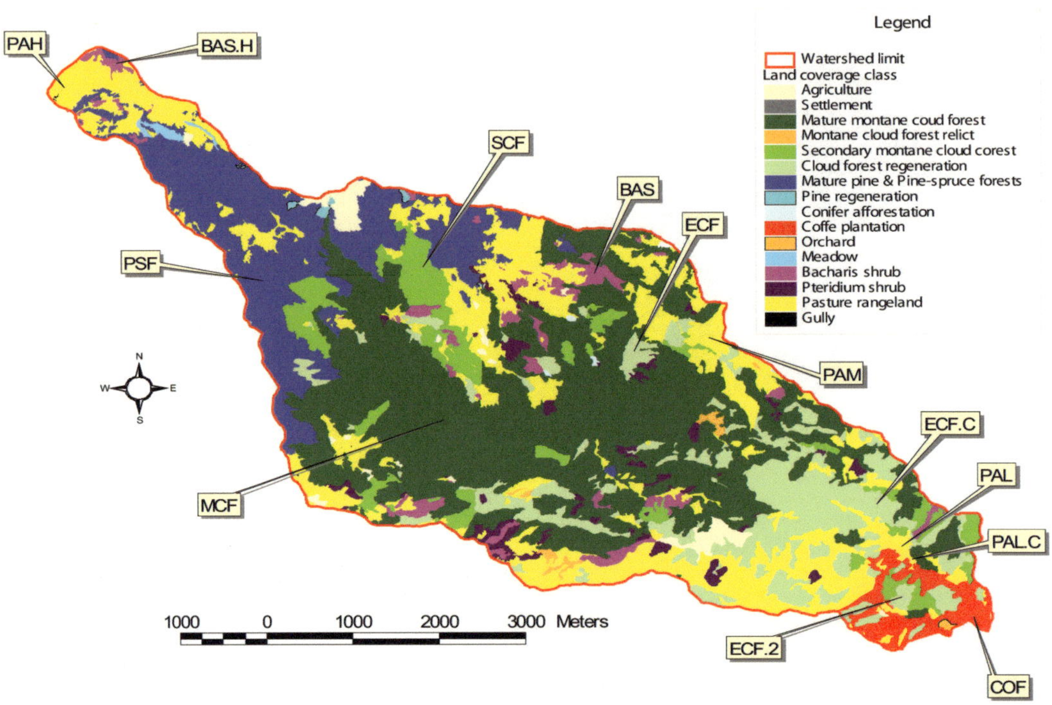

Fig. 3. Land use coverage map of Gavilanes watershed for 2009 (Gómez-Tagle and Geissert, unpublished map).

the high portion has only three (March-April and December). The budget figures for the middle portion indicate only one month of deficit (March) (Figure 2). All water budget estimates are positive indicating total annual surplus in the whole watershed.

2.2 Sampling procedures

Sampling took place between October 2008 and November 2009, following a stratified model design that included each major portion of the watershed (high, middle, low) and the land use/coverage class (Figure 1). A summary of sites and measurements is presented in Table 1.

Unsaturated infiltration measurements were conducted in the field using INDI-INECOL tension infiltrometers based on the design of Špongrová (Kechavarzi, et al., 2009; Špongrová, 2006; Špongrová, et al., 2009) with 10.0 cm diameter and an effective contact surface of 0.00785 m^2 (Figure 4). Commercially available grounded sand (92% sand) was used as contact material. Water height in the infiltrometers was recorded every two minutes with a Campbell Scientific CR1000 datalogger and several pre-calibrated Motorola Free-Scale MPX2010DP differential pressure transducers with 0 to 10 kPa pressure range (Motorola, 2002). Recording was carried out until unsaturated flow reached a steady state condition which usually occurred before two hours of elapsed time at a particular tension.

Land use & coverage	Watershed portion	Key	Actual Payment	Altitude (masl)	Observations	Infiltration measurements
Mature cloud forest	Middle	MCF	Yes	2100-2160	-	16
Secondary cloud forest	Middle	SCF	Yes	2090-2150	-	16
Early regeneration cloud forest	Middle	ECF	Yes	2200-2240	-	30
	Low	ECF2		1360-1420	-	30
	Low	ECF.C		2140-2180	From coffee plantation	30
Coffee plantation	Low	COF	No	1350-1400	Shade coffee plantation	30
Shrub	Middle	BAS	No	2280-2300	*Bacharis spp.* y *Pteridium sp.* Shrub	30
	High	BAS.H	No	2940-2950	*Bacharis spp.-Muhlembergia sp.* Shrub	12
Pasture	Middle	PAM	No	2050-2120	Degraded pasture	17
	Low	PAL	No	1370-1420	Degraded pasture	30
	Low	PAL.C	No	1360-1390	Well managed pasture	10
	High	PAH	No	2930-2940	Well managed pasture	18
Pine-spruce forest	High	PSF	Yes	2870-2890	Pine-spruce (pine-*Abies religiosa*) forest	34

Table 1. Unsaturated infiltration sampling measurements for each land use/coverage class and watershed portion.

The exponential Gardner equation (Equation 1) describes the relationship between tension and hydraulic conductivity, $K(h)$ is the unsaturated conductivity, K_s is the saturated conductivity, a is the inverse of capillary length in the soil and h corresponds to tension (Gardner, 1958). Using this model it is possible to estimate K_s using tension infiltrometer data.

The Logsdon and Jaynes (1993) non-linear simultaneous solution (Equation 2) allows the estimation of K_s and a parameter using the infiltrometer radius r, the π constant (3.1416), $q(h)$ or steady state infiltration flow for the applied tension h. This approach requires the

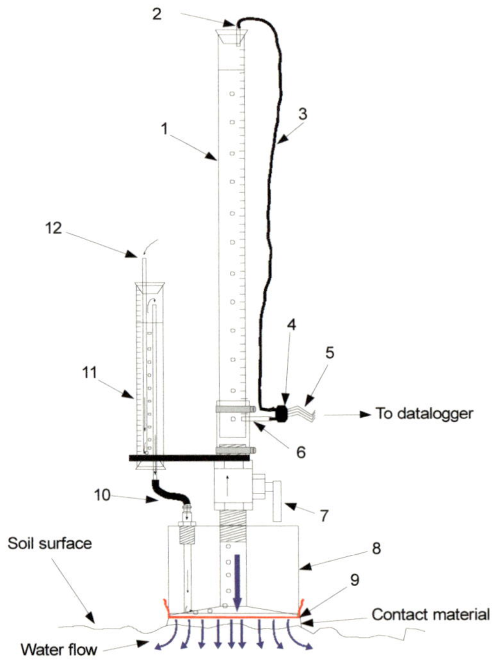

Fig. 4. Schematics of the INDI-INECOL tension infiltrometer used in this study. Main reservoir (1), top rubber lid with tubing connection (2), pressure transducer tubing to air chamber in main reservoir (3), MPX2010DP differential pressure transducer (4), four line cable to datalogger (5), reservoir to transducer tubing connection (6), two way valve (7), machined aluminum base (8), 15 µm nylon mesh (9), mariotte to base flexible connection (10), mariotte reservoir (11), air inlet (12). Arrows depict air and water flow in and out the infiltrometer.

measurement of steady state flow rates at two or more tensions (h). In our case we used three tensions applied in a decreasing order (-0.882, -0.343 and -0.049 kPa).

$$K(h) = K_s \cdot e^{\alpha h} \tag{1}$$

$$q(h) = [1 + (4 / \pi r \cdot \alpha)] \cdot K_s e^{\alpha h} \tag{2}$$

The method proposed by Watson and Luxmoore (1986) allows the estimation of number of pores of the effective conductive porosity, this is the interconnected porosity that actually conducts water at certain tension. Number of pores, percentage of pore space and proportion of infiltration flow was estimated for three apparent diameter intervals of effective pores ≥ 1.0 mm (-0.0145 kPa), between 0.3 and 1.0 mm (-0.049 to -0.0145 kPa), and between 0.01 and 0.3 mm (-1.47 to -0.049 kPa). The largest apparent diameter interval corresponds to macropores while the other size classes represent mesopores. In order to compare pore size contribution to total infiltration flow, proportion of infiltration flow in percentage was determined from $K(h)$ defined by the Gardner equation (Equation 1) (Gardner, 1958), and applying Equation 3, where FP_d is the flow proportion for certain diameter interval d, K_d is the unsaturated hydraulic conductivity for the pore diameter interval and K_s is the saturated hydraulic conductivity. The proportion of infiltration flow was calculated for the apparent pore diameter categories listed above but also for < 0.01 mm (> -1.47 kPa) micropores according to Watson and Luxmoore (1986).

$$FP_d = (K_d / K_s) \cdot 100 \tag{3}$$

In addition to the unsaturated infiltration measurements in the field, soil samples were obtained by carefully removing the contact material from the soil surface. Bulk density was determined by the core method (cylinder 5.5 cm diameter and 4.0 cm length) (Miller & Donahue, 1990). Particle size distribution was performed by the Bouyoucos Hydrometer Method (Gee & Or, 2002), organic carbon by means of dry combustion on a CN Truspec LECO analyzer and final moisture content by gravimetry. Coarse sand and fine sand fractions were separated by sieving.

2.3 Data analysis

Data analysis included Shapiro-Wilk normality tests, Pearson´s correlation analysis between physicochemical properties and target hydrophysical variables (K_s, α), analysis of variance (ANOVA) and mean comparison test applying the Tukey Honest Significant Differences (Bates, 2006).
Given the hierarchical structure of sampling in which data were collected at different spatial scales, a mixed-effect model was conducted in order to elucidate if the current payment scheme classes depicted statistical differences in K_s. This model was constructed as follows: saturated conductivity was log transformed (Log K_s) and analyzed as the response variable. A categorical variable named Payment, with two levels (yes and no) was generated and associated with each land use type depending on the possibility that a determinate land use was considered within the payments for hydrological ecosystem services or not. This follows the municipality ranking for ecosystem services payment scheme, namely forested and not forested, practically defined as the presence or absence of trees. This scheme does not differentiate between cover classes like mature cloud forest and pine plantations which may differ substantially in K_s and infiltration. The model was analyzed using lmer function released by Bates (2006) and specified according to Crawley (2007). Thus random effects were listed from largest to smallest spatial scale as follows: zone/payment/land uses/site.
All data analysis were performed using standard mathematical and statistical methods within the R language and environment for statistical computing (R version 2.10.1) (R_Development_Core_Team, 2004) and the packages for R *lmer4* (Bates & Maechler, 2010) and *lattice* (Sarkar, 2008).

3. Results

3.1 Hydraulic conductivity (K_s)

Two hundred and sixty five unsaturated infiltration measurements were successfully conducted and processed. K_s values obtained with the Logsdon and Jaynes method (1993) ranged from 2.8 x 10^{-7} to 2.42 x 10^{-5} m s^{-1}, while the α parameter was between 0.1 and 7.96 m^{-1}. The overall K_s data set did not exhibit a normal distribution either directly or through log-transformation. However, Shapiro-Wilk test of normality on original K_s and log-transformed K_s of watershed portions confirmed the log-normal distribution at this scale (Table 2). At the individual land use/cover scale the log-normal distribution was confirmed for eleven out of thirteen classes (Table 3). This concurs with earlier published results concerning the probabilistic distribution of K_s and other flow related soil properties (Esteves, et al., 2005; McIntyre & Tanner, 1959; Rogowski, 1972; Russo & Bresler, 1981; Soil Survey Staff, 1993).

	Original K_s data (m·s^{-1})							
Watershed Zone	min	median	mean	max	sd	W	p Value	Normality
High	2.49E-06	7.24E-06	7.84E-06	1.87E-05	3.42E-06	0.916	0.001	No
Low	2.80E-07	4.49E-06	5.40E-06	2.42E-05	4.68E-06	0.842	<0.001	No
Middle	8.33E-07	3.11E-06	3.98E-06	1.68E-05	2.61E-06	0.846	<0.001	No
	Log transformed K_s (ln m·s^{-1})							
Watershed Zone	min	median	mean	max	sd	W	p Value	Normality
High	-1.29E+01	-1.18E+01	-1.18E+01	-1.09E+01	3.42E-06	0.991	0.939	Yes
Low	-1.51E+01	-1.23E+01	-1.25E+01	-1.06E+01	4.68E-06	0.982	0.124	Yes
Middle	-1.40E+01	-1.27E+01	-1.26E+01	-1.10E+01	2.61E-06	0.992	0.882	Yes

Table 2. Descriptive statistics and normality (Shapiro-Wilk) analysis for K_s of different watershed portions (α=0.95).

The Tukey HSD test revealed the existence of five groups for natural log- transformed hydraulic conductivity values according to land use/cover classes (a to e in Figure 5A) and two groups according to watershed portion (Figure 5B). In Figure 5A the low case letters indicate membership to the groups. Because of the in-class dispersion many land use/cover classes belong to different groups, except for four classes that belong only to group d. Red lines indicate K_s threshold values that separate low, intermediate and high K_s values within the data set, while blue dashed lines indicate Soil Survey Staff (1993) class limits; low (1 x 10^{-7} m·s^{-1}), moderately low (1 x 10^{-6} m·s^{-1}) and moderately high (1 x 10^{-5} m·s^{-1}).

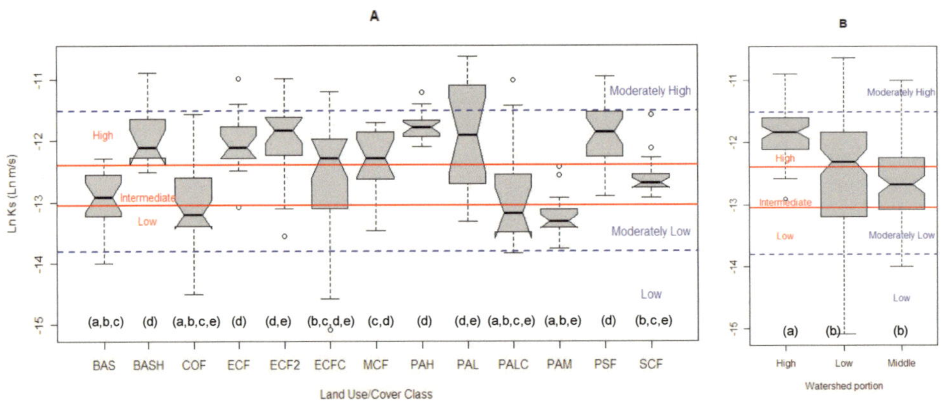

Fig. 5. Boxplots of log transformed K_s for all land use/coverage classes (A) and watershed portions (B). Land use/coverage classes and watershed portions with same letters are no different from each other, different letters indicate significant difference α=0.95 (TukeyHSD test). Red lines represent the threshold values for K_s values for different land uses/cover within the dataset, blue dashed lines correspond to K_s class thresholds according to Soil Survey Staff (1993). Bacharis-Pteridium shrub middle (BAS), Bacharis shrub high (BASH), coffee plantation low (COF), early cloud forest regeneration middle (ECF), early cloud forest regeneration low (ECF2), early cloud forest regeneration from coffee plantation low (ECFC), mature cloud forest middle (MCF), pasture high (PAH), pasture low (PAL), well managed pasture low (PALC), pasture middle (PAM), pine-spruce forest (PSF), secondary cloud forest (SCF).

Land use/cover	Original K_s data ($m \cdot s^{-1}$)							
	min	median	mean	max	sd	W	p Value	Normality
BAS	8.3E-07	2.5E-06	2.7E-06	4.6E-06	1.2E-06	0.944	0.2797	Yes
BASH	3.7E-06	5.5E-06	7.4E-06	1.9E-05	4.3E-06	0.777	0.0052	No
COF	5.0E-07	1.9E-06	2.7E-06	9.5E-06	2.2E-06	0.795	0.00007	No
ECF	2.1E-06	5.5E-06	6.5E-06	1.7E-05	3.2E-06	0.852	0.0057	No
ECF2	1.3E-06	7.2E-06	7.7E-06	1.7E-05	4.0E-06	0.942	0.1151	Yes
ECFC	2.8E-07	4.6E-06	4.7E-06	1.4E-05	3.1E-06	0.94	0.1124	Yes
MCF	1.4E-06	4.7E-06	5.0E-06	8.2E-06	2.1E-06	0.942	0.3699	Yes
PAH	5.5E-06	7.6E-06	8.0E-06	1.4E-05	2.0E-06	0.889	0.03694	No
PAL	1.6E-06	6.7E-06	9.1E-06	2.4E-05	7.2E-06	0.878	0.01594	No
PALC	9.8E-07	1.9E-06	3.9E-06	1.7E-05	4.7E-06	0.582	0.00003	No
PAM	1.1E-06	1.6E-06	1.9E-06	4.0E-06	7.8E-07	0.795	0.00176	No
PSF	2.5E-06	7.1E-06	7.9E-06	1.8E-05	3.8E-06	0.934	0.06263	Yes
SCF	2.4E-06	3.1E-06	3.6E-06	9.4E-06	1.7E-06	0.614	0.00002	No
Land use/cover	Log transformed K_s ($\ln m \cdot s^{-1}$)							
	min	median	mean	max	sd	W	p Value	Normality
BAS	-14.0	-12.9	-12.9	-12.3	0.489	0.952	0.3899	Yes
BASH	-12.5	-12.1	-11.9	-10.9	0.470	0.911	0.2201	Yes
COF	-14.5	-13.2	-13.1	-11.6	0.721	0.964	0.4153	Yes
ECF	-13.1	-12.1	-12.0	-11.0	0.453	0.974	0.8385	Yes
ECF2	-13.6	-11.8	-11.9	-11.0	0.602	0.963	0.3873	Yes
ECFC	-15.1	-12.3	-12.6	-11.2	0.894	0.926	0.0488	No
MCF	-13.5	-12.3	-12.3	-11.7	0.497	0.929	0.2351	Yes
PAH	-12.1	-11.8	-11.8	-11.2	0.231	0.953	0.4752	Yes
PAL	-13.3	-11.9	-11.9	-10.6	0.845	0.952	0.4032	Yes
PALC	-13.8	-13.2	-12.9	-11.0	0.897	0.85	0.0573	Yes
PAM	-13.8	-13.3	-13.2	-12.4	0.346	0.92	0.1468	Yes
PSF	-12.9	-11.9	-11.9	-11.0	0.487	0.98	0.8347	Yes
SCF	-12.9	-12.7	-12.6	-11.6	0.3	0.753	0.0006	No

Table 3. K_s descriptive statistics and normality analysis; *Bacharis-Pteridium* shrub middle (BAS), *Bacharis* shrub high (BASH), coffee plantation low (COF), early cloud forest regeneration middle (ECF), early cloud forest regeneration low (ECF2), early cloud forest regeneration (ECF), coffee plantation low (ECFC), mature cloud forest middle (MCF), pasture high (PAH), pasture low (PAL), well managed pasture low (PALC), pasture middle (PAM), pine-spruce forest (PSF), secondary cloud forest (SCF), (α=0.95).

According to this land uses/coverages COF, PALC and PAM have low K_s, while BAS, ECFC and SCF have intermediate K_s and BASH, ECF, ECF2, MCF, PAH, PAL and PSF have high K_s.

In the case of watershed portion, Figure 5B shows a larger variation in K_s in the low portion of the watershed than in the two other portions. The K_s in the high portion of the watershed is statistically different to the middle and lower portions, which are not statistically different from each other.

3.2 Effective porosity and infiltration flow proportions

Total porosity ranged from 54.5 to 91.9 %, descriptive overall statistics for total porosity and effective pore numbers are shown in Table 4. The span between maximum and minimum was two orders of magnitude (≥ 1.0 mm apparent pore diameter) and three orders of magnitude for the other two classes. Total porosity did not correlate with the number of effective pores for any of the three diameter classes analyzed.

	Percentage				
Total porosity	min	median	mean	max	sd
	54.47	74.09	74.47	91.91	9.50
Apparent pore diameter (mm)	Number of effective pores per m²				
≥ 1.0	1.28E+00	7.39E+01	1.26E+02	6.83E+02	1.82E+02
0.3 - 1.0	1.58E+02	9.12E+03	1.55E+04	8.43E+04	2.24E+04
0.01 - 0.3	1.28E+04	7.39E+05	1.26E+06	6.83E+06	1.82E+06

Table 4. Overall total porosity statistics and number of effective pores for three apparent diameter classes.

Test for normality (Shapiro-Wilk) revealed that for the three apparent diameter pore size intervals nine out of thirteen land use/cover classes had log-normal distribution.

Flow proportion results indicated that, in general, macropores (≥ 1.0 mm apparent diameter) contributed about 3% (0.15 - 11%) to total infiltration flow (Table 5). Large mesopores (0.3 - 1.0 mm apparent diameter) contributed nearly in 7% (0.3 - 21%) of flow, while small size mesopores (0.01 – 0.3 mm apparent diameter) accounted for nearly 74% (13 - 86%) of total infiltration flow. Micropore (< 0.01 mm apparent diameter) contribution was extremely variable and ranged between being insignificant (0.003%) and contributing almost to all of the infiltration flow (99.27%) in specific cases, while on average, micropores contribute up to 40% of the total infiltration flow. These results contrasted with those of Watson and Luxmoore (1986) in forested soils where macropore flow accounted nearly 73% of the infiltration flow. In this specific setting texture and soil particle distribution related mesoporosity have a major role in infiltration than structure related macropores, this will be discussed and analyzed further in the text.

Apparent pore diameter (mm)	Flow proportion (%)				
	min	median	mean	max	sd
≥ 1.0	0.147	2.656	3.093	11.050	2.051
0.3 - 1.0	0.342	5.921	6.689	21.250	4.167
0.01 - 0.3	13.190	81.490	74.940	85.980	15.311
< 0.01	0.003	25.840	39.570	99.270	36.042

Table 5. Infiltration flow proportion statistics of four apparent pore diameter sizes.

Flow proportions in all three watershed portions were highly variable. Infiltration flow contribution from macropore (≥ 1.0 mm, Figure 6A) and large mesopores size (0.3 - 1.0 mm, Figure 6B) in the low portion of the watershed was significantly lower than in the Middle and High portions. The contribution of small mesopores flow contribution (0.01 - 0.3 mm, Figure 6C) is significantly higher for the middle portion of the watershed.

Figure 6 also shows that the infiltration flow through micropores, was statistically different for the three portions of the watershed even when it had very wide ranges (Figure 6D). The land use/cover level reflects the flow trends mentioned above in which the largest flow proportion occurred through small size mesopores (0.01 - 0.3 mm). For certain land

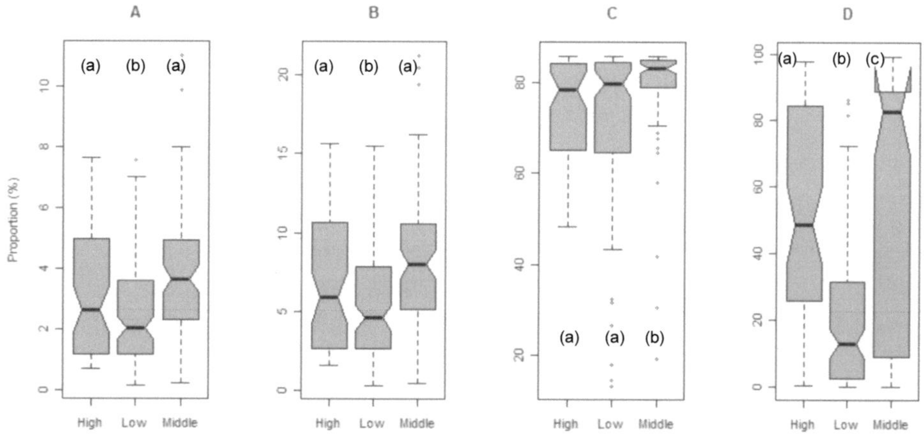

Fig. 6. Boxplots of flow proportion for four apparent pore diameter sizes at the watershed portion level; apparent pore diameter ≥ 1.0 mm (A), 0.3 - 1.0 mm (B), 0.01 - 0.3 mm (C) and < 0.01 mm (D). Different low case letters indicate Tukey HSD statistical difference (α=0.95).

use/cover classes (PAL) the infiltration flow proportion departs from the general mean (≈75%) and goes as low as 61.1% while mean micropore flow proportion increases up to 34.3%. In other land use/cover classes e.g. PSF, the 0.3 - 1.0 mm pore flow proportion increases near 10% and the micropore flow proportion decreases under 10% indicating a shift in infiltration flow to larger mesopore use and lower micropore use.

3.3 Physicochemical soil properties

Two hundred and thirty six samples were analyzed in the laboratory for nine physicochemical soil properties. Overall descriptive statistics of these soil properties are presented in Table 6.

Soil Property	min	median	mean	max	sd
Bulk density (g/cm^3)	0.1	0.5	0.5	1.1	0.09
Final moisture content (% wt)	21.4	52.9	67.3	182.2	42.0
Total porosity (%)	54.47	74.09	74.47	91.91	9.50
Clay %	4.5	26.9	29.7	54.4	5.24
Silt %	12.0	36.0	36.5	71.4	7.17
Sand %	9.6	30.3	33.8	68.4	9.19
Fine Sand %	2.5	14.8	14.5	50.5	4.78
Coarse Sand%	0.4	19.9	19.3	46.4	7.30
Organic Carbon%	1.3	16.0	15.2	44.8	3.40

Table 6. Descriptive statistics for the analyzed soil properties.

Final moisture content varied from 21.4 to 182.23%; soils in the low watershed soils had overall lower final water content, while the high and middle watershed soils had larger moisture contents (not shown). Bulk density was low, concurring with previously reported values for volcanic Andisols in the area (0.2 - 1.3 g/cm^3; (Meza-Pérez & Geissert-Kientz, 2006) and elsewhere (Alvarado & Forsythe, 2005; Porta, et al., 1999). The lower portion of the watershed has the highest values of bulk density, while the higher and middle portions have lower values (Figure 7A).

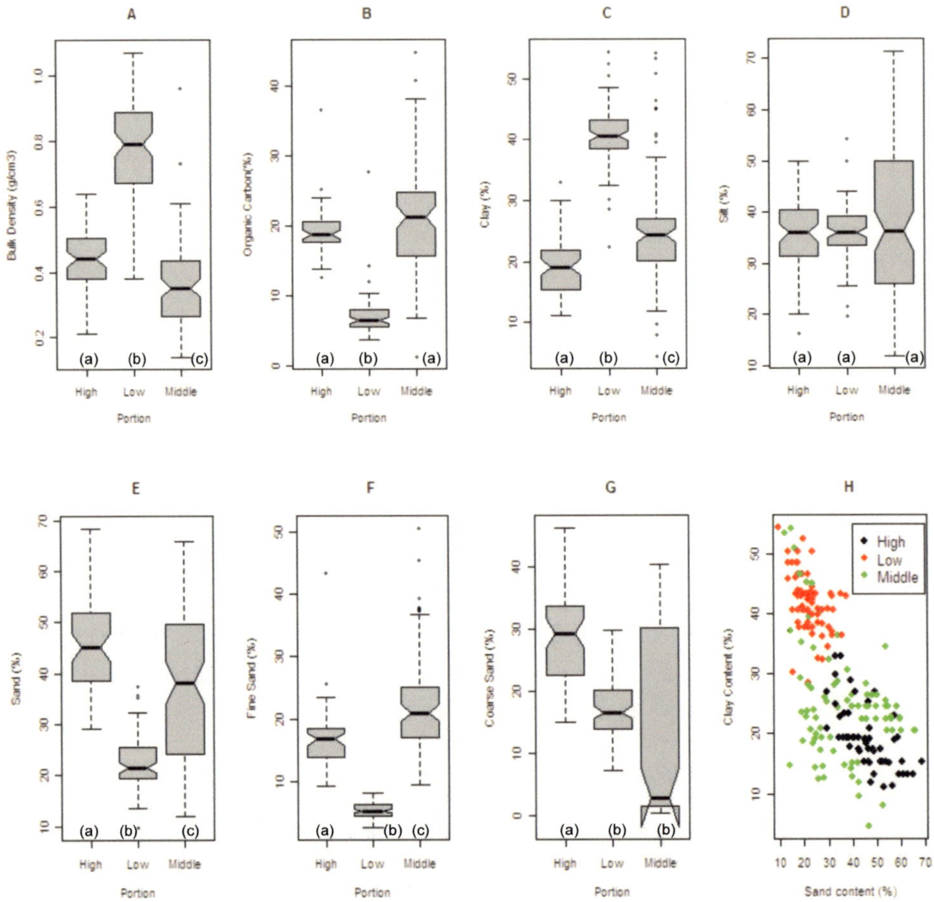

Fig. 7. Boxplot of soil properties by portion of the watershed: Bulk density (A), organic carbon (B), clay content (C), silt content (D), sand content (E), fine (F) and coarse sand content (G), overall sand *vs.* clay relationship (H). Different lower case letters indicate statistical difference between portions of the watershed (Tukey HSD), α=0.95.

Total porosity ranged between 54.5% and 94.0%. Andisols from areas nearby are reported to have porosities between 44 and 91% (Meza-Pérez & Geissert-Kientz, 2006). A wide range was found in the organic carbon content throughout the watershed (between 1.3 and 44.8%, Table 6) although a trend was found relating the carbon content with the portion of the watershed. The lower portion had significant lower values than the middle and higher portions, while the middle portion had the largest variation (Figure 7B). Meza-Pérez and Geissert-Kientz (2006) reported organic carbon contents between 0.4 and 8.0% in forested and agricultural surface soil from an area near the higher portion of the watershed, while Marín-Castro (2010) reported mean contents between 22 and 49% in forest (Mature and secondary cloud forest) and pasture land from the middle portion of the watershed.

Since silt content is mostly constant, clay and sand contents show an opposite trend (Figure 7H). The lower portion of the watershed had significantly higher clay contents and lower sand contents, while the high and middle portions resemble one another, although clay is significantly higher and sand lower in the middle than in the higher portion (Figure 7C, E). Thus for these two particle sizes every portion of the watershed was different.

There were no significant differences in silt content in the three portions of the watershed (Figure 7D). Fine sand content (Figure 7F) shows an opposite trend to that of clay (Figure 7H) and resembles the sand content (Figure 7E). These boxplots indicate that the middle portion of the watershed had the largest variation.

Nine of the 12 USDA soil texture classes (Soil Survey Staff, 1993) are present in surface soils of the study area. The most frequent classes were Clay (27.7 % of the samples), followed by Loam (26.9 %), Clay Loam (13.4 %), Sandy Clay Loam (10.1%) and Silty Loam (9.7%). The other five classes represented less than six percent each (Figure 8A). Sandy Clay was the only texture class with one sample. Sand, Loamy Sand and Silt texture classes are not present in surface soils.

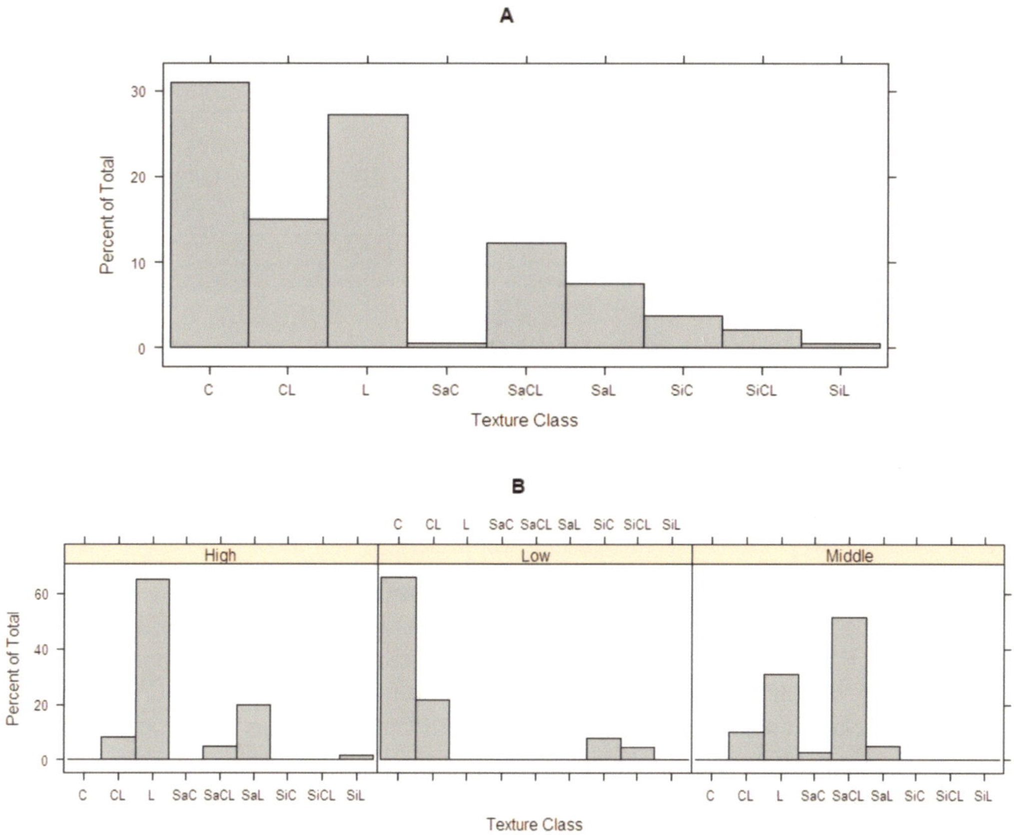

Fig. 8. Texture classes. Overall sample percentage by texture class (A), texture class by watershed portion (B), see Figure 9 for texture key.

In terms of texture class, the study area is quite diverse, because it has nine out of twelve classes. The texture differed depending on the portion of the watershed. The higher portion is dominated by Loam, while the lower is dominated by Clay and Clay Loam. The middle portion is dominated by Sand Clay Loam and Loam textures (Figure 8B). The middle portion of the watershed had the greatest number of texture classes (8 of 9) is the middle portion with eight out of the nine classes identified in the study area (Figure 8B).

Further, land use/covers are not homogeneous regarding texture. Most of the land use/cover classes include three or four texture classes (Figure 9).

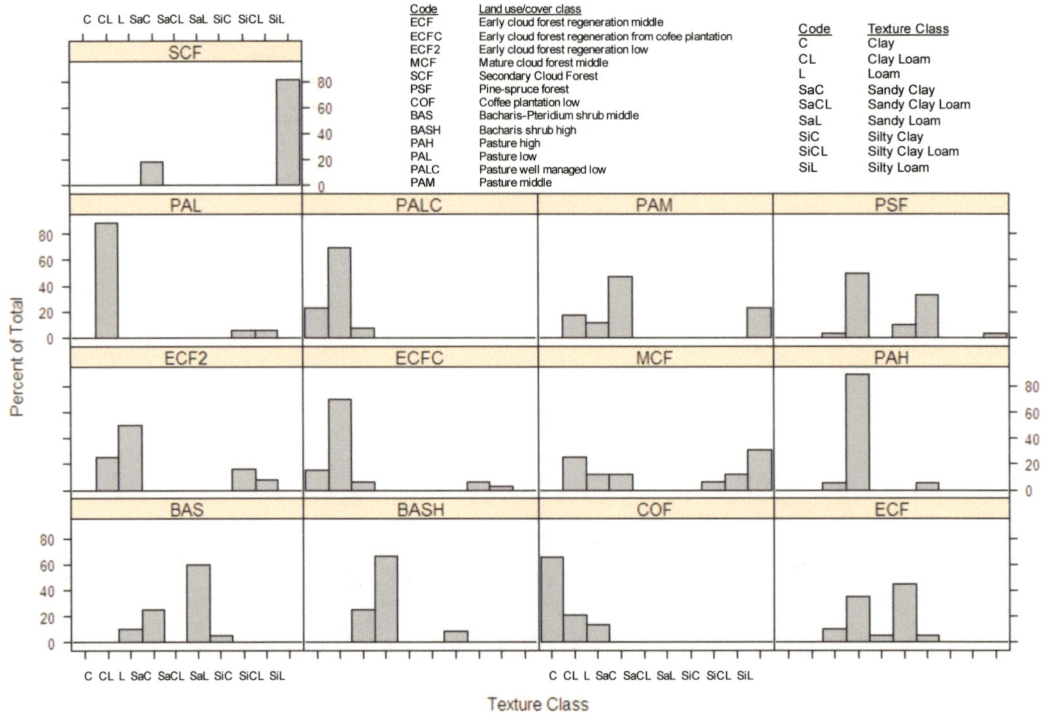

Fig. 9. Texture classes by land use/cover, texture classes follow the watershed portion relationship. There is no cause-effect relationship between land use/cover and texture.

3.4 K_s, porosity and flow proportion relationships with soil properties

At the watershed level correlation analysis with the whole dataset (Pearson´s product-moment correlation) indicated that Log K_s had a significant negative overall correlation with clay content, as well as a minor positive correlation with coarse sand fraction. The highest significant correlation score for Log K_s occurred with Clay/Silt ratio (Table 7). None of the other soil properties showed a significant correlation.

The alpha parameter (α) of the Gardner exponential model (Gardner, 1958) that expresses the inverse of the average capillary length within the soil matrix had a significantly negative relationship with bulk density (t= -5.738, p<0.001, r= -0.3513), clay content (t= -2.837, p=0.005, r= -0.1816) and coarse-sand/fine-sand ratio (t= -4.646, p<0.001, r= -0.2895). This parameter was also positively correlated with organic carbon content (t= 5.061, p<0.001, r= 0.3129), sand content (t= 2.634, p= 0.009, r= 0.1690) and fine sand content (t= 5.08, p<0.001, r= 0.3139). Increasing values in bulk density are related with surface soil compaction or heavier textures, and therefore the reduction in α.

A negative correlation was also found between α and clay and coarse-sand/fine/sand ratios. Low bulk density values are related to loose granular and micrugranular structured soils with high α values.

Similar to the α parameter, the number of macropores per square meter were positively correlated with organic carbon content, sand and fine-sand content and negatively with bulk density, clay content and coarse-sand/fine-sand and clay/silt ratios. Bulk density scored highest of all tested physicochemical properties ($t= -3.9034$, $p<0.001$, $r= -0.2472$). The same trends were found for the number of pores in the other two diameter intervals studied. From the analyzed physicochemical properties, bulk density had always the top scores.

Flow proportions for micropores (<0.01 mm in diameter) had positive significant correlations with bulk density, clay content and coarse-sand/fine-sand ratio and negative correlations with organic carbon, sand, and fine-sand fraction.

Watershed portion	Variable	r	p-value
High	Organic carbon%	-0.293	0.023
Low	Clay %	-0.238	0.025
Low	Organic carbon%	-0.273	0.010
Low	Coarse sand%	-0.285	0.007
Low	Fine sand %	-0.290	0.006
Middle	Bulk Density (g/cm^3)	-0.393	< 0.001

Table 7. Significant correlations between Log K_s and physicochemical properties at the watershed portion level.

While correlations were generally low in the whole watershed, within watershed portions correlations suggest distinct relationships between K_s and physicochemical variables (Table 7). At this scale, the tightest correlation was found for bulk density and K_s in the middle portion.

During the field work it was observed that in the middle portion, all land uses except of BAS and PAM presented high root density in the first fifteen centimeters. Compacted surface soil by cattle trampling and overgrazing was observed at PAM sites, nevetheless bulk density values for this site were rather lower than expected (mean 0.47 g/cm^3) and not statistically different from those in the SCF (mean 0.45 g/cm^3). However bulk density values were different from those in the MCF (mean 0.36 g/cm^3).

Significant correlations between Log K_s and specific physicochemical variables at the land use/cover scale occurred only in five out of the thirteen land uses analyzed (Table 8). Bulk density correlated with Log K_s in PSF, ECFC and BASH, the first two showed positive relationships while the latter had a negative relationship. Clay percentage was negatively correlated at the ECF site. Organic carbon content was negatively correlated with Log K_s at PSF and ECFC sites. Silt percentage showed a positive correlation at the ECFC and PAM sites. At the land use/cover scale only a few land use/cover classes showed significant correlations. For the ECFC class six out of the eight analyzed physicochemical variables had significant correlations with Log K_s.

Correlations indicate different controls across scales. For some physicochemical properties certain land use/cover class may influence the relationship at the watershed portion scale, e.g. bulk density in PSF class ($r= -0.498$) influences the overall correlation for the high portion of the watershed ($r= -0.293$) (Figure 10B, C).

Land use/cover class	Watershed Portion	Variable	r	p-value
BASH	High	Bulk Density (g/cm^3)	-0.838	<0.001
ECF	Middle	Clay (%)	-0.492	0.028
ECFC	Low	Bulk Density (g/cm^3)	0.442	0.019
ECFC	Low	Coarse Sand%	-0.630	<0.001
ECFC	Low	Fine Sand%	-0.699	<0.001
ECFC	Low	Organic Carbon%	-0.697	<0.001
ECFC	Low	Sand %	-0.681	<0.001
ECFC	Low	Silt %	0.740	<0.001
PAM	Middle	Silt %	0.495	0.043
PSF	High	Bulk Density (g/cm^3)	0.511	0.004
PSF	High	Organic Carbon%	-0.498	0.005

Table 8. Significant correlations between Log K_s and physicochemical properties at the land use/cover level. *Bacharis* shrub (BASH), early cloud forest regeneration (ECF), early cloud forest regeneration from coffee plantation (ECFC), pasture (PAM) and pine-spruce forest (PSF).

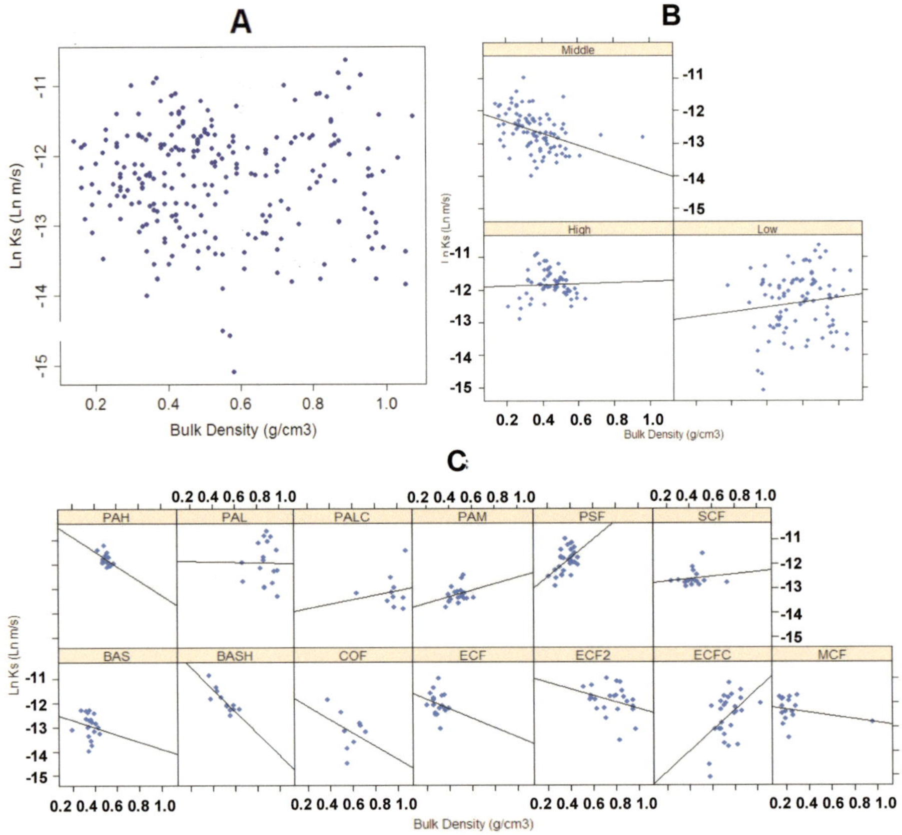

Fig. 10. Correlation between bulk density and Log K_s at three scales levels, whole watershed (A), watershed portion (B) and land use/coverage class (C).

3.5 Mixed-effect model

A large percentage of the variation of conductivity measurements was not explained by the factors tested in the mixed-effect model (residual = 40%, in Table 9). Despite the above, 19% of the variation was observed for land uses; 18 % between sites; 16%within watershed portions and only 4 % between payments (Table 9). Firstly these results indicate that there was not significant variation in conductivity values between those areas potentially considered within payments programs and those that are not; secondly, that land use did have an important effect on the overall variance in the data. According to this, Akaike's information criterion and p values showed that most of variation in K_s could be explained by land use and site (Table 10).

Groups	Name	Variance	Variance components (%)	Std.Dev.
Site:(LUC:(Payments:Portion)	(Intercept)	0.125908	18.10	0.35484
LUC:(Payments:Portion)	(Intercept)	0.138129	19.86	0.37166
Payments:Portion	(Intercept)	0.032249	4.635	0.17958
Portion	(Intercept)	0.114736	16.50	0.33873
Residual		0.284224	40.88	0.53313
Total		0.695246	100	

Table 9. Summary of random effects for the full model "A" (Log K_s ~1+(1 | Portion/Payments/LUC/Site). LUC= Land use/cover. Number of obs: 265, groups: LUC:(Payments: PortionZ), 14; Payments: Portion, 6; Portion,3: AIC= 513.4; BIC= 534.9; logLik=-250.7; deviance=500.4; REMLdev= 501.4.

Models	Df	AIC	BIC	logLik	Chisq	Chi DF	Pr(>Chisq)
modelB	5	510.61	528.51	-250.31			
modelA	6	512.13	533.61	-250.06	0.4822	1	0.4874
modelC	4	509.01	523.33	-250.51			
modelB	5	510.61	528.51	-250.31	0.3999	1	0.5271
modelD	3	520.94	531.68	-257.47			
modelC	4	509.01	523.33	-250.51	13.924	1	0.0001904 ***

Model A (full moldel) = (Log K_s b~1+(1 | Portion/Payments*/LUC/Site));
Model B = (Log K_s b~1+(1 | newzone*/LUC/Site)):
Model C= (Log K_s b~1+(1 | newpayment*/Site)) and
Model D = (Log K_s b~1+(1 | newLUC*))
*Factors were re-coded according to Crawley (2007).

Table 10. Summary of Model simplification.

4. Discussion

4.1 Hydraulic conductivity

In this study, we found an important variation in K_s within different portions of the watershed. The low and middle-elevation portions were not statistically different from each other while the high portion had higher K_s. We found that differences in K_s and effective porosity are affected by soil properties, pedogenesis and land management.

Our research questions addressed the relationship between land use/cover and K_s. At this level, K_s values ranged between low and moderately high according to the Soil Survey Staff infiltration classes (1993). The local classification derived from this study separates the obtained values into three classes: low, intermediate and high. This classification used key threshold values at 2.15 x 10-6 $m \cdot s^{-1}$ and 4.12 x 10-6 $m \cdot s^{-1}$.An interesting result was that at this level an increase in plant cover was not related to an increase in K_s values in all cases. The increase in plant cover after anthropogenic disturbance has been reported to be related to an increase in K_s values in certain environments (Li & Shao, 2006; Ziegler, et al., 2004). In other environments, it seems to be an age threshold for plant cover recovery, under this threshold there is a clear relationship between plant cover recovery and K_s values, but over it there are no differences in K_s values between different ages of recovery and plant cover (Zimmermann, 2007; Zimmermann, et al., 2006).

The widely reported negative impact of land use change from forested to pasture lands on K_s and other hydrophysical variables (e.g. Singleton, et al., 2000; Tobón, et al., 2004; Zimmermann, et al., 2006) was not completely verified by our results. In our case the lowest K_s values corresponded to a pasture site (PAM) but there were also pasture sites with high K_s values (e.g. PAH, PALC, PAL). Moreover, the highest variation in K_s for a specific land use/cover class occurred in another pasture site (PAL) (Table 2; Figure 5A). The well-managed pasture site (PALC) had both high and low K_s values while the poorly managed pasture site (PAL) had some of the highest K_s values of the study. It seems that pasture land use/cover classes do not reflect differences in K_s or effective porosity behaviour. Some ideas about this facts are mentioned in the following lines.

Field observations helped to elucidate the pattern of K_s in different pasture sites. While the PAM, PAL and PALC sites occurred in very steep slopes (> 40º), PAH occurred in gentle slope terrain (3-8 º) in the high portion of the watershed. Also PAL, PAM and PALC sites were heavily grazed by cattle (cows), horses, donkeys and occasionally goats and sheep. In the PAH site grazing was done mainly by young animals, calves mostly. In the earlier pasture sites cattle was aimed to local dairy production and in the latter to commercial ranching of calves. This implies significant differences in management and also in cattle´s effect on hydrophysical properties as K_s. In gentle slope small and young animals have less impact than larger animals on steep slopes.

Further, soil particle distribution is also crucial in the explanation of these differences in K_s since coarser textures have larger pores and therefore higher K_s. The high portion of the watershed with higher K_s has also lower clay content (Figure 7C) and larger sand and coarse sand contents (Figure 7E, G), while the low portion of the watershed has higher clay and lower sand and fine sand contents than the other two portions (Figure 7C, E, F). The middle portion has characteristics of both the lower and higher portions e.g. (sand and clay content; Figure 7H). The difference in clay and sand proportions seems to be related to volcanic ash sources and to pedogenetic factors. First, the high portion of the watershed is close to several monogenetic cinder and scoria cones and received considerable amounts of fresh ash materials during the cone´s eruptions, while the middle and low portions received lower inputs of these materials. Second, both the middle and low portions are older reliefs in which pedogenetic processes have occurred longer than in the high portion and thus the original volcanic materials had been already weathered into secondary clay minerals, which explains the higher clay contents in the lower portion.

The low K_s values for the coffee plantation in production (COF) may be explained by repeated traffic of plantation workers. These low K_s values grouped COF site with sites depicting the lowest K_s values in our study (ECFC, PALC and PAM). Even when coffee plantations have a dense tree cover that provides shading, there is no direct relationship between such a tree cover and high K_s values.

The formation of soils in this watershed (i.e. Andisols) are strongly controlled by climatic factors, especially temperature and rainfall. The rate of chemical weathering increases remarkably with increasing temperature and soil moisture. As the degree of weathering increases, metastable non crystalline materials are gradually consumed by transformation to more stable crystalline minerals (i. g. halloysite, kaolinite, gibbsite) (Dahlgren, et al., 2004).

Changes in temperature and rainfall induced by the elevation gradient across the watershed have important impacts on soil properties. Surface horizons of the Andisols from the upper portion, which developed under dryer and colder conditions have sand shifting textures, while the Andisols from the warmer lower portion with contrasting wet/dry conditions have higher clay content caused by more intense weathering. These transitions of Andisols in the tropics caused by changes in soil moisture regime have been documented by several authors (e.g. Dubroeucq, et al., 1998; Nizeyimana, 1997; Zehetner, et al., 2003). Since the central portion of the watershed consists of a highly dissected landscape, accelerated erosion and redeposition on steep slopes may have created unstable landscape positions where the andic topsoil materials have been locally removed creating intermixed soil patterns with a more variable texture.

4.2 Effective porosity and flow proportions

Flow proportions for different pore sizes indicate that infiltration processes in this system are controlled mainly by particle size related porosity rather than bypass structural related macropores. Our results contrast greatly with earlier data for forest soils where 73% of infiltration flow was transmitted by 0.04% of soil volume (Watson & Luxmoore, 1986) or where macropores flow constitutes 85% of ponded flow (Wilson & Luxmoore, 1988).

In our case macropore related flow (mean 3.09 %) was conducted through approximately 0.009 % of soil volume. The highest value in our study was for an early cloud forest site in the middle portion of the watershed where 11.05% of infiltration flow was conducted by only 0.04% of soil volume. This discrepancy between our results and earlier results in forested soils may be attributed to soil type related structure. Andisols are known for their granular and microgranular structure in natural conditions, and even when preferential flow has been reported in this type of soil (Eguchi & Hasegawa, 2008), our results indicate that in this particular setting most of the water flow occurs through fine mesopores and micropores of the soil matrix and not through macropores.

4.3 Hydrological PES schemes

The mixed effect model analysis indicated no significant statistical difference in K_s between land uses under the economical compensation scheme and those land uses not considered in the compensation scheme. Field observations and informal interviews with land owners allowed a larger understanding of the land use/cover - K_s relationship. Some land use/coverages correspond to different transitional stages of a natural successional process. In some of the cases K_s figures are not statistically different from those of a mature cloud forest, which indicates that either part of the original infiltration capacity of the soil had been recovered or that it was not severely affected by land use change.

Land management practices also are a major driving force in the modification of infiltration patterns. Since Log K_s values from early regeneration of cloud forest were not statistically different from that of secondary or mature cloud forests, we conclude that the early regeneration patches may be infiltrating just as much as the mature or secondary cloud forests. This should be taken in account in the PES program since early regeneration cloud forests, if protected through monetary incentives, may evolve into secondary or mature cloud forests. However, in transformed and intensively managed environments such as pastures, land use/cover may not be a good criterion to assess hydraulic functionality of soil.

Further, agroecosystems such as the coffee plantations may not be suited for hydrological PES schemes. Perhaps they could be included in other PES schemes such as carbon sequestration or biodiversity conservation. But it is yet to be addressed if coffee plantations in the Gavilanes watershed represent adequately the coffee plantations within the region. Further research may include different coffee plantation configuration and similar agroecosystems.

5. Conclusions

The main aim of this chapter was to examine the relationships between land use/cover and key hydrophysical variables (K_s), in order to strengthen and aid policy making related to Payment for Ecosystem Services initiatives.

The relationship between land use/cover and hydraulic conductivity is very complex in the study area. Although land use/cover is a major driving force in K_s behavior, soil properties also contribute to hydraulic conductivity through the parent material differences (closeness to ejecta sources), pedogenetical processes (secondary mineral clay formation) and geomorphological processes (relief formation and erosion).

Successional stages such as early cloud forests have no statistical difference in K_s with mature forest formations indicating that the hydraulic functionality of the soil had been recovered or that it was not severely affected by land use change. Including these successional stages within the local and federal PES schemes may promote their recovery and transition into secondary or mature forest.

The high variability of K_s in pasture use is related to local cattle management strategies, thus a general pasture land use classification may not be a suficient criterion (or an efficient criterion) for spatial mapping of K_s.

6. Acknowledgments

This research was conducted with financial support of the following grants: Mexican National Council for Science and Technology (CONACyT), grant number 106788, postdoctoral fellowship CONACyT grant number CVU-168649 and SEP-PROMEP grant number 103-5/09/4106. Publication was funded with UMSNH-CA-195-CAMBIO

CLIMÁTICO resources. We are grateful to Adán Hernández, Sergio Cruz, Rafael Morales, Mario Guevara, Ricardo Vázquez, Jethro Meza, Saraí Sánchez, Gloria Reyes and Gabriela Guerra for valuable field assistance. We thank Luis Martínez, Lourdes Cabrera, Ninfa Portilla and Sandra Rocha for soil properties analysis. The climate data from de middle portion of the watershed were kindly supplied by Dr. Friso Holwerda. We also want to thank Dr. Sybil Gotsch for the comments to the manuscript.

7. References

Alvarado, A. & Forsythe, W. (2005). Variación de la densidad aparente en órdenes de suelos de Costa Rica. *Agronomía Costarricense*, 29. 001. 85-94, 0377-9424.

Bates, D. (2006). Linear mixed model implementation in lme4, Department of Statistics, University of Wisconsin-Madison 32.

Bates, D. & M. Maechler (2010). Package lme4 v 0.999375-37. CRAN, R project. http://lme4.r-forge.r-project.org/ .

Buytaert, W., Wyseure, G., De Bièvre, B. & Deckers, J. (2005). The effect of land-use changes on the hydrological behaviour of Histic Andosols in south Ecuador. *Hydrological Processes*, 19. 3985–3997, 1099-1085.

Cerdà, A. (1995). Soil moisture regime under simulated rainfall in a three years abandoned filed in southeast Spain. *Physics and Chemistry of The Earth*, 20. 3-4. (1995/0) 271-279, 1474-7065.

Cerdà, A. & Doerr, S. H. (2005). Influence of vegetation recovery on soil hydrology and erodibility following fire: an 11-year investigation. *Int. J. Wildland Fire*, 14. 4. 423-437, 1049-8001.

Conafor. (2008). México tiene el mayor programa de servicios ambientales del mundo: Banco Mundial, Secretaría de Medio Ambiente y Recursos Naturales, B139-2008.

Dahlgren, R. A., Saigusa, M. & Ugolini, F. C. (2004). The nature, properties and management of volcanic soils. *Advances in Agronomy*, 82. 113-182. 978-0-12-385040-9.

Dubroeucq, D., Geissert, D. & Quantin, P. (1998). Weathering and soil forming processes under semi-arid conditions in two Mexican volcanic ash soils. *Geoderma*, 86. 1-2. 99-122, 0016-7061.

Eguchi, S. & Hasegawa, S. (2008). Determination and characterization of preferential water flow in unsaturated subsoil of andisol. *Soil Sci Soc Am J*, 72. 320-330, 0361-5995.

Esteves, M., Descroix, L., Mathys, N. & Marc Lapetite, J. (2005). Soil hydraulic properties in a marly gully catchment (Draix, France). *CATENA Gully Erosion: A Global Issue*, 63. 2-3. (2005/10/31) 282-298, 0341-8162.

Ferrer, J. M., Estrela Monreal, T., Sanchez Del Corral Jimenez, A. & Garcia Melendez, E. (2004). Constructing a saturated hydraulic conductivity map of Spain using pedotransfer functions and spatial prediction. *Geoderma*, 123. 3-4. (2004/12) 257-277, 0016-7061.

García, E. (2004). Modificaciones al sistema de clasificación climática de Köppen, Instituto de Geografía UNAM, México, D.F.

Gardner, W. R. (1958). Some steady-state solutions of the unsaturated moisture flow equation with applications to evaporation from a water table. *Soil Science*, 85. 228-232, 1538-9243.

Gee, G. W. & Or, D. (2002). Particle-Size analysis. In: *Methods of Soil Analysis: Part 4 Physical Methods*, 4. 255-293, 089118841X.

Gómez-Tagle, A. (2009). Linking hydropedology and ecosystem services: differential controls of surface field saturated hydraulic conductivity in a volcanic setting in central Mexico. *Hydrol. Earth Syst. Sci. Discuss*, 6. 2499-2536, 18122108.

Holwerda, F., Bruijnzeel, L. A., Muñoz-Villers, L. E., Equihua, M. & Asbjornsen, H. (2010). Rainfall and cloud water interception in mature and secondary lower montane cloud forests of central Veracruz, Mexico. *Journal of Hydrology*, 384. 1-2. 84-96, 0022-1694.

Karlsen, R. H. (2010). Stormflow processes in a mature tropical montane cloud forest catchment, Coatepec, Veracruz, Mexico, Vrije Universiteit Amsterdam, Amsterdam

Kechavarzi, C., Spongrová, K., Dresser, M., Matula, S. & Godwin, R. J. (2009). Laboratory and field testing of an automated tension infiltrometer. *Biosystems Engineering*, 104. 2. 266-277, 1537-5110.

Lassen, L. & Lull, H. (1951). Infiltration and the use of the infiltrometer. *Association Internationale d´Hidrologie Scientifique, Assemblée générale de Bruxelles*, pp. 310-315, Bruxelles, Association Internationale d´Hidrologie Scientifique.

Li, Y. Y. & Shao, M. A. (2006). Change of soil physical properties under long-term natural vegetation restoration in the Loess Plateau of China. *Journal of Arid Environments*, 64. 1. (2006/1) 77-96, 0140-1963.

Logsdon, S. D. & Jaynes, D. B. (1993). Methodology for Determining Hydraulic Conductivity with Tension Infiltrometers. *Soil Sci Soc Am J*, 57. 6. 1426-1431.

Marín-Castro, B. E. (2010). Variación espacial de la conductividad hidráulica saturada en suelos de origen volcánico bajo tres usos de suelo en el centro de Veracruz, México, INECOL, A.C., Xalapa.

Mcintyre, D. S. & Tanner, C. B. (1959). Anormally distribuited soil physical measurements and nonparametric statistics. *Soil Science*, 88. 133-137, 0038-075X.

Meza-Pérez, E. & Geissert-Kientz, D. (2006). Estabilidad de estructura en andisoles de uso forestal y cultivados. *Terra Latinoamericana*, 24. 2. 163-170, 0187-5779.

Miller, R. W. & Donahue, R. L. (1990). Soils, An introduction to Soils and Plant Growth, Prentice-Hall Inc., 0138203334, Englewood Cliffs, New Jersey.

Motorola, I. (2002). MPX2010, MPXV2010G Series, Motorola, Inc., Motorola Freescale Semiconductor, Inc., Semiconductor Technical Data.

Naranjo, J. & Murgueitio, E. (2006). Propuesta de PSA hídrico en el municipio de Marinilla, Antioquía. *Congreso Internacional Servicios Ecosistémicos en los Neotrópicos*, 13-19 Nov. Valdivia, Chile, Bosque 20.2. 172.

Nizeyimana, E. (1997). A toposequence of soils derived from volcanic materials in Rwanda: morphological, chemical, and physical properties. *Soil Science*, 162. 2. 350-360, 0038-075X Online ISSN: 1538-9243.

Ordoñez, J. A. B., De Jong, B. H. J., García-Oliva, F., Aviña, F. L., J.V.Pérez, Guerrero, G., Martínez, R. & Masera, O. (2008). Carbon content in vegetation, litter, and soil under 10 different land-use and land-cover classes in the Central Highlands of Michoacan, Mexico. *Forest Ecology and Management*, 225. 7. 2074-2084, 0378-1127.

Pérez-Maqueo, O., Delfín, C., Fregoso, A., Cotler, H. & Equihua, M. (2005). Modelos de simulación para la elaboración y evaluación de los programas de servicios ambientales hídricos. *Gaceta INE*, 78. 47-66, 1405-2849.

Porta, J., López-Acevedo, M. & Roquero, C. (1999). Edafología para la agricultura y el medio ambiente, Mundi-Prensa, 84-8476-148-7, Bilbao, España.

R_Development_Core_Team. (2004). R: A language and environment for statistical computing, R Foundation for Statistical Computing, Vienna, Austria.

Rogowski, A. S. (1972). Water physics: Soil variability criteria. *Water Resources Research*, 8. 1015-1023, 0043-1397.

Russo, D. & Bresler, E. (1981). Soil hydraulic properties as stochastic processes: I. An analysis of field spatial variability. *Soil Sci Soc Am J*, 45. 4. 682-687, 0361-5995.

Sarkar, D. (2008). Lattice: Multivatiate Data Visualization with R, Springer, 978-0-387-75968-5, New York.

Singleton, P. L., Boyes, M. & Addison, B. (2000). Effect of treading by dairy cattle on topsoil physical conditions for six contrasting soil types in Waikato and Northland, New Zealand, with implications for monitoring. *New Zealand Journal of Agricultural Research*, 43. 4. 559-567, 0028-8255.

Soil Survey Staff. (1993). Soil Survey Manual, Soil Conservation Service, United States Department of Agriculture, 8172331134, Washington, D.C., USA.

Špongrová, K. (2006). Design of an automated tension infiltrometer for unsaturated hydraulic conductivity measurement, Cranfield University at Silsoe. 92.

Špongrová, K., Kechavarzi, C., Dresser, M., Matula, S. & Godwin, R. J. (2009). Development of an Automated Tension Infiltrometer for Field Use. *Vadose Zone J*, 8. 3. 810-817.

Thornthwaite, C. W. (1948). An approach toward a rational classification of climate. *Geographical Review*, 38. 1. 55-94, 00167428.

Tobón, C., Bruijnzeel, L. A., Frumau, A. & Calvo, J. C. (2004). Changes in soil physical properties after conversion of tropical montane cloud forest to pasture in northern Costa Rica. *Second International Symposium Mountains in the Mist*, pp. 39, Waimea, Hawai´i, July 27- August 2, 2004.

Watson, K. W. & Luxmoore, R. J. (1986). Estimating Macroporosity in a Forest Watershed by use of a Tension Infiltrometer. *Soil Sci Soc Am J*, 50. 3. 578-582, 0361-5995.

Wilson, G. V. & Luxmoore, R. J. (1988). Infiltration, Macroporosity, and Mesoporosity Distributions on Two Forested Watersheds. *Soil Sci Soc Am J*, 52. 2. (March 1, 1988) 329-335, 0361-5995.

Wit, A. M. W. (2001). Runoff controlling factors in various sized catchments in a semi-arid Mediterranean environment in Spain, Universiteit Utrecht, Utrecht, The Nederlands.

Zehetner, F., Miller, W. P. & West, L. T. (2003). Pedogenesis of volcanic ash soils in andean Ecuador. *Soil Sci Soc Am J*, 67. 1797-1809, 0361-5995.

Ziegler, A. D., Giambelluca, T. W., Tran, L. T., Vana, T. T., Nullet, M. A., Fox, J., Vien, T. D., Pinthong, J., Maxwell, J. F. & Evett, S. (2004). Hydrological consequences of landscape fragmentation in mountainous northern Vietnam: evidence of accelerated overland flow generation. *Journal of Hydrology*, 287. 1-4. 124-146, 0022-1694.

Zimmermann, B. (2007). Spatial and temporal variability of the soil saturated hydraulic conductivity in gradients of disturbance, Doktogrades, der Mathematisch - Naturwissenschaftlichen Fakultät der Universität Potsdam, Postdam. 106.

Zimmermann, B. & Elsenbeer, H. (2008). Spatial and temporal variability of soil saturated hydraulic conductivity in gradients of disturbance. *Journal of Hydrology*, 361. 1-2. 78-95, 0022-1694.

Zimmermann, B., Elsenbeer, H. & De Moraes, J. M. (2006). The influence of land-use changes on soil hydraulic properties: Implications for runoff generation. *Forest Ecology and Management*, 222. 1-3. (2006/2/15) 29-38, 0378-1127.

11

Hydraulic Conductivity and Landfill Construction

Witold Stępniewski[1], Marcin K. Widomski[1] and Rainer Horn[2]
[1]*Lublin University of Technology, Faculty of Environmental Engineering,*
Dept of Land Surface Protection,
[2]*Christian Albrechts University, Institute for Plant Nutrition and Soil Science,*
[1]*Poland*
[2]*Germany*

1. Introduction

Landfills are important engineered constructions spread all over the world. Their number is calculated in thousands as the production of wastes in Europe only, reaches each year 3000 million tones of which 14% (about 415kg per capita) is municipal waste (EEA, 2004). Of this in 1999 about 57% was landfilled, 16% was incinerated, 20% recycled and composted and 7% was treated in other way. There are numerous types of landfills from simple dumping sites to rather sophisticated constructions constituting real bioreactors. Due to uncountable biochemical reactions occurring within the waste body, landfills produce biogas and leachates which threaten the pollution of air, water and soil. The environmental impact of landfills depends, to a high extent, on a bottom liner and top capping isolating the landfill from the surrounding. The quality of this isolation is determined by the water permeability as, in fact, no constructions are completely impermeable.

There are two essential types of liners i.e. mineral clay liners and synthetic liners of different geomembranes (or combination of both). As durability of synthetic liners is limited in time the mineral clay liners, which can persist thousands of years, if managed in a sophisticated way as it was proved by the countless layered natural soils worldwide, are preferred as a long term impermeable and rigid system. It is necessary to emphasize that landfill should preferably have a bottom liner and top capping. The function of bottom liner is to prevent the deeper soil layers and the groundwater from contamination with soluble substrates and irreversible pollution of the future drinking water reservoirs. The function of the top capping is to avoid infiltration of the precipitation water (from rain and from snow melting) and migration of methane and odors from the biogas to the atmosphere. However, the top capping system also has to guarantee optimal (or at least satisfactory) conditions for plant growth while the deep rooting of plants must be prevented. Thus, these conflicting requirements can be only fulfilled by special mineral soil systems which, if they are adjust, will preserve their properties for ever.

2. Materials appropriate for mineral liner construction

The EU Landfill Directive (1999/31/EC) distinguishes three types of landfills i.e. landfills for hazardous waste, landfills for non hazardous waste, and landfills for inert waste. This directive, among others, says that the landfill must be situated and designed in a way ensuring the prevention of pollution of atmosphere, groundwater, surface water and soil. It

can be achieved by combination of bottom liner and geological barrier during operation phase, and by combination of geological barrier and top liner during the aftercare phase.
The directive determines that landfill base and sides should consist of mineral layer with the following requirements:

- landfill for hazardous waste – the layer should be characterized by the hydraulic permeability k equal or lower than 10^{-9} m·s^{-1} and thickness equal at least 5 m,
- landfill for non hazardous waste – the same permeability and thickness equal or higher than 1m,
- landfill for inert waste –hydraulic permeability of 10^{-7} m·s^{-1} or less and thickness of at least 1 m.

In case of lack of a natural geological barrier it can be prepared artificially. The minimum thickness of artificially established barrier is 0.5 m. For non hazardous and hazardous landfill categories an artificial sealing liner and a drainage layer (≥0.5 m) is required.
Often local soils can be used for construction of landfill bottom liner, after application of external loads leading to their compaction. Usually recommended soil properties to achieve hydraulic permeabilities of order 10^{-9} m·s^{-1} by compaction are: percentage of fines (<0.075 mm) ≥ 30%, plasticity index between 20 and 30 and percentage of gravel (5 to 50 mm) ≤ 20% (Roehl et al., 2009). The ranges of grain size - distributions providing hydraulic conductivities k ≤ 10^{-7} m s^{-1} appropriate for landfill liner constructions are presented in Fig. 1.

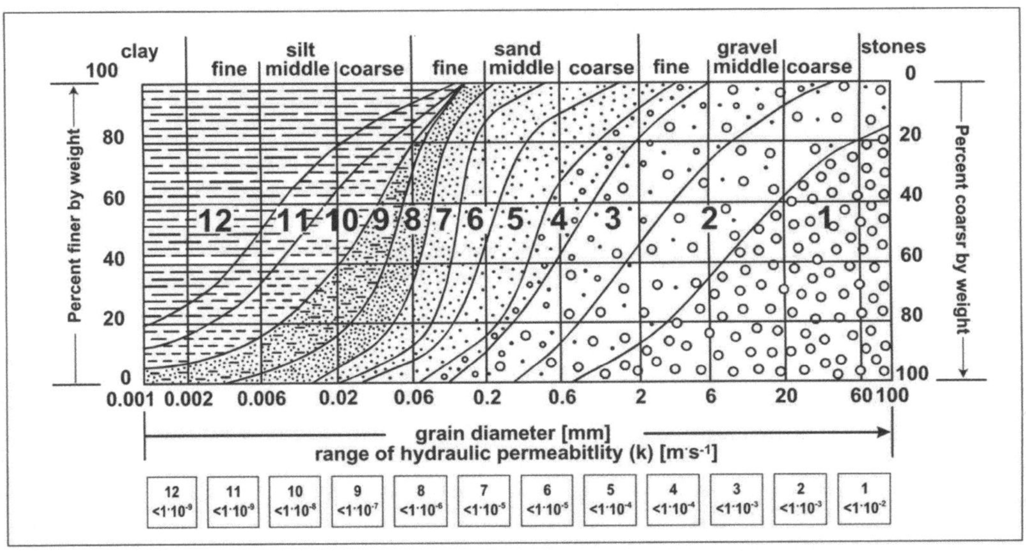

Fig. 1. Ranges of hydraulic permeability (expressed as k coefficient in m·s^{-1}) as related to grain-size distribution areas. The range of k ≤ 10^{-7} m·s^{-1} is considered as an important barrier feature for mineral sealing in most of national regulations. Grain size distribution areas 10, 11 and 12 refer to this range. After Alamgir et al. (2005).

3. Factors affecting hydraulic permeability

3.1 Theoretical background

Laminar flow rate of a liquid through a cylindrical capillary, according to Hagen-Poiseuille equation is proportional to the 4th power of the capillary diameter. The other factors are the gradient of pressure constituting the external driving force of the flow and the dynamic

viscosity being the property of the liquid itself. In case of porous materials (such as eg. soil) characterized by a very complicated pore structure the laminar flow rate, as described by equation of Darcy (1856) is proportional to the pressure gradient and hydraulic permeability k characterizing the properties of the material in which the flow takes place.
Hydraulic conductivity k is related to soil and permeating fluid according to Kozeny - Carman equation (Mitchel & Soga, 2005):

$$k = \frac{\rho g}{\mu} \frac{1}{K_n T^2 S_0^2} \left(\frac{e^3}{1+e} \right) S^3 \qquad (1)$$

Where the particular symbols have the following meaning:
k - hydraulic conductivity, g - acceleration of gravity, ρ - fluid mass density, μ - fluid viscosity, T - tortuosity, K_n - pore shape factor, S_0 - wetted surface area per unit volume of particles, e - void ratio, S - degree of saturation.
Thus, any factors affecting the above properties should result in altering the water permeability of porous material applied to municipal waste landfill bottom liner.
It was confirmed experimentally that soil hydraulic conductivity depends, among others, on its particle size distribution and specific surface area as well as on void ratio, swelling and ion exchange capabilities (Alamgir et al., 2005; Baumann, 1999; Benson & Trast, 1995; Egloffstein, 2001; Foged & Baumann 1999; Mitchell & Jaber 1990; Vukovic & Soro, 1992). Hydraulic conductivity usually decreases with the increase of the content of fine particles (Alamgir et al., 2005; Sivapullaiah et al., 2000), as shown in Fig. 1.

3.2 Compaction effects on hydraulic conductivity of soil materials

The key question for solid municipal wastes landfill constructors and operators is how to reduce the negative effects of the waste body like landslides, leachate infiltration to ground water and soil, odors, rain and wind erosion on the surrounding environment. An appropriate isolation of waste body can be achieved by the construction of bottom and top liners limiting the leachates outflow and infiltration of water (e.g. Bagchi, 2004; Horn & Stępniewski, 2004; Tatsi & Zouboulis 2002; Wysocka et al., 2007) while allowing at the same time the gas emission and oxygen inflow in the top capping. According to literature reports and engineering practice two different approaches of liner construction may be observed: application of polymer membranes and usage of frequently local mineral materials containing significant amounts of clay (Bagchi, 2004). Both approaches have their benefits and limits but in some cases, especially in developing countries application of mineral clay liners, despite risk of cracking, sometimes supported by simple membranes is welcome by the local authorities (e.g. Ahmed, 2008; Gunarathna et al., 2007;). Such attempts were noted not only in e.g. Asian less developed countries but also in Europe.
Water permeability of natural soils is often higher than the required values described by national and international standards for bottom liner construction (in most countries, as it was mentioned earlier the minimum required saturated conductivity for bottom liner should be no more than $1 \cdot 10^{-9}$ $m \cdot s^{-1}$) and unsuitable even for top capping constructions. Consequently mechanical compaction approaches can be used to decrease the hydraulic conductivity. However, such mechanical compaction caused by external loads generating static and dynamic forces leads to increased bulk density, decreased porosity as well as shifts in pore shapes and size distributions but reduces the strength of the system because of an anthropogenically created positive pore water pressure due to dynamic kneading. (e.g. Flowers & Lal, 1998; Radford at al., 2000; Horn, 2004; Yavuzcan et al., 2005; Zhang et al., 2006).

The influence of compaction process on hydraulic properties of soil can be easily explained by the Hagen - Poiseuille law and was proved amongst others by Kooistra and Tovey (1994) who found out that a voids reduction in size and shape caused by passing wheeling machines resulted in smaller macroporosity (pore diameter > 100 μm) by approx. 3 % .
Irrespective of cracking risks - which can be prevented by compaction at the water content lower than the water content at Proctor density - such compacted liners made of local soils or other materials (by-products, bentonite-soil mixtures etc.) are commonly applied in construction of municipal solid waste landfills. According to numerous reports, they appear successful in limiting infiltration of leachates to soil and groundwater environment as well as reducing infiltration of surface water into waste body (e.g. Ahn & Jo 2009; Bagchi, 2004; Gunarathna, 2007; Horn & Stepniewski, 2004; Islam et al., 2008; Wysocka et al., 2007).
Figure 2 presents the reported effects of compaction on hydraulic conductivity of selected different porous materials (local soils, industrial by-products and bentonite mixtures) applied in construction of European or Asian landfills bottom and top liners. The degree of compaction is reflected by soils bulk density changed after stress application.

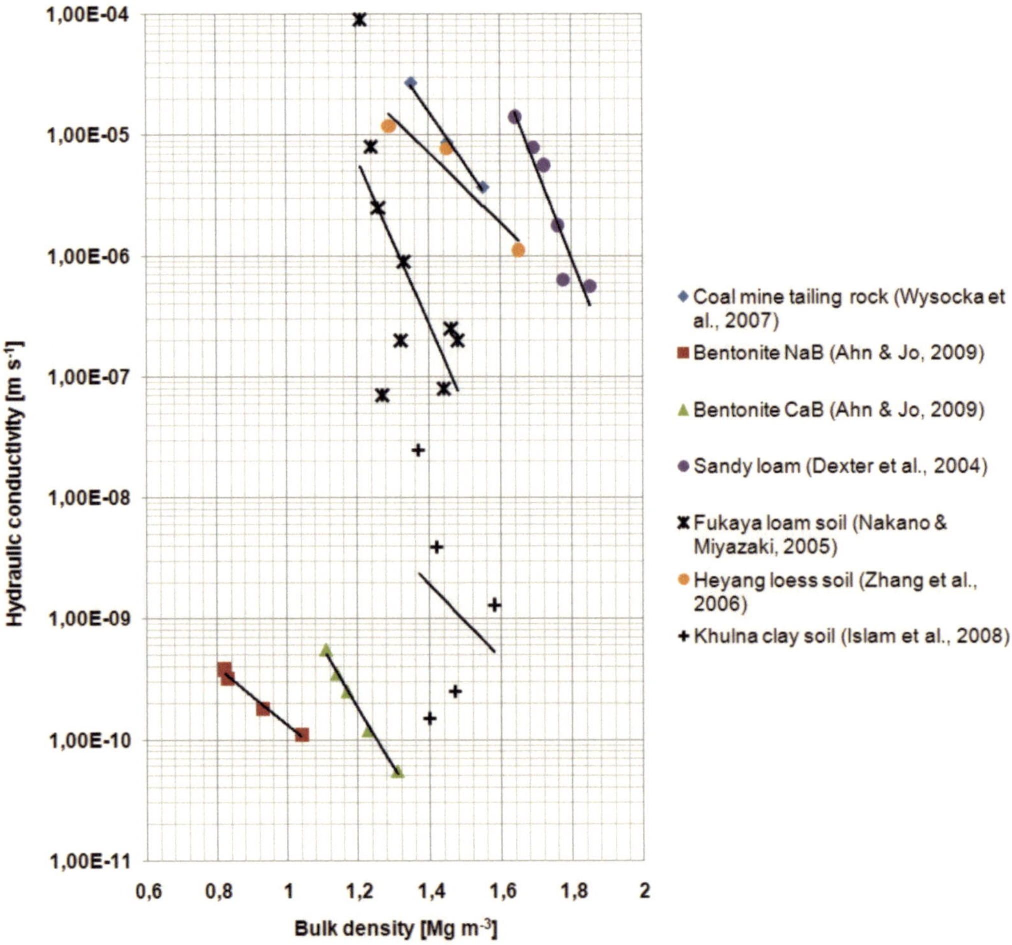

Fig. 2. Effect of bulk density on hydraulic conductivity of various porous materials (compiled from different sources)

Presented results of literature studies show that several local soils and other porous materials after compaction fulfill the requirements of solid waste landfills' liners construction. Reports state that loess, sandy loam and loam soils as well as coal mining tailing rock are applicable to construction of landfills' top liners presenting saturated water conductivity reaching the value of 10^{-6} - 10^{-7} $m{\cdot}s^{-1}$. Observed value of tested porous materials' hydraulic conductivity significantly may reduce the infiltration of surface water to waste body through landfill top capping.
The other compacted materials presented in Fig. 2 such as clay soil and bentonite mixtures, may be useful in construction of landfill bottom liners – the application of compaction process resulted in reduction of saturated hydraulic conductivity below 10^{-9} $m{\cdot}s^{-1}$, even in case of bentonite below 10^{-10} $m{\cdot}s^{-1}$.
Changes in pore size and continuity, however, alter the hydraulic conductivity and water retention characteristics, which initially may result in reduced infiltration abilities and limited storage capacity but - on the long run - because of swelling and shrinkage which coincides with a non rigid pore system, we even determine higher values of the hydraulic conductivity. Thus, the impacts of soils compaction depend, among others, on soil type, soil moisture during compacting, intensity and kind of loading as well as frequency. Junge et al. (2000) proved that in the course of soil compaction and re-drying these weak substrates crack and result in the formation of new macropores with a very high hydraulic conductivity. The results presented by Islam et al. (2008) on compaction of clay soil applied to construction of bottom liner of an experimental municipal solid waste landfill in Khulna, Bangladesh showed that molding water content during the compaction process affects the value of hydraulic conductivity of compacted soils irrespective of the obtained bulk density value (Fig. 3).

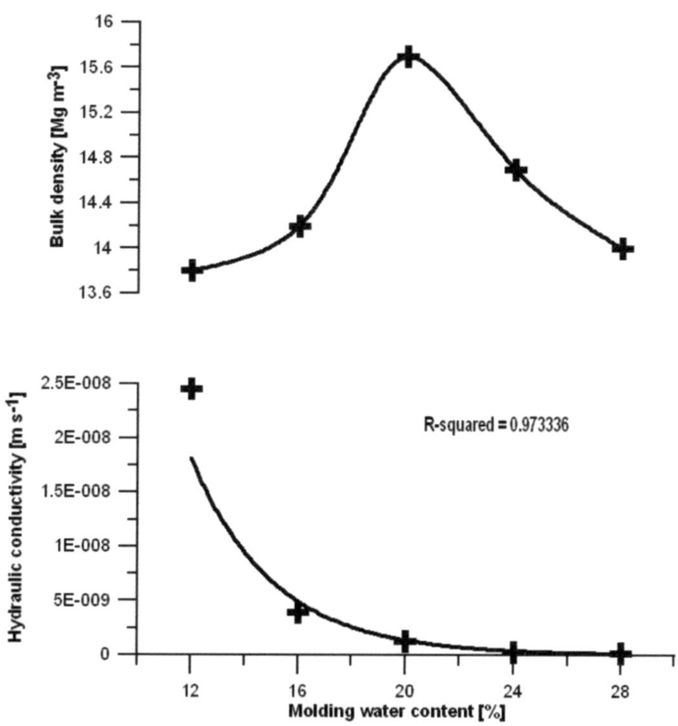

Fig. 3. Proctor curve and hydraulic conductivity of compacted clay soil, versus molding water content. Based on Islam et al. (2008).

As it can be seen, the same bulk density can be achieved at two different levels of molding water content during the process of compaction both on the left and right side of Proctor's curve. Increase of molding water content resulted in a decrease of hydraulic conductivity down to $2.5 \cdot 10^{-10}$ $m \cdot s^{-1}$ at molding water content of 28%. Nonetheless this situation is not reflected in changes of saturated hydraulic conductivity of the tested soil and only focus on the short time effect while the shrinkage induced crack formation thereafter enhances the hydraulic conductivity by many orders of magnitude (Junge et al., 1997).

3.3 Influence of leachates on water permeability

Municipal landfill leachates, generated during infiltration of surface (rain or melted snow) water through the waste body, are commonly considered as one of the most dangerous types of wastewater, significantly influencing environmental conditions as containing high concentrations of ammonium, salts, organic matter, etc. (Di Iaconi et al., 2010). The volume of leachates and their composition depend on the amount of water infiltrating the waste body, and on chemical reactions occurring between the solid and liquid phases – dissolution, ion exchange and biochemical processes (Francisca & Glatstein, 2010). The reported full composition of leachates of different municipal solid waste landfills all over the world were presented by e.g. Ehrig (1989); Fatta et al. (1998), Kjeldsen & Christoffersen (2001), Kylefors (2003), Kulikowska & Klimiuk (2008); Tatsi & Zouboulis (2002) and Ziyang et al. (2009).

Migration of leachates generated inside the waste body of municipal landfill to soil and groundwater is prevented by bottom liners of different construction based on porous materials of low permeability. As it was mentioned, bottom liners usually have multilayer layout and consist of natural or compacted clay or mixtures of clayey soils, granular filters and geosyntetics (e.g.: Francisca & Glatstein, 2010; Ozcoban et al., 2006; Petrov & Rove 1997; Touse- Foltz et al., 2006).

Particle size, specific surface area, void ratio and fluid properties as well as soil fabric, compaction energy and thixotropy are the main factors limiting the water and contaminants movement in compacted porous materials of landfill liners (Benson & Trast 1995; Vukovic & Soro, 1992). According to reported numerous studies (e.g. Mitchell & Jaber, 1990, Sivapullaiah 2000; Schmitz 2006) evaluating soil and liquid properties controlling the saturated hydraulic conductivity in liners, hydraulic conductivity decreases along with increased content of fine particles. The increased mechanical stress observed in compacted soils results in reduction of electrical forces effect on soil behavior but the soil fabric is affected by the chemical properties of the flowing liquid (Mitchell & Soga, 2005). The other important factor influencing soil behavior is its retention capacity depending on adsorption mechanisms delaying the transport of contaminants through the soil (Francisca & Glatstein, 2010); ions present in permeating liquid are absorbed by mineral phase surface, in the rate and amount controlled by surface charge density, ion concentration and valence, and pH. So, according to Schmitz (2006) landfill leachate containing high ionic concentration should increase the hydraulic conductivity as increased ionic concentration should decrease the double-layer thickness. But, as it was reported (e.g. Francisca & Glatstein, 2010; Mitchell & Soga, 2005) this mechanism has sometimes a negligible effect on the experiment field test since it is relevant only in case of high porosity or freshly compacted soils.

Another factor influencing changes of hydraulic conditions of liner porous material treated with leachates is bioactivity causing pore clogging (e. g. Brovelli et al., 2009). Nutrients load present in leachates is responsible for increased formation and development of bacteria and yeast colonies resulting in partial or permanent soil pore blocking (Francisca & Glatstein, 2010; Rebata-Landa & Santamarina, 2006). Decrease of porous media hydraulic conductivity may be in this case related to the presence of biofilm covering surface of mineral particles, thus significantly reducing sizes of micropores and increasing the resistance of fluid flow.

Research concerning effects of leachate on hydraulic conductivity of natural clay was conducted by Ozcoban et al., (2006). Natural clay soils applied as liner in municipal solid waste landfill in Kemerburgaz, Turkey, were tested. Soil samples containing kaolinite were permeated with distilled water and leachate in a vertical reactor - constant head permeameter (each test lasted 3-4 weeks). Tests conducted by Ozcoban et al., (2006) confirmed that clay soils, under laboratory conditions show a very little increase of hydraulic conductivity after being permeated with leachate: $9.848 \cdot 10^{-10}$ $m \cdot s^{-1}$ for water vs. $10.8 \cdot 10^{-10}$ $m \cdot s^{-1}$ for leachate.

The hydraulic and compaction characteristics of leachate-contaminated lateric Indian soil were presented by Nayak et al. (2007). The soil was sampled at local open waste dump where municipal solid wastes were deposited without shredding and segregation. Four different levels of leachate concentration were tested: 0%, 5%, 10% and 20%. The increase of hydraulic conductivity of soil due to leachate addition was observed (see Table 1).

Leachate content	k 10^{-7} [$m \cdot s^{-1}$]	Increase %
0%	3.07	-
5%	3.698	20.46
10%	4.542	22.82
20%	5.792	27.52

Table 1. Effect of leachate concentration on hydraulic permeability of lateric soil (Nayak et al., 2007)

These observations fully support the earlier mentioned thesis of Mitchell & Soga (2005) or Schmitz (2006) and prove increase of hydraulic conductivity of leachate treated soils.

Not numerous investigations were conducted to define the influence of leachates presence on hydraulic conductivity of porous materials applied to bottom liners of municipal solid waste landfill. Studies of Francisca & Glatstein (2010) focused on long term hydraulic conductivity of compacted silt soils of Chaco-Pampean plain, Argentina, also with 5% and 10% bentonite addition. Permeability tests were conducted for distilled water and filtered landfill leachate. The compaction liquid content was at the constant level of 20%. Hydraulic conductivity was measured weekly during the period of 15 months by the standard falling head procedure according to ASTM D5856 (Francisca & Glatstein, 2010). Observations showed a decrease of pore volume of flow after 15 months treating soil samples with leachate - Table 2.

	Pore volume of flow	
Sample	Distilled water	Leachate
0% of bentonite	3.6	1.37
5% of bentonite	2.3	1.51
10% of bentonite	2.7	2.0

Table 2. Changes in pore volume of flow after 15 months of distilled water and leachate permeation (Francisca & Glatstein, 2010)

Results of hydraulic conductivity test after long term permeation of the tested samples with distilled water and leachate conducted by Francisca & Glatstein (2010) are presented in Fig 4. The decrease of water permeability of the tested soils after 15 months of leachate treatment is visible in case of 0% and 5% bentonite content. The reported changes in soil water conductivity, according to the paper, could have resulted from expansion/shrinking of expansive minerals, mineral clogging and bioactivity – the mechanism of ion exchange and reduction of double-layer thickness which should increase the water permeability has in this case a negligible effect.

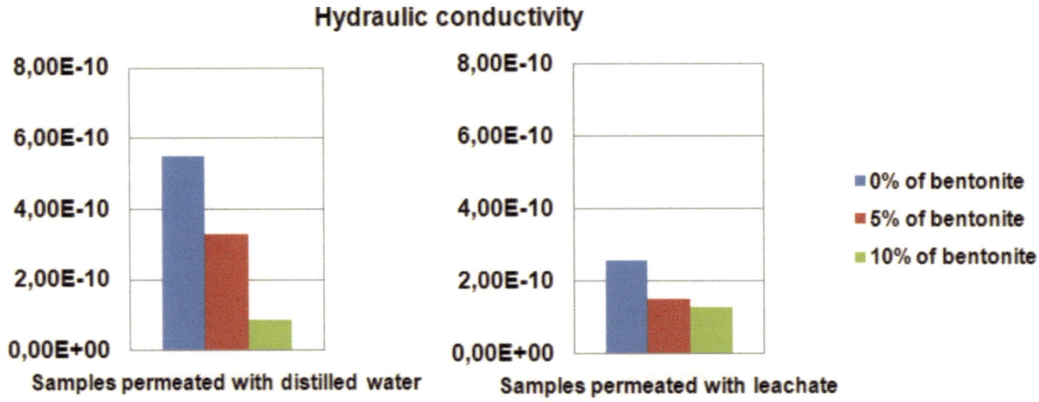

Fig. 4. Hydraulic conductivity k ($m \cdot s^{-1}$) of compacted silt with bentonite amendments (based on data of Francisca & Glatstein, 2010)

Several studies were focused on determination of leachate effect on geosynthetic materials applied in construction of landfill bottom liner – usually geosynthetic clay liners (GCLs) used as hydraulic barrier in landfills, remediation sites or other contamination systems. Shan and Lai (2002) tested the hydraulic conductivity of two different geosynthetic liners: Bentomat ST and Claymax 200R, CETCO, USA, using different liquids as penetrating medium. Both tested GCLs were approx. 6 mm thick and both contained bentonite in the amount of 3.6 kg/m^2. The hydraulic conductivity tests were conducted according to standard ASTM D5887 procedure at effective pressure of 34.5 kPa with typical time of hydration equal to 48 hours (7 days for tap water). The trials of sequential permeation by water and then by leachate were also conduced. The results of the measurements conducted by Shan and Lai (2002) are presented in Table 3.

Permeate	Hydraulic conductivity [$m \cdot s^{-1}$]	
	Bentomat ST	Claymax 200R
Deionized distilled water	$2.7 \cdot 10^{-11}$	$2.7 \cdot 10^{-11}$
Tap water	$4.4 \cdot 10^{-11}$	$4.8 \cdot 10^{-11}$
Landfill leachate	$3.0 \cdot 10^{-11}$	$2.6 \cdot 10^{-11}$
Tap water→leachate	$3.7 \cdot 10^{-11}$	$1.9 \cdot 10^{-11}$

Table 3. Hydraulic conductivity of two types of geosynthetic clay liners (Shan & Lai, 2002).

Both materials showed the same value of water permeability for deionized distilled water and higher values for tap water; the observed increase reaching approx. 70%. Application of landfill leachate as a permeate caused a decrease of GCLs water conductivity for both materials of approx. 31,8% and 84.6% in comparison to water conductivity for tap water. Then, sequential tests of permeation by tap water and then by leachate showed a decrease of hydraulic conductivity of the studied geosynthetics by, respectively, 15,9% and 60,4% in comparison to results obtained for tap water only.

Material	Direction of changes of permeability	Reference liquid	Source
Natural clay of a landfill	Little increase, 3-4 weeks test	water	Ozcoban et al., 2006
Lateric Indian soil of an open damp	Up to 50% increase of permeability with 5 - 20% leachate in water	water	Nayak et al., 2007
Silt soil with 5 and 10% of bentonite	Decrease, 15 month test	distilled water	Francisca & Glatstein, 2010
Two geosynthetic liners	Decrease	tap water	Shan & Lai, 2002

Table 4. A summary of the effect of landfill leachate on hydraulic permeability of selected materials (as compared to water) according to different sources.

Summing up we may state that cited, exemplary reports show different effects of leachate on saturated conductivity of landfill liner materials (Table 4). It should be emphasized that there no data related to the action time of many years, as the longest test did not exceed 15 months (Francisca & Glatstein, 2010). It should be added that hydraulic conductivity may change due to modification of soil water repellency by leachates (cf. Hartman et al., 2010).

4. Amendments used to improve hydromechanical properties of liners

In many cases hydraulic permeability of local soils may be insufficient even after application of external loads leading to compaction (e.g. Bogchi, 2004). An example of such situation is a silty soil from Chaco - Pampean plain in the center and north- east of Argentina covering 600 000 km^2 characterized by mean hydraulic conductivity 10^{-8} $m{\cdot}s^{-1}$ after compaction (Francisca & Glatstein, 2010). Thus this material requires modification in order to be useful for landfill liner construction.

Numerous researches, presented in Table 5, reported different attempts of decreasing water hydraulic permeability of various materials by application of series of amendments to meet the required threshold values. This Table shows that among many materials tested, bentonite shows high popularity. Bentonite is a natural clay characterized by a very high swelling capacity, high ion exchange capacity and very low value of water permeability. The most important characteristics of bentonites are high montmorillonite content (60-90%), high water absorption capacity (200-700% weight), swelling volume of 7-30 ml, pH suspension value 9-10.5, plasticity 140-380%, and cation exchange capacity 0.60-0.90 mol/kg (Egloffstein, 2001).

From other materials we should mention claystones, natural zeolites, fly ashes, water glass, silica fume, cement and some other waste materials (see below). A special attention deserves quick lime (CaO) which can be used to reduce water content of the material during compaction and to stabilize the liner structure (eg. Wiśniewska & Stępniewski, 2007).

Material	Description	Minimal reported k: $m \cdot s^{-1}$	Sources
Fly ash	Ash from incineration plant, sewed < 4 mm, mixed with Freidland clay	No data	Travar et al., 2009
	C type fly ashes generated in Columbia Generating Station Unit II, Portage, Wisconsin added in 28% to sand and bottom ash, compacted at 18% water content	$3.1 \cdot 10^{-10}$	Palmer et al., 2000
Quicklime and water glass	Waste rock of coal mine (Bogdanka, Poland), mixed with 2% (by mass) of quicklime (CaO) and 6% by mass of water-glass.	$1 \cdot 10^{-10}$	Wiśniewska & Stępniewski, 2007
Silica fume	Silica fume from Ferro-Chromite Factory in Antalya, Turkey mixed in different proportion with natural clays of clay pit in Oltu, Turkey.	$9{,}03 \cdot 10^{-10}$ for 25% of silica fume*	Kalkan & Akbulut, 2004
Bentonite	Japanese commercial bentonite Kunigel-V1 extracted from Tsukinuno Mine, Japan, mixed with sand	$1 \cdot 10^{-11}$-$1 \cdot 10^{-12}$ for bentonite content 5-50%.	Komine 2004, 2010
	Bentonite of 92% sodium montmorillonite (by Minarmco SA) added to Chaco-Pampean silt in the amount of 5 and 10%.	$3.3 \cdot 10^{-10}$ – 5% bentonite $8.5 \cdot 10^{-11}$ – 10% bentonite	Francisca & Glatstein, 2010
	Compacted sodium and calcium exchanged bentonite, Gyungsang, Korea	$5.4 \cdot 10^{-12}$ for Ca bentonite $9.9 \cdot 10^{-12}$ for Na bentonite	Ahn & Jo, 2009
	Bentonite compacted (intermediate and modified by Proctor test) different shapes and sizes commercial gravel particles by AquaBlok, Ltd.	$6.08 \cdot 10^{-12}$ (intermediate Proctor) $5.98 \cdot 10^{-12}$ (modified Proctor)	Roberts & Shimanoka, 2008
	Commercial Na-bentonite and Ca-bentonite (Concarde Mining) mixed with crushed, natural zeolites (Etibank-Bigadic, Turkey) at different proportions	$5 \cdot 10^{-10}$-$8 \cdot 10^{-10}$	Kaya & Dudukan, 2004
Geosynthetic clay liners (GCL)	A thin layer of sodium or calcium bentonite bonded to a layer or layers of geosynthetic	$2 \cdot 10^{-10}$-$2 \cdot 10^{-12}$ depending on confining stress (general info)	Bouazza, 2002
	Bentonite based medium-heavy and heavy GCL, after ion exchange in situ for 1-3 years	$1 \cdot 10^{-10}$-$1 \cdot 10^{-11}$ permeability increased during observation period	Egloffstein, 2001

Material	Description	Minimal reported k: $m \cdot s^{-1}$	Sources
Geosynthetic clay liners (GCL)	Modification of commercial GCL Claymax 200R (CETCO, USA: layer of bentonite between two polypropylene geotextiles) - GCL was modified by standard bentonite with HDTMA-bentonite or BTEA bentonite sprayed instead upper geotextile.	$6.8 \cdot 10^{-11}$ - unaltered Claymax 200R $1.2 \cdot 10^{-10}$ - 30% BTEA-bentonite $3.4 \cdot 10^{-11}$ - 30% HDTMA-bentonite	Lorenzetti et al., 2005
Cement	Askale Cement Factory, Erzurum, Turkey, mixed with Oltu - Erzurum (Turkey) clay.	$8.53 \cdot 10^{-10}$	Kalkan, 2006
Claystone	Northpatagonian smectite rich claystones mixed with sand - 15 % of claystones	$5.34 \cdot 10^{-12}$	Musso et al., 2010
Red mud	By product of the caustic leaching of bauxite to produce alumina, redish-brown color, superfine particle-size distribution, mixed with Oltu - Erzurum, Turkey clay and Askale Cement Factory, Erzurum, Turkey cement in different proportions.	$3.73 \cdot 10^{-10}$	Kalkan, 2006
Rubber	Pulverized form of tires rubber added to mixture of C type fly ash (90%) by Soma thermal power plant, Turke and bentonite.	$9.5 \cdot 10^{-12}$ after 0 days $2.7 \cdot 10^{-11}$ after 28 days Both for 90 % fly ash, 3% rubber and 7% of bentonite content.	Cokca & Yilmaz, 2004

*Normalized permeability = k_{exp} (ρ_{std}/γ_{dmax}) – where: ρ_{std} is the standard value of specific gravity adopted by the Authors as 2.65; γ_{dmax} - maximum dry bulk density of the sample.

Table 5. Various amendments applied to lower the water permeability of different materials likely to be used for construction of landfill liners (according to different sources).

5. Capillary barrier concept and landfill liner construction

5.1 Introduction

Recently the capillary barrier system is more often applied because the natural soil behaviour concerning water fluxes and direction of fluxes underlines the long term efficiency of layered systems for multidimensional water transport. It is well known, that soils are highly heterogeneous and anisotropic materials because a myriad of processes influence the formation of physical structure with time. A major impact influencing the soil structure during the preparation of waste deposit capping systems occurs due to machinery application through compaction, mixing and a degeneration of processes that thereafter again promote aggregation (Ahuja et al., 1984). One property that is highly sensitive to all changes in particle arrangements and structure formation due to physical, chemical and anthropogenic processes is the conductivity of pores, which as Bear (1972) mentioned is strongly influenced by pore-geometric factors, like total porosity, pore-size distribution, shape of the pore system, continuity and tortuosity. Normal swelling and shrinkage processes, especially in clayey, silty, and loamy soils result in the formation of direction dependent secondary pores and aggregate shapes. The same is also true when soils are loaded (as can be also proved e.g. by the tillage-induced plough pan), which alters pore geometries affecting the air and water-filled pores and, consequently their functionality. Thus, the development of soil structure can be evaluated through the presence of a direction-dependent behavior of hydraulic properties. These properties present anisotropy if they are direction dependent, otherwise they would be considered as isotropic.

The development or the preparation of soil structure or various layers most often reveals anisotropy. Stratified soils, consisting of fine layers parallel to the surface, exhibit a dominant horizontal component of the saturated hydraulic conductivity (k_{sh}) greater than its vertical component (k_{sv}) (Mualem, 1984; Tigges, 2000). Under unsaturated conditions, the direction dependent hydraulic fluxes can either be also anisotropic in horizontal or vertical direction or they can also for a certain matrix potential range becoming isotropic. Single soil horizons also present anisotropy, which can be also used to define the aggregate formation theory and to quantify the consequences for the 3 d effects on water fluxes. Consequently these functions can be furthermore broadened if during the construction of capping systems a defined layering is achieved in order to e.g. support long-term impermeability of such capping systems (Hartmann et al., 2009; Horn et al., 2001). In the following the consequences of direction-dependent behavior on mass transport at the scale of soil horizons or soil layers will be defined in order to also evaluate the consequences for water movement on the waste deposit scale. However, it must be also stated that the anisotropy or isotropy depends on the drying history, too, which underlines that in order to really predict the 3D water fluxes a very detailed analysis of the hydraulic properties must be done. Thus, the anisotropic behavior of hydraulic conductivity plays an important role in the analysis and modeling of transport processes in soils, especially in heterogeneous soils conducting water or chemicals in 2 and 3 dimensions.

5.2 Physical principles

Irrespective of the material, any kind of mass movement always occurs in the direction of the steepest slope and depends on the conductivity and gradient dependent flux. Considering the theoretical background, the hydraulic conductivity/matrix potential functions are to be defined as vectors which are described in the complete form of the Laplace equation for multidimensional water flux where the hydraulic conductivity /matrix potential relation has to be included for the x,y,z, directions as soon as $k_x \neq k_y \neq k_z$. This boundary condition is in general to be accepted if we deal with non equally sized spheres, which in addition have not reached the smallest possible pore volume or maximum amount of contact points. It therefore also requires more consideration for sampling and, in the consequence, it also has to be included in all discussions dealing with mass movement in systems under a given slope as well as it affects the construction of layered capping sealing systems concerning the mass flow of water and sediments along the slope line. Additionally it also has implications for the reformation of structure elements if especially swell/shrink processes and gravitational forces result in crack formation in the freshly compacted soil layers at a water content higher than that at the Proctor density.

5.3 Hydraulic properties of layered soils

5.3.1 Effect of pore size distribution, water saturation, and pore connectivity

Water content /matrix potential functions define the pore size distribution which gives an insight in the ratio of the volume of pores related to the volume of the bulk soil. The coarser the soil the better aerated, the finer the pores the less aerated and the longer remains the water in these pores. Aggregate formation, anthropogenic effects result in an intense change in the pore diameter as well as in the total porosity, which also causes interruptions between different soil layers and can also cause stagnant water (=stagnic soil properties) or altered water flow intensity and directions (for more detailed info see Gräsle et al., 1995; Nielsen & Kutilek 1995).

5.3.2 Effect of hydraulic conductivity/matrix potential ratio

Apart from the pore size distribution we need a better insight in the water flux properties because they primarily define the transport within the soils. It is well known that under saturated conditions coarse textured soils with the dominance of coarse pores have a high saturated flow rate (=saturated hydraulic conductivity), while in soils with dominating medium and fine pores the saturated hydraulic conductivity gets smaller. Under unsaturated conditions however, samples with more medium and fine pores keep their unsaturated hydraulic conductivity the longer the smaller the pores are while coarse pores are emptied very quickly. Consequently the remaining hydraulic conductivity is intensely reduced (Fig. 5). Between the curves of 2 materials we can define the cross over suction value, which defines the pF value at which both materials have an identical hydraulic conductivity. If the soil dries out more intensely the finer textured soil with the higher amount of finer pores has a higher hydraulic conductivity, while left from this point more water flows in the coarser textured material.

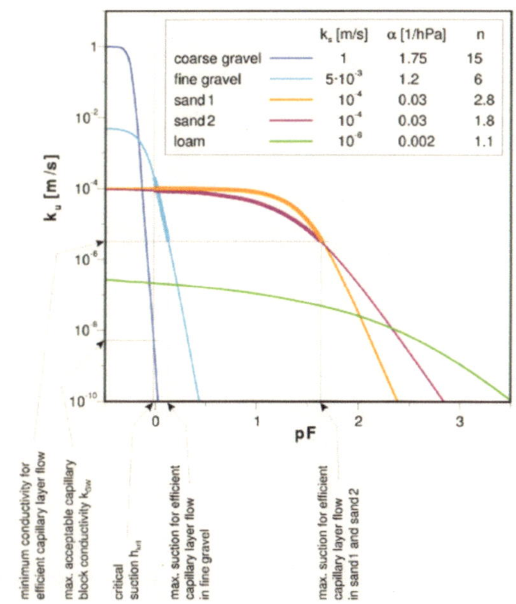

Fig. 5. Hydraulic conductivity versus matrix potential (defined as pF value) for very coarse up to finer textured soils.

The cross over suction value (Hillel, 1998) defines the matrix potential or pF value where the hydraulic properties of the various materials are identical. Thus, if we now come back to a very old picture of W. Gardner, *(Washington State University, USA)* the consequences for the water flow in layered systems can be easily understood. In this example (Fig. 6) he prepared a layered sample with fine over coarse sand and proved that at t1 the water mostly flows downwards, at t2 the water front concentrates at the boundary (=coarser) layer until the water content is increased sufficiently and the matrix potential declined. At t3 the matrix potential in the top layer gets sufficiently high, which results in a pronounced vertical infiltration in the coarser lower layer.

t1

t2

t3

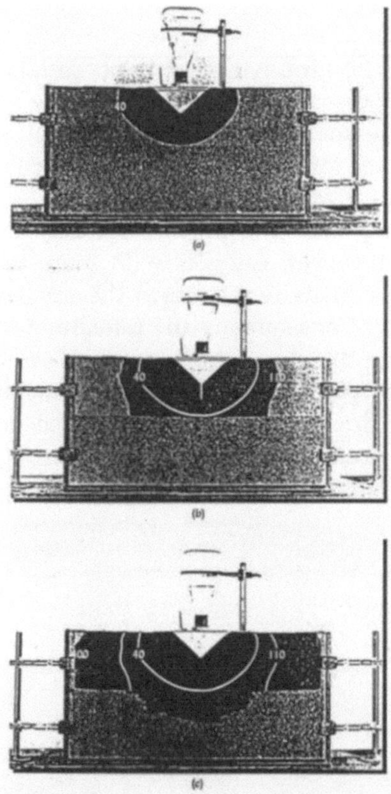

Fig. 6. Downward water movement in a layered sand tank as a function of time: the top layer consists of finer sand and the bottom one of coarser sand. (Picture originates from W.H. Gardner, Washington State University, USA). At t1 the water mostly flows downwards, at t2 the water front concentrates at the boundary (=coarser) layer until the water content is increased sufficiently and the matrix potential declined. At t3 the matrix potential in the top layer gets sufficiently high, which results in a pronounced vertical infiltration in the coarser lower layer.

5.4 Application of the anisotropy principle to the capillary barrier concept

The capillary barrier concept requests the defined construction of the various layers in dependence of the expected climatic conditions and on the later land use (Fig. 7). On top of the waste body and the compensation layer, the capillary block contains coarse textured material (e.g. gravel) with very coarse pores which are emptied immediately when even only a very small unsaturation degree is reached. Thereafter follows the capillary layer which consists of finer material like fine sand. Finally, a recultivation soil layer ensures the storage of sufficient plant available water, nutrient storage and rootability and guarantees a mostly rigid i.e. long term stable topsoil.

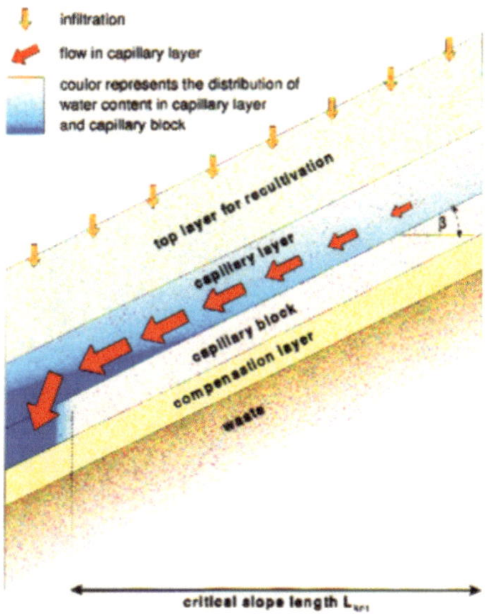

Fig. 7. Construction of waste capping systems applying the concept of capillary barrier

Consequently, the capillary barrier system can be ranked as an waste deposit sealing system, which facilitates not only the reuse of soils available in the region if their physical properties are known. If the above defined principles are agreed it becomes obvious that in order to also quantify the lateral fluxes the tensorial functions of the hydraulic conductivity can be used to construct an impermeable and long- term stable capping system (Horn, 2002; Baumgartl et al., 2004).

5.5 Example: two dimensional hydraulic fluxes in a layered waste deposit capping system (Rastorf, Germany)

Based on the hydraulic conductivity and continuous matrix potential measurements the 2 dimensional fluxes within the topsoil waste deposit capping system (Figs. 8-10) could be verified (Hartmann et al., 2009).

It become obvious that the flow direction as well as the dynamics of the changes between vertical upwards, downwards or lateral flow can be quantified and always related to the present situation of the matrix potential dependent hydraulic conductivity for the various layers. In case of the upwards flow, the drying intensity of the topsoil layer was high and caused the capillary rise (Fig 8) while in Fig. 9 the hydraulic conductivity of the topsoil was much higher than that of the coarser layer below. However, even if the re-saturation results in less negative matrix potential values in the finer topsoil and the coarser subsoil layer is left of the cross over suction value and a vertical downward water movement occurs it is still reversing as soon as the soil layers dry out again. Thus, such layering can be classified as a long- term stable system with a self „reparing= reversing" flux system.

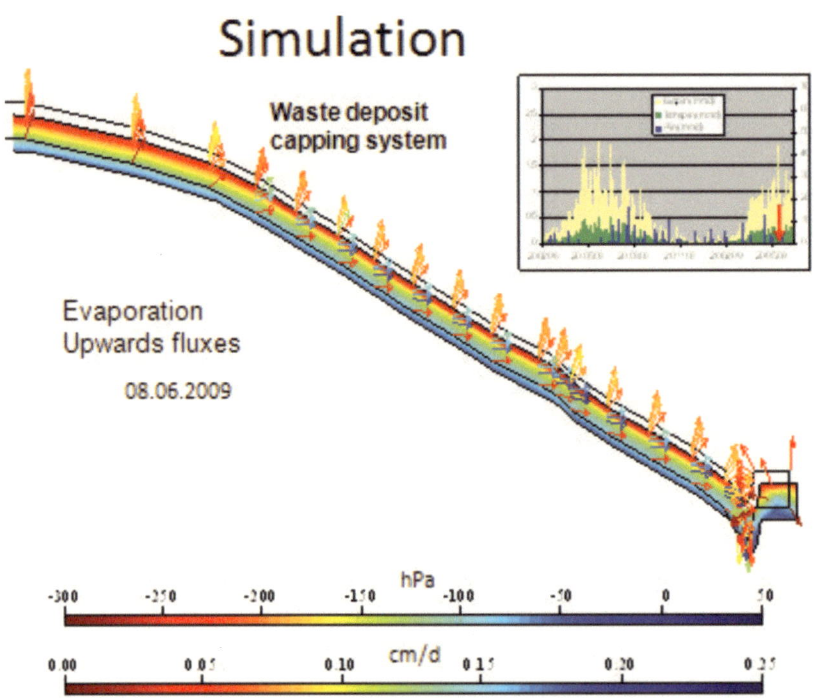

Fig. 8. Simulated hydraulic fluxes in a waste deposit capping system: the case presents situation with dominant upwards flow.

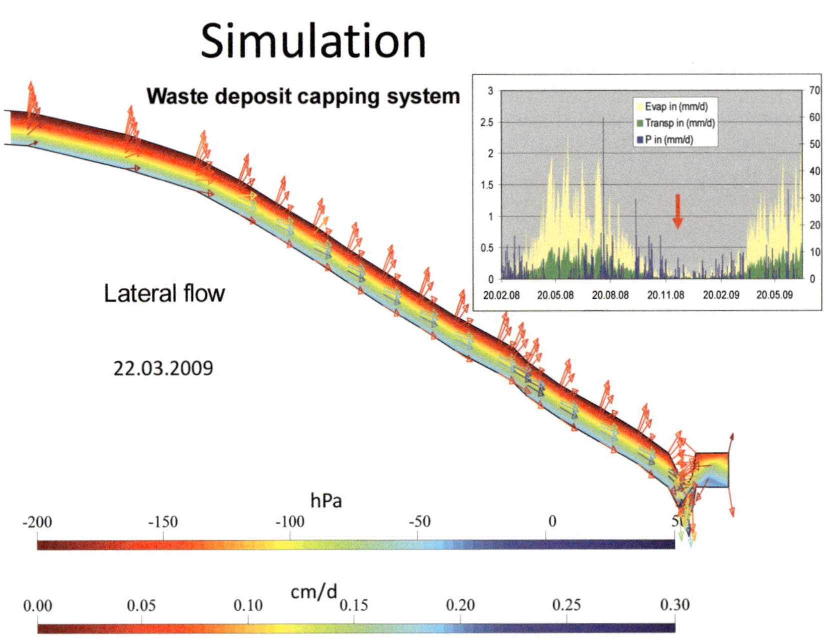

Fig. 9. Simulated hydraulic fluxes in a waste deposit capping system with the lateral flow situation

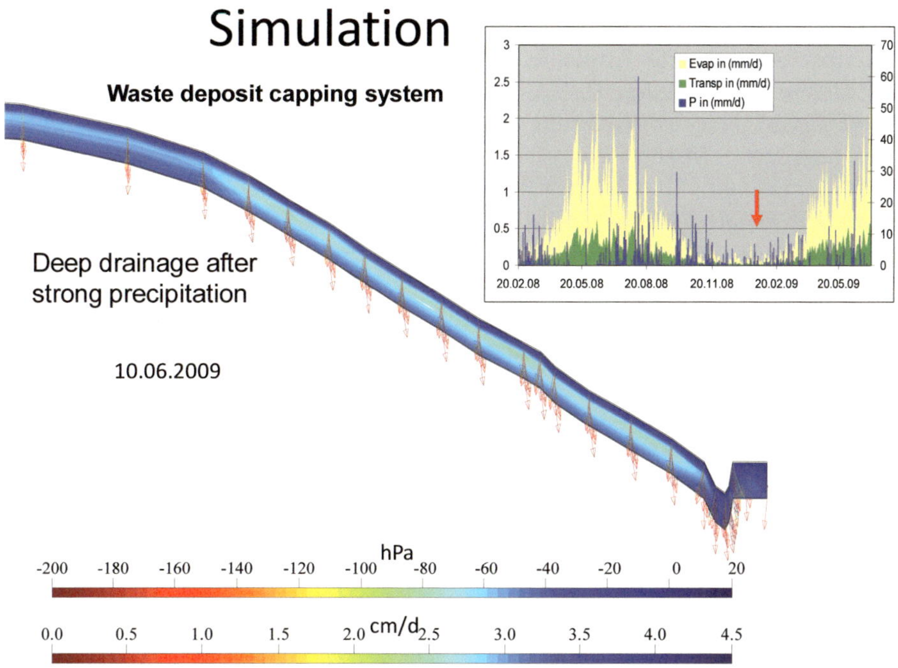

Fig. 10. Simulated hydraulic fluxes in a waste deposit capping system: an example of short term deep drainage in case of a heavy rainfall event.

6. Conclusions

Hydraulic conductivity is a key factor for landfill construction. In the case of bottom liner it is the matter of sufficiently low saturated hydraulic conductivity and long - term stability in time. In case of top capping the situation is much more complicated as problem of removal of infiltrating rain water and presence of soil recultivation layer are involved.

In this case the impermeability is always proved as long as the cross over suction value between the hydraulic conductivity/matric potential relationships of the two layers under consideration is exceeded in all directions for the underlying soil layer. Under those boundary conditions the lateral movement of water is guaranteed also in structured soils. It must be underlined that the anisotropy depends always on the mechanical or hydraulic prestresses which coincides with a strong control need of these hydraulic or mechanical stresses.

Anisotropy of hydraulic conductivity is also proved for the unsaturated state and its consideration results in a better validation of modeled versus measured water fluxes on all scales.

7. References

Ahmed, M., (2008). Improvement of solid waste disposal facilities at Matuail in Dhaka City. Proceedings of the National Seminar on Solid Waste Management - WasteSafe, Khulna, Bangladesh 1-12

Ahn, H.-S. & Jo, H.Y., (2009). Influence of exchangeable cations on hydraulic conductivity of compacted bentonite. Applied Clay Science 44, 144–150

Ahuja, L.R., Naney, J.W., Green, R.E. & Nielsen, D.R. (1984): Macroporosity to characterize spatial variability of hydraulic conductivity and effects of land management. Soil Sci. Soc. Am. J. 48, 699 - 702.

Alamgir, M., McDonald, Ch., Roehl, K.E., Ahsan A. (Eds.) (2005). Integrated Management and Safe Disposal of Municipal Solid Waste in Least Deveoped Asian Countries. A Feasibility Study. p 83. Khulna University of Engineering and Technology, Khulna, Bangladesh.

Albrecht, K.A. & Cartwright, K., (1989). Infiltration and hydraulic conductivity of a compacted earthen liner. Groundwater 27 (1), 14-19.

Bagchi, A., (2004). Design of landfills and integrated solid waste management. John Wiley and Sons Inc. Hoboken. New Jersey.

Baumgartl, T., Horn, R., Richards, B.G., (2004). Soil physical processes in capping systems of landfills - possibilities of capillary break capping. 277 - 288. in: Juhasz,A.L., G.Magesan, R.Naidu (eds) Waste Management, Science Publ. ISBN: 1-57808-323-0

Bear, J., (1972). Dynamics of Fluids in Porous Media. Elsevier, New York, p. 764. [4]

Benson, C.H., Trast, J.M., (1995). Hydraulic conductivity of thirteen compacted clays. Clays and Clay Minerals 43 (6), 669–681

Bouazza, A., (2002). Geosynthetic clay liners. Geotextiles and Geomembranes 20. 3–17

Brovelli, A., Malagureea, & Barry D.A., (2009). Bioclogging in porous media: Model development and sensitivity to initial conditions. Environmental Modelling & Software, 24, 611-626.

Cokca, E. & Yilmaz, Z., (2004). Use of rubber and bentonite added fly ash as a liner material. Waste Management 24. 153–164

Darcy, H., (1856). Les fontaines publiques de la vill de Dijon. Paris, Dalmont.

Dexter, A.R., Czyz, E.A. & Gate, O.P., (2004). Soil structure and the saturated hydraulic conductivity of subsoils. Soil & Tillage Research 79. 185–189

Di Iaconi C., Pagano M, Ramadori R., & Lopez A. (2010). Nitrogen recovery from a stabilized municipal landfill leachate. Bioresource Technology 101, 1732–1736

EEA (2004). Waste and Material Flows 2004. European Topic Centre on Waste and Material Flows. European Environmental Agency, Copenhagen, 76 p.

EU (1999). Landfill directive: Council Directive 1999/31/EC of 26 April 1999 on the landfill of waste. Official Journal of the European Communities. L182, 1-19.

Egloffstein, T.A., (2001). Natural bentonites–influence of the ion exchange capacity and partial desiccation on permeability and self-healing capacity of bentonites used in GCLs. Geotextiles and Geomembranes, 19, 427- 444.

Ehrig, H., (1989). Leachate quality. In: Christensen, T.H., Cossu, R.,Stegmann, R. (Eds.), Sanitary Landfilling: Process, Technology, and Environmental Impact, Academic Press, London, pp. 213–229

Fatta, D., Voscos, C., Papadopoulos, A., Loizidou, M.,(1998). Leachate quality of a MSW landfill. Journal Environmental Science and Health Part A A33 (5), 749–763

Flowers, M.D. & Lal, R., (1998). Axle load and tillage effects on soil physical properties and soybean grain yield on a mollic ochraqualf in northwest Ohio. Soil Till. Res. 48, 21–35

Foged, N. & Baumann, J., (1999). Clay membrane made of natural plasticity clay: leachate migration due to advection and diffusion. Engieneering Geology, 54, 129 - 137.

Francisca, F.M. & Glatstein, D.A. (2010). Long term hydraulic conductivity of compacted soils permeated with landfill leachate. Applied Clay Science 49. 187–193

Gardner, W.H. Washington State University. USA. Private communication.

Gräsle, W., Richards, B.G., Baumgartl, T. & Horn R., (1995): Interaction between soil mechanical properties of structured soils and hydraulic processes- theoretical fundamentals of a model. 719 - 725, in: Alonso E.E. & Delage P. (eds.) Unsaturated soils. Balkema Verlag ISBN: 9054105844

Gunarathna, H.A.Y.R., Kapukotuwa, A.S.B., Karunarathna, A.K. & Basnayake, B.F.A. (2007). Construction and Evaluation of a Low Cost, Waste Polyethylene Based Liner for Small Landfills. Proceedings of the International Conference on Sustainable Solid Waste Management, Chennai, India. pp.226-233

Hartmann ,P., Fleige, H., Horn, R. (2010). Water repellency of fly ash- enriched forest soil from eastern Germany. Europ. J. Soil Sci. 61,107-1078.

Hartmann, P., Gebhardt, S., Janßen,I., Fleige,H., Horn,R. (2009). Auswirkungen des Klimawandels auf die Gestaltung von geschichteten mineralischen Abdeckungssystemen von Deponien. Abschluss und Rekultivierung von Deponien und Altlasten: Egloffstein und Burghardt (Hrsg) : 17, 207-217, ISBN: 978-3-939662-06-8, ICP Eigenverlag Bauen und Umwelt

Hillel, D., (1998). Environmental Soil Physics. Lewis Publ.

Horn, R., (2002). Analysis of drying processes of mineral capping material. (in German, with English summary and captures) Analyse des Austrocknungsverhaltens von mineralischem Bodenmaterial anhand von Laboruntersuchungen. Ed.: Ramke W.-G.e.a., Höxter, Höxteraner Berichte zu angewandten Umweltwissenschaften, 3, 269-278

Horn, R., (2004). Recommendations for the preparation of no shrinking mineral liner systems (in German with English summary and captions) Empfehlungen zur Herstellung nicht schrumpfanfälliger mineralischer Dichtungen mit zusätzlicher tonmineralischer Wasserspeicherschicht für Deponie- ober flächen abdichtung systeme. Müll und Abfall 36, 67-71

Horn R., Baumgartl, T., Possel, H., Koselowske, M., Albiker, B., (2001). Waste deposit sealing systems with mineral materials- a long term impermeable alternative? (in German, with English summary and captures) Deponieabdichtungen aus mineralischem Material - eine undurchlässige Langzeitperspektive? Wasser and Abfall, 3,38 - 41,

Horn, R. & Stępniewski, W., (2004). Modification of mineral liner to improve its long term stability. Int. Agrophysics 18 317-323

Islam, M.R., Alamgir, M., Mohiuddin, K.M. & Hasan, K.M.M. (2008). Investigation of physical properties of a selected soil to use as a compacted clay liner in sanitary landfill. Proceedings of the National Seminar on Solid Waste Management - WasteSafe 2008, Khulna, Bangladesh 167-174

Junge, T., Thienemann, J., Gräsle, W., Baumgartl, T. & Horn R., (1997): Effect of mechanical and hydraulic stresses on the long term stability of waste deposits. (in German, with an English abstract and figure captures). Müll and Abfall 8, 475 - 478

Junge, T., Gräsle, W., Bensel, G. & Horn, R., (2000). Effect of pore water pressure on tensile strength. Z. Pflanzenernährung und Bodenkunde, 163, 21 - 27

Kalkan, E., (2006). Utilization of red mud as a stabilization material for the preparation of clay liners. Engineering Geology 87. 220–229

Kalkan, E. & Akbulut S., (2004). The positive effects of silica fume on the permeability, swelling pressure and compressive strength of natural clay liners. Engineering Geology 73. 145–156

Kaya, A. & Durukan, S. (2004). Utilization of bentonite-embedded zeolite as clay liner. Applied Clay Science 25. 83– 91.

Kjeldsen, P., Christoffersen, M., (2001). Composition of leachate from old landfills in Denmark. Waste Management and Research 19, 249–256.

Komine, H., (2004). Simplified evaluation on hydraulic conductivities of sand–bentonite mixture backfill. Applied Clay Science 26. 13– 1.

Komine, H., (2010). Predicting hydraulic conductivity of sand–bentonite mixture backfill before and after swelling deformation for underground disposal of radioactive wastes. Engineering Geology 114. 123–134.

Kooistra, J.M., Tovey, N.K., (1994). Effects of compaction on soil microstructure. In: Soane, B.D., van Quwerkerk, C. (Eds.), Soil Compaction in Crop Production. Elsevier Science (Chapter 5), pp. 91–111

Kulikowska, D. & Klimiuk, E. (2008). The effect of landfill age on municipal leachate composition. Bioresource Technology 99. 5981–5985

Kylefors, K., (2003). Evaluation of leachate composition by multivariate data analysis (MVDA). Journal of Environmental Management 68. 367–376

Lorenzetti, R.J., Bartelt-Hunt, S.L., Burns, S.E. & Smith, J.A., (2005). Hydraulic conductivities and effective diffusion coefficients of geosynthetic clay liners with organobentonite amendments. Geotextiles and Geomembranes 23.385–400

Mitchell, J.K., Jaber, M., (1990). Factors controlling the long-term properties of clay liners. Waste containment systems: construction, regulation and performance. ASCE. Geotechnical Special Publication 26, 84–105

Mitchell, J.K., Soga, K., (2005). Fundamentals of Soil Behavior, 3rd Edition. John Wiley & Sons, Hoboken.

Mualem, Y., (1984): Anisotropy of unsaturated soils. Soil Sci. Soc. Am. J. 48, 505–509.

Musso, T.B., Roehl, K.E., Pettinari, G. & Vallés, J.M., (2010). Assessment of smectite-rich claystones from Northpatagonia for their use as liner materials in landfills. Applied Clay Science 48 (2010) 438–445

Nakano, K. & Miyazaki, T., (2005). Predicting the saturated hydraulic conductivity of compacted subsoils using the non-similar media concept. Soil & Tillage Research 84, 145–153.

Nayak, S., Sunil, B.M., Shrihari, S., (2007). Hydraulic and compaction characteristics of leachate-contaminated lateritic soil. Engineering Geology 94, 137–144

Nielsen, D. & Kutilek, M., (1995). Soil Hydrology, Catena.

Ozcoban, M.S., Tufekci, N., Tutus, S., Sahib, U. & Celik, O,. (2006). Leachate removal rate and effect of leachate on the hydraulic conductivity of natural (undisturbed) clay. Journal of Scientific and Industrial Research 65. 264-269

Palmer, B.G., Edil, T.B. & Benson, C. H., (2000). Liners for waste containment constructed with class F and C fly ashes. Journal of Hazardous Materials 76. 193–216.

Petrov, R., Rowe, R., (1997). Geosynthetic Clay Liner (GCL) – chemical compatibility by hydraulic conductivity testing and factors impacting its performance. Canadian Geotechnical Journal 34, 863–885.

Radford, B.J., Bridge, B.J., Davis, R.J., McGarry, D., Pillai, U.P., Rickman, J.F., Walsh, P.A. & Yule, D.F., (2000). Changes in the properties of a vertisol and responses of wheat after compaction with harvester traffic. Soil Till. Res. 54, 155–170.

Rebata-Landa, V., Santamarina, J.C., (2006). Mechanical limits to microbial activity in deep sediments. Geochemistry, Geophysics, Geosystems 7, Q11006.

Roberts, A.A., Shimaoka, T. (2008). Analytical study on the suitability of using bentonite coated gravel as a landfill liner material. Waste Management 28, 2635–2644.

Roehl, K.E., Alamgir, M., Razzaque, M.A., Sarkar, G., (2009). Geoenvironmental properties of clays for landfill construction in Bangladesh. Proceedings of the Internat. Conf. on Solid Waste Manag. Waste Safe 2009, vol 2, pp 503-511, 9-10 Nov. 2009, Khulna, Bangladesh. (Eds. Alamgir et al.). ISBN 978 984 33-0761-3.

Schmitz, R.M., (2006). Can the diffuse double layer theory describe changes in hydraulic conductivity of compacted clay? Geotechnical and Geological Engineering 24. 1835–1844

Shan, H.-Y., Lai, Y.-J., (2002). Effect of hydrating liquid on the hydraulic properties of geosynthetic Clay lines. Geotextiles and Geomembranes 20. 19-38

Sivapullaiah, P.V., Sridharan, A., Stalin, V.K., (2000). Hydraulic conductivity of bentonite-sand mixtures. Canadian Geotechnical Journal 37 (2), 406–413

Tatsi, A.A. & Zouboulis, A.I., (2002). A field investigation of the quantity and quality of leachate from a municipal solid waste landfill in a Mediterranean climate (Thessaloniki, Greece). Ž. Advances in Environmental Research 6, 207 - 219.

Tigges, U., (2000). Untersuchungen zum mehrdimensionalen Wassertransport unter besonderer Berücksichtigung der Anisotropie der hydraulischen Leitfähigkeit. Ph. D. Thesis. Schriftenreihe des Instituts für Pflanzenernährung und Bodenkunde, Christian Albrechts University, Kiel, Germany, Nr. 56, 145 S. ISBN: 0933-680 X.

Touze-Foltz, N., Duquennoi, C. & Gaget, C.E., (2006). Hydraulic and mechanical behavior of GCLs in contact with leachate as part of a composite liner. Geotextiles and Geomembranes 24. 188–197

Travar, I., Lidelöwa, S., Andreas L., Thamb, G. & Lagerkvist, A., (2009). Assessing the environmental impact of ashes used in a landfill cover construction. Waste Management 29.1336–1346.

Vuković, M. & Soro, A., (1992). Determination of hydraulic conductivity of porous media from grain size distribution. Water Resources Publications, Littleton, Colorado

Wiśniewska, M. & Stępniewski, W. (2007). The influence of lime, water-glass and clay addition of sealing properties of waste rock from Bogdanka. pp 271-275. In Environmental Engineering (Pawlowski et al. - Eds), Francis and Taylor Group, London, ISBN13 978-0-415-40818-9

Wysocka, A., Stępniewski, W. & Horn, R., (2007). Swelling-shrinkage properties and hydraulic conductivity of a compacted coal mine tailing rock likely to be used for landfill capping. Int. Agrophysics 21. 405-408

Yavuzcan, H.G, Matthies, D. & Auernhammer, H., (2005). Vulnerability of Bavarian silty loam to compaction under heavy trafic: impacts of tillage method nad soil water content. Soil & Tillage Research 84, 200-215

Zhang, S., Grip, H. & Lovdahl, L., (2006). Effect of soil compaction on hydraulic properties of two loess soils in China. Soil & Tillage Research 90, 117–125

Ziyang, L., Youcai, Z., Toa, Y., Yu, S., Huili, C., Nanwen, Z. & Renhua, H., (2009). Natural attenuation and characterization of contaminants composition in landfill leachate under different disposing ages. Science of the Total Environment 407. 3385–3391.